Sustainable Agriculture Management in Semi-Arid Climates

A. K. Singh • Pavan Kumar • S. S. Singh
Editors

Sustainable Agriculture Management in Semi-Arid Climates

Volume 1

Springer

Editors
A. K. Singh
Rani Lakshmi Bai Central Agricultural University
Jhansi, Uttar Pradesh, India

S. S. Singh
Rani Lakshmi Bai Central Agricultural University
Jhansi, Uttar Pradesh, India

Pavan Kumar
Forest Resource Management, College of Horticulture and Forestry
Rani Lakshmi Bai Central Agricultural University
Jhansi, Uttar Pradesh, India

ISBN 978-3-031-94061-3 ISBN 978-3-031-94062-0 (eBook)
https://doi.org/10.1007/978-3-031-94062-0

This Springer imprint is published by the registered company Springer Nature Switzerland AG
The registered company address is: Gewerbestrasse 11, 6330 Cham, Switzerland

Foreword

The semi-arid regions of the world represent unique landscapes where agriculture is both a challenge and an opportunity. These regions face extreme climatic variability, water scarcity, and soil degradation, but they also hold vast potential for innovation and sustainable growth. As global populations grow and the impacts of climate change intensify, addressing the intricate challenges of sustainable agriculture in semi-arid climates becomes critical to ensuring food security, enhancing livelihoods, and promoting environmental resilience.

This book *Sustainable Agriculture Management in Semi-Arid Climates: Opportunities and Threats*, Vol. I offers a comprehensive exploration of these critical issues. It brings together pioneering research and innovative practices designed to address the unique challenges of semi-arid climates while unlocking opportunities for sustainable development. This book delves into a wide range of topics, including climate-resilient farming techniques, the role of biochar in modern agriculture, soil and water management strategies, and the use of advanced technologies, such as remote sensing and GIS. It also emphasizes the importance of livestock production, value addition in agricultural products, and sustainable aquaculture, providing a holistic perspective on agricultural sustainability.

This volume reflects the combined expertise of leading researchers and practitioners who present practical solutions grounded in science, technology, and traditional knowledge. The insights offered here go beyond theoretical discussions, providing actionable strategies to transform semi-arid agriculture into a resilient and thriving sector.

By addressing the interconnected dimensions of climate adaptation, resource management, and value addition, this book serves as a vital resource for researchers, policymakers, agricultural professionals, and students. It not only enriches our understanding of sustainable agricultural practices but also inspires innovation and collaboration to create a more sustainable future for semi-arid regions and beyond.

I commend the editors and contributors for their dedication to producing such a significant and timely work. This volume will undoubtedly contribute to advancing the global agenda for sustainable agriculture and climate resilience

Panjab Singh
Rani Lakshmi Bai Central
Agricultural University
Jhansi, India

Preface

Sustainable agriculture is the cornerstone of food security, economic stability, and environmental health, especially in semi-arid regions where climatic uncertainties and resource constraints pose unique challenges. Semi-arid climates, characterized by limited and erratic rainfall, are particularly vulnerable to the adverse effects of climate change, necessitating the adoption of innovative and resilient agricultural practices. This book, *Sustainable Agriculture Management in Semi-Arid Climates: Opportunities and Threats*, Vol. I, offers a comprehensive exploration of the multifaceted approaches required to sustain agricultural productivity while preserving ecological balance in these fragile environments.

The book is divided into thematic sections that address the key aspects of sustainable agriculture in semi-arid regions. It begins by examining adaptive strategies under changing climate scenarios, highlighting the integration of climate-smart agricultural techniques, agroforestry practices, and water-efficient technologies. These strategies are crucial in mitigating the impacts of climate change and ensuring the long-term viability of farming systems.

Soil and water management emerge as critical pillars of sustainability, with a focus on innovative techniques for soil conservation, nutrient management, and the use of modern tools like remote sensing and GIS to monitor and optimize resources. By addressing issues such as soil salinity, irrigation water quality, and environmental variability, the book underscores the importance of preserving these fundamental resources for future generations.

Livestock nutrition and production, a vital component of agriculture in semi-arid regions, are explored with an emphasis on farmer-friendly technologies and capacity-building initiatives. Special attention is given to sustainable livestock practices that enhance rural livelihoods while minimizing ecological footprints.

Additionally, the value addition of agricultural products is discussed as a pathway to enhance economic opportunities and improve nutrition. The chapters on meat and fish processing and the utilization of aquatic resources for sustainable aquaculture provide actionable insights into diversifying income sources and increasing food security.

The accelerating pace of climate change poses significant challenges to agricultural sustainability, particularly in semi-arid regions where climatic variability directly affects soil health, water availability, crop production, and livestock management. This book, organized into four thematic sections, explores innovative strategies and sustainable practices designed to address these challenges while promoting resilience and productivity in agriculture.

In part "**Adaptation Strategies Under Climate Change Scenario**", the chapters focus on current approaches and techniques for climate-resilient agriculture in semi-arid regions. From assessing the impacts of irrigation water quality to adopting agroforestry and other climate-smart agricultural practices, these chapters provide critical insights into adaptive mechanisms for sustaining agriculture under changing climatic conditions.

Part "**Soil and Water Management for Agricultural Sustainability**" delves into the foundational aspects of soil and water conservation, offering strategies for maintaining soil health and ensuring water quality. Through case studies and advanced methodologies such as remote sensing indices and GIS-based analytical tools, the chapters emphasize the importance of innovative approaches to monitor and manage salinity, nutrient levels, and environmental variability.

The role of livestock in agricultural sustainability is examined in part "**Livestock Nutrition and Production**, which highlights the potential of farmer-friendly technologies and capacity development initiatives. With a special focus on pig farming in semi-arid regions, this section outlines strategies for enhancing livelihood opportunities and achieving sustainable development.

Finally, part "**Value Addition in Meat and Fish for Health and Nutrition**" underscores the importance of leveraging value-added products to improve human health and nutrition. Innovative uses of aquatic macrophytes in aquaculture and strategies for enhancing the nutritional value of meat and fish products are explored, offering practical solutions for sustainable production systems.

The contributions in this book collectively address the critical intersections of climate resilience, agricultural sustainability, and rural development. It aims to serve as a valuable resource for researchers, practitioners, policymakers, and students seeking to understand and implement sustainable agricultural practices in the face of a rapidly changing climate. This book serves as a vital resource for researchers, policymakers, agricultural practitioners, and students. It not only identifies the threats posed by climate variability but also highlights opportunities to innovate and transform agricultural systems in semi-arid regions. By combining scientific advancements with traditional knowledge, this work contributes to building resilient, sustainable, and equitable food systems. We hope this book inspires new ideas and fosters collaborative efforts toward achieving sustainable agriculture in semi-arid climates worldwide.

Jhansi, Uttar Pradesh, India

A. K. Singh
Pavan Kumar
S. S. Singh

Contents

About the Editors

A. K. Singh Vice Chancellor, Rani Lakshmi Bai Central Agricultural University, Jhansi. He has been engaged in teaching and research along with a wide national and international exposure in extension management and training. He looked after the extension services of Chandra Shekhar Azad University of Agriculture and Technology, Kanpur, before joining ICAR services as Zonal Coordinator, Zone IV and presently guiding and managing entire frontline extension system and Krishi Vigyan Kendras of the country as Deputy Director General (Agricultural Extension), Indian Council of Agricultural Research. Major initiatives pursued by him are Farmer FIRST Programme, Attracting and Retaining Youth in Agriculture (ARYA), Mera Gaon Mera Gaurav (MGMG), Extramural Research Grant, Coordinated and network research in extension, cluster demonstration of pulses and oilseeds, launched four new initiatives like Value Addition and Technology Incubation Centre Agriculture (VATICA), Knowledge System and Homestead Agriculture Management in Tribal Areas (KSHAMTA), Nutri-sensitive Agricultural Resources and Innovations (NARI) and Farm Innovation and Resource Management, ICT initiatives including KVK, and mobile-based agro-advisory through Kisan portal covering one crore farmers. Initiated first virtual extension through 2.0 portal with both read and write options and online monitoring of KVKs, pioneering and experimenting district Strategic Research and Action Plan (SREP) in association with World Bank experts and MANAGE, Hyderabad. Developed KVK guideline. He has published more than 113 research papers, 55 books and book chapters, and about 50 popular articles. He is recipient of 15 awards/recognitions including Swami Sahajanand Saraswati Best Extension Scientist Award of ICAR, Leadership Award of Banaras Hindu University (BHU), and Excellence in Science Award of CSAUA&T, Kanpur, Fellow of National and International Academic Societies. He is Chief Editor and Editor of NAAS rated Refereed Journals of National repute, Reviewer of Current Science Journal. President and Vice President of National Professional Societies and organized 20 national and international seminars/conferences.

Pavan Kumar is a Faculty of Forest Resource Management, College of Horticulture and Forestry, Rani Lakshmi Bai Central Agricultural University, Jhansi, U.P., India. He obtained his Ph.D. degrees from Faculty of Natural Sciences, Jamia Millia Islamia, New Delhi. He did B.Sc. (Botany) and M.Sc. (Environmental Science) from Banaras Hindu University, Varanasi, India, and subsequently obtained Master's degree in Remote Sensing (M. Tech) from Birla Institute of Technology, Mesra Ranchi, India. His current research interests include Resilient Agriculture and Climate Change studies. He is recipient of Innovation China National academy award for Remote Sensing. Dr. Kumar has published more than 75 research papers in international journals and authored several books (Springer, Elsevier, Taylor & Francis, and Wiley). He has visited countries like USA, France, the Netherlands, Italy, China, Indonesia, Brazil, West Africa, and Malaysia for various academic/scientific assignments, workshop, and conferences. Dr. Kumar is member of the International Associations for Vegetation Science, USA, and the Institution of Geospatial and Remote Sensing, Malaysia.

S. S. Singh Director of Extension, Rani Lakshmi Bai Central Agricultural University, Jhansi, U.P., India. His present research interests include environmental management, sustainable development, agriculture management, and applications of remote sensing and GIS. He has four books to his credit. He has published more than hundred research papers in journals of repute. Dr. Singh has presented 50 research papers at national and international conferences including at Sapienza University of Rome, Italy, University of British Columbia, Canada, University of Western Cape, Bellville, South Africa, and University of Brighton, U.K. He has delivered invited talks at various universities. Ten doctoral degrees to the scholars have been awarded under his supervision. He has chaired academic sessions at various conferences. He is the reviewer of many scientific research journals and member of scientific bodies.

Adaptation Strategies Under Climate Change Scenario

Current Approaches Scenario of Semi-Arid Climate

A. K. Singh, Pavan Kumar, and S. S. Singh

Abstract Semi-arid climates pose a significant challenge for sustainable development due to their fragile ecosystems, limited water resources, and susceptibility to climate change. This chapter explores the current approaches to managing semi-arid regions, focusing on water management, soil conservation, biodiversity preservation, and the impact of climate change. Through case studies and global examples, we delve into technological innovations, policy frameworks, and sustainable land management strategies employed to build resilience in semi-arid areas. Future scenarios concerning the effects of climate change and population growth are also discussed, with an emphasis on sustainable solutions, including water-efficient technologies, climate-smart agriculture, and ecosystem-based adaptation. The integration of traditional knowledge and modern techniques provides a holistic approach to combating the challenges faced by semi-arid regions, ensuring food security, water conservation, and biodiversity protection.

Keywords Climate change adaptation · Food security · Renewable energy · Semi-arid climate · Sustainable agriculture · Sustainable land use practices

1 Introduction

1.1 Characteristics of Semi-Arid Climates

Semi-arid climates are typically defined by receiving between 250 and 500 mm of rainfall annually. These regions experience long dry seasons, short rainfall periods, and high evaporation rates due to extreme heat. Semi-arid climates exist across

A. K. Singh · S. S. Singh
Rani Lakshmi Bai Central Agricultural University, Jhansi, Uttar Pradesh, India

P. Kumar (✉)
Forest Resource Management, College of Horticulture and Forestry, Rani Lakshmi Bai Central Agricultural University, Jhansi, Uttar Pradesh, India

A. K. Singh et al. (eds.), *Sustainable Agriculture Management in Semi-Arid Climates*, https://doi.org/10.1007/978-3-031-94062-0_1

globe, consisting of Sub-Saharan Africa, the Middle East, parts of the United States, Australia, and Central Asia. One of the critical challenges in semi-arid regions is the variability in rainfall, both in terms of amount and timing (Lal, 2004a, 2004b; Trenberth et al., 2014; UNCCD, 2017; Reij et al., 2009; Agarwal & Narain, 1997). This unpredictability complicates agricultural planning and water management, creating a constant struggle for resources. Additionally, the ecosystems in these regions are finely tuned to cope with water scarcity, leading to unique vegetation and wildlife that are highly vulnerable to human activity and climate change. Semi-arid climates are also characterized by extreme temperature fluctuations and prolonged dry seasons, making them one of the most challenging environments for both human habitation and natural ecosystems. According to the Köppen climate classification, semi-arid regions typically receive between 250 and 500 mm of annual rainfall, which is often erratic and unevenly distributed throughout the year (Peel et al., 2007a, 2007b). The combination of low precipitation and high evaporation rates creates a water deficit, exacerbating the challenges of maintaining water resources and supporting agriculture. One of the primary features of semi-arid climates is the marked contrast between wet and dry seasons. The majority of the rainfall occurs during a short, often unpredictable rainy season, followed by long dry periods where temperatures can soar, particularly in subtropical semi-arid zones. For example, regions like the Sahel in Africa experience extremely high temperatures during the dry season, with average high temperatures exceeding 40 °C (Nicholson, 2013). These extreme conditions limit the types of crops that can be grown and place significant stress on both ecosystems and human settlements.

Temperature variability is another hallmark of semi-arid climates. In addition to extreme heat during the dry season, semi-arid regions often experience cold winters, particularly in temperate semi-arid areas. For example, semi-arid regions in Central Asia, such as parts of Kazakhstan and Mongolia, are subject to hot summers and freezing winters, with temperatures ranging from over 35 °C in summer to well below freezing in winter (Chen et al., 2019). This significant temperature variation requires adaptive strategies for agriculture, water management, and infrastructure. The unpredictable nature of rainfall in semi-arid regions often leads to periodic droughts, which can have devastating effects on local agriculture and water availability. Drought events in semi-arid climates are often exacerbated by climate change, as rising global temperatures are expected to increase the frequency and intensity of droughts (Trenberth et al., 2014).

Another important characteristic of semi-arid climates is the vulnerability of their soils. Semi-arid regions often have thin, nutrient-poor soils that are susceptible to erosion. When combined with overgrazing, deforestation, and unsustainable agricultural practices, this leads to soil degradation and desertification—a process that has affected vast areas of the world, particularly in regions like Sub-Saharan Africa and Central Asia (UNCCD, 2017). As vegetation cover is lost, the ability of the soil to retain moisture decreases, further compounding the challenges of water management in these fragile ecosystems. Despite these challenges, semi-arid regions are home to millions of people who depend on agriculture, pastoralism, and natural resources for their livelihoods. These regions often support unique ecosystems, with

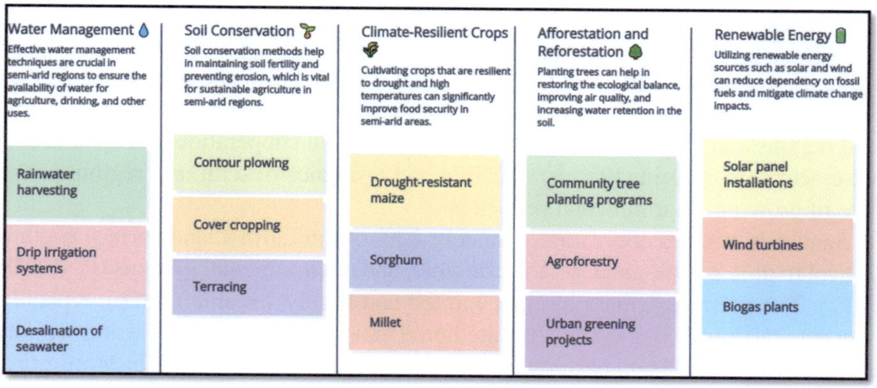

Fig. 1 Analysis of various strategies and methodologies currently being employed to manage and adapt to the semi-arid climate conditions

species that have adapted to survive in harsh conditions. Plants and animals in semi-arid areas have developed remarkable adaptations, such as deep root systems in plants and drought tolerance in animals, to cope with the variability in water availability (Schwinning & Sala, 2004) (Fig. 1).

1.2 Global Importance of Semi-Arid Regions

Semi-arid regions cover approximately 40% of the Earth's land surface, supporting nearly 2 billion people. These areas are essential for agricultural production, pastoralism, and biodiversity conservation. However, the fragile ecosystems in semi-arid climates face mounting pressure due to climate change, unsustainable land use, and population growth. Effective management of these regions is critical to mitigating the impacts of climate change and ensuring the sustainability of natural resources. Semi-arid regions are essential for global biodiversity and agriculture, serving as the home to millions of people. Agriculture, livestock grazing, and small-scale farming dominate the economies in these areas. However, the precarious nature of water and soil resources creates vulnerability to environmental degradation.

The global importance of semi-arid regions extends to international environmental and policy discussions, particularly in relation to climate change, desertification, and sustainable development. The United Nations Convention to Combat Desertification (UNCCD) has highlighted the need for global action to address the degradation of land in dry lands, including semi-arid regions, through sustainable land management practices (UNCCD, 2017). Global initiatives, such as the Sustainable Development Goals (SDGs), particularly Goal 15 (Life on Land) and Goal 13 (Climate Action), emphasize the importance of preserving ecosystems and combating desertification in semi-arid regions to ensure a sustainable future (UN, 2015). International frameworks such as the Paris Agreement and the Convention on Biological Diversity (CBD) recognize the critical role that semi-arid regions play in climate regulation, biodiversity conservation, and food security. By focusing on sustainable land use, climate adaptation, and ecosystem restoration, these frameworks aim to mitigate the impacts of climate change and desertification on semi-arid regions and their inhabitants (FAO, 2019). Global cooperation and local action are essential to ensuring the sustainability and resilience of semi-arid regions in the face of growing environmental pressures.

Semi-arid regions cover approximately 40% of the Earth's land surface and are critical to global ecological, socio-economic, and environmental systems (UNCCD, 2017). These areas, defined by their limited and highly variable rainfall, typically between 250 and 500 mm annually, are home to nearly 2 billion people and support essential agricultural and pastoral economies (Peel et al., 2007a, 2007b; Mortimore, 2009). Despite the challenges posed by low rainfall, water scarcity, and frequent droughts, semi-arid regions contribute significantly to global food security, biodiversity conservation, and climate regulation. However, they also face increased vulnerability due to land degradation, desertification, and climate change. As global attention increasingly focuses on sustainable development and resilience, understanding the importance of semi-arid regions is essential for addressing some of the world's most pressing environmental and socio-economic issues.

1.3 Contribution to Global Food Security

Semi-arid regions play a crucial role in global food production, particularly in dryland agriculture and livestock systems. Many of the world's staple crops, including millet, sorghum, maize, and wheat, are cultivated in these regions due to their drought-resistant characteristics (Mugalavai et al., 2008). For example, in Sub-Saharan Africa, dryland agriculture provides livelihoods for millions, with small-scale farmers relying on rain-fed crops and livestock (Otte et al., 2012). The livestock sector is also significant, as semi-arid regions support large populations of cattle, goats, and sheep, providing essential food products such as meat, milk, and wool (Thornton et al., 2009).

While these regions are pivotal to global food supply chains, they are highly susceptible to climate variability, making food security precarious (FAO, 2011). Droughts, land degradation, and over-reliance on rainfed agriculture heighten their vulnerability. The unpredictability of rainfall, which can result in both droughts and floods, exacerbates the challenges of managing agricultural systems in semi-arid climates (Wheeler & von Braun, 2013). Climate change is expected to intensify these challenges, threatening food security not only in semi-arid regions but globally, as these areas are key suppliers of staple crops and livestock (Vermeulen et al., 2012). Semi-arid regions are biodiversity hotspots, home to a wide range of endemic species uniquely adapted to the extreme environmental conditions. These regions often contain critical ecosystems, including savannas, grasslands, and shrublands, which host species that have evolved to survive in conditions of water scarcity and temperature extremes (Schwinning & Sala, 2004). For example, the Cerrado in Brazil and the Horn of Africa are rich in endemic plant and animal species, many of which are not found anywhere else in the world (Myers et al., 2000). The biodiversity of semi-arid regions provides essential ecosystem services that support both local communities and global environmental health. These services include water filtration, soil stabilization, and carbon sequestration, which are critical for mitigating the impacts of climate change and maintaining the resilience of ecosystems (Chapin et al., 2000; Lal, 2004a, 2004b). However, human activities, such as overgrazing, deforestation, and unsustainable agricultural practices, have led to the degradation of these ecosystems, threatening biodiversity and the services they provide (Millennium Ecosystem Assessment, 2005; Pfenninger & Staffell, 2016; UNEP, 2011).

This chapter aims to provide a comprehensive analysis of current strategies used to manage the challenges posed by semi-arid climates. It will highlight water management approaches, soil conservation practices, biodiversity protection, and the role of climate-smart agriculture. The chapter also discusses the role of technology and policy in shaping future scenarios for these regions, emphasizing sustainable and adaptive solutions.

2 Challenges of Semi-Arid Climates

2.1 *Water Scarcity*

Water scarcity is one of the most pressing issues in semi-arid regions. The limited rainfall, combined with high rates of evaporation, means that water resources are constantly in short supply. Rivers and reservoirs dry up quickly during prolonged dry seasons, leaving communities without reliable access to water. Agriculture, which is highly water-dependent, suffers in these regions. Irrigation systems often cannot keep pace with demand, leading to crop failure, reduced yields, and food insecurity. Additionally, the competition for water between agricultural, industrial, and domestic uses places further strain on already depleted water supplies.

2.2 *Soil Degradation and Desertification*

Semi-arid regions are highly prone to soil degradation and desertification. Unsustainable agricultural practices, deforestation, and overgrazing contribute to soil erosion and loss of fertility. Once the protective vegetative cover is removed, wind and water erosion strip away the topsoil, leaving the land barren and unproductive. Desertification, the process by which fertile land turns into desert, is a critical issue in semi-arid regions. According to the United Nations, desertification affects around 2 billion people worldwide, and semi-arid regions are particularly vulnerable due to their already fragile ecosystems (Lal, 2004a, 2004b; Ghassemi et al., 1995).

Soil degradation and desertification present serious threats to the sustainability of semi-arid regions, with profound implications for food security, biodiversity, water resources, and livelihoods. While the climatic conditions of semi-arid regions contribute to soil vulnerability, human activities such as unsustainable agriculture, overgrazing, and deforestation have accelerated the degradation process. Global and regional efforts, including the UNCCD and sustainable land management practices, offer hope for reversing desertification and restoring the health of semi-arid ecosystems (Reij et al., 2009). As the impacts of climate change intensify, addressing soil degradation in semi-arid regions will be critical for ensuring the long-term sustainability of these environments and the communities that depend on them.

The semi-arid climate itself plays a crucial role in accelerating soil degradation. Low and variable rainfall leads to prolonged dry periods during which the land is more susceptible to erosion by wind and water (Nicholson, 2013; Peel et al., 2007a, 2007b; Eswaran et al., 2001). During intense rainfall events, the lack of protective vegetation and the structure of semi-arid soils, which are often shallow and poorly developed, contribute to surface runoff and soil erosion (Lal, 2001; FAO, 2011; Ghassemi et al., 1995). The erratic nature of precipitation, combined with high evaporation rates, further depletes soil moisture, leaving the land vulnerable to degradation (Reij et al., 2009; UNCCD, 2017). Soil degradation also affects the availability and quality of water resources in semi-arid regions. Eroded soil particles often end up in rivers and reservoirs, reducing water quality and storage capacity (Millennium Ecosystem Assessment, 2005; Reij et al., 2009).

2.3 *Biodiversity Threats*

Semi-arid regions are biodiversity-rich ecosystems that support unique species adapted to extreme environmental conditions, such as low and variable rainfall, high temperatures, and frequent droughts. Despite their ecological significance, biodiversity in semi-arid regions faces mounting threats due to habitat loss, land degradation, climate change, and unsustainable human activities. Overgrazing, deforestation, and agricultural expansion have significantly reduced vegetation cover, leading to habitat fragmentation and the decline of both plant and animal species. Climate change exacerbates these threats by increasing the frequency and intensity of droughts, altering precipitation patterns, and stressing already fragile ecosystems. Desertification, driven by unsustainable land use and climatic changes, further accelerates the loss of biodiversity by degrading soils and reducing the productivity of ecosystems (Eswaran et al., 2001; MEA, 2005; UNCCD, 2017). Efforts to mitigate biodiversity loss in semi-arid regions must prioritize sustainable land management, ecosystem restoration, and the integration of indigenous knowledge into conservation strategies. Given the critical role that biodiversity plays in maintaining ecosystem services and supporting human livelihoods, it is essential to address these threats to ensure the long-term sustainability of semi-arid ecosystems. The semi-arid ecosystem supports a variety of plant and animal species that have adapted to the harsh conditions of limited water and extreme heat (Schwinning & Sala, 2004). However, habitat loss, over-extraction of natural resources, and climate change are threatening the biodiversity in these regions. Many species in semi-arid zones are specialized and unable to migrate to new habitats, making them highly vulnerable to extinction.

2.4 Economic and Social Vulnerabilities

Semi-arid regions are often home to marginalized populations that depend heavily on agriculture and pastoralism for their livelihoods. These communities are particularly vulnerable to the impacts of climate variability, as they lack the financial resources and infrastructure needed to adapt to environmental changes. Prolonged droughts can lead to food insecurity, displacement, and conflict over resources.

3 Current Approaches to Water Resource Management

Water resource management in semi-arid regions is a critical challenge due to the limited and highly variable rainfall, frequent droughts, and increasing pressures from population growth and climate change. Effective management strategies are essential to ensure the sustainable use of water resources for agriculture, human consumption, and ecosystem maintenance. Various approaches, both traditional and modern, have been developed to address the unique water management needs of semi-arid regions. These approaches include traditional water harvesting techniques, modern technologies for water conservation, integrated water resource management (IWRM), and community-led initiatives (Fig. 2).

3.1 Traditional Water Harvesting Techniques

In many semi-arid regions, communities have long relied on traditional water harvesting techniques to cope with water scarcity. These systems, often based on indigenous knowledge, have sustained livelihoods for centuries by capturing and storing water during periods of rainfall for use during dry seasons. For example, qanats in the Middle East, jessours in North Africa, and rainwater harvesting systems in India are highly effective in semi-arid climates (Agarwal & Narain, 1997). These systems are designed to make the most of sporadic rainfall by diverting, storing, and managing runoff water in natural or artificial reservoirs. Traditional water harvesting

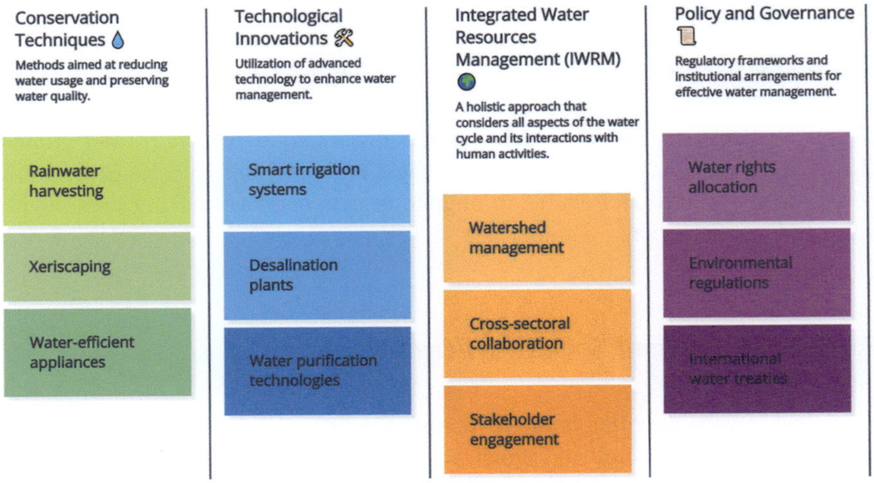

Fig. 2 Analysis delves into the modern strategies and methodologies employed in the management of water resources, highlighting the significance, challenges, and innovative solutions

methods remain relevant today and are often integrated with modern techniques to increase their efficiency. For example, rainwater harvesting is being promoted in India's semi-arid regions, not only as a means of water storage but also as a way to recharge groundwater, which is critical during periods of drought (Falkenmark & Rockström, 2004).

3.2 Modern Water Conservation Techniques

Advances in technology have led to the development of several modern water conservation techniques that are now widely used in semi-arid regions to maximize the efficiency of water use. One such technique is drip irrigation, which delivers water directly to the roots of crops, minimizing water loss through evaporation and ensuring that water is used efficiently (Postel et al., 2001). Drip irrigation systems have been particularly successful in Israel, a semi-arid country that has become a global leader in water-efficient agriculture (Keller & Keller, 2003). These systems allow farmers to maintain productivity even in regions with very low rainfall. Graywater recycling and desalination are other modern methods used in semi-arid regions. Graywater recycling involves the treatment and reuse of domestic wastewater for agricultural or landscape irrigation, thereby reducing the demand on freshwater resources. Desalination, which involves removing salt from seawater to produce fresh water, is increasingly being used in coastal semi-arid regions, such as the Middle East and North Africa, to supplement freshwater supplies (Gleick, 2003). Although desalination is energy-intensive, advances in renewable energy technologies are making it a more viable option in water-stressed regions.

3.3 Integrated Water Resource Management (IWRM)

Integrated Water Resource Management (IWRM) is an approach that promotes the coordinated development and management of water, land, and related resources to maximize economic and social welfare without compromising the sustainability of ecosystems (Global Water Partnership, 2000). IWRM is particularly important in semi-arid regions, where water resources are scarce and must be carefully managed to balance the needs of agriculture, industry, and domestic use. IWRM strategies often involve collaboration among stakeholders, including governments, local communities, and the private sector, to develop comprehensive water management plans. These plans typically include measures for improving water-use efficiency, protecting water quality, and enhancing the resilience of water systems to climate variability (Molle, 2008). For example, in the semi-arid regions of Southern Africa, IWRM approaches have been used to manage shared river basins, ensuring equitable water distribution among countries and communities (SADC, 2012).

3.4 Groundwater Management and Artificial Recharge

Groundwater is a crucial resource in many semi-arid regions, but over-extraction of groundwater has led to the depletion of aquifers, threatening long-term water availability. To address this, groundwater management practices are being implemented to monitor extraction rates and ensure sustainable use. Additionally, artificial groundwater recharge techniques are being used to replenish aquifers. These techniques involve diverting surface water, such as floodwater or treated wastewater, into underground aquifers through recharge wells or infiltration basins (Bouwer, 2002).

In India, for example, groundwater recharge has been promoted in semi-arid regions through the construction of check dams, percolation tanks, and other structures designed to capture rainwater and allow it to infiltrate the soil, thereby replenishing groundwater supplies (Kumar et al., 2008). This approach not only addresses water scarcity but also helps to mitigate the impacts of drought.

3.5 Community-Led Water Management Initiatives

Community involvement is essential for the success of water resource management in semi-arid regions, particularly in rural areas where local knowledge and participation are key to managing scarce resources effectively. Participatory water management approaches, which involve communities in decision-making processes, have proven successful in many semi-arid regions. In Africa's Sahel region, for example, community-led initiatives such as farmer-managed natural regeneration (FMNR) have been used to restore degraded landscapes and improve water retention (Reij et al., 2009). By working together to manage natural resources, local communities have been able to rehabilitate millions of hectares of land, increasing agricultural productivity and resilience to drought. Similarly, in Latin America, participatory watershed management has been employed to improve water governance and increase water availability in semi-arid regions (Cruz-Angón et al., 2011). These initiatives often involve training local farmers in sustainable water use practices, improving irrigation infrastructure, and promoting the use of water-efficient crops.

4 Approaches to Soil Conservation and Sustainable Agriculture

4.1 Contour Plowing and Terracing

Contour plowing and terracing are widely used soil conservation techniques in semi-arid regions that help to reduce soil erosion. Contour plowing involves plowing along the natural contours of the land, which helps to slow down water runoff and prevent soil from being washed away during heavy rains (Morgan et al., 2005). This technique is particularly effective on slopes, where water erosion is most severe. In many semi-arid regions, contour plowing has been adapted by local farmers to retain moisture and improve soil fertility (Pimentel, 2006).

4.2 Agroforestry

Agroforestry, the integration of trees and shrubs into agricultural systems, plays a crucial role in soil conservation in semi-arid regions. Trees help to stabilize soils, prevent erosion, and improve soil fertility through the decomposition of leaf litter and organic matter (Nair, 1993). Additionally, trees provide shade, reduce evaporation, and enhance water retention in the soil, making agroforestry an effective strategy for semi-arid climates where water is a limiting factor. For example, in the Sahel region of Africa, agroforestry systems using *Faidherbia albida* (a nitrogen-fixing tree) have been shown to improve crop yields and restore degraded soils (Reij et al., 2009). The deep root systems of these trees also help to stabilize soils, reduce erosion, and increase the availability of water for crops.

4.3 Conservation Tillage

Conservation tillage is a sustainable agricultural practice that minimizes soil disturbance, thereby reducing erosion and improving soil structure. Unlike traditional plowing, conservation tillage leaves crop residues on the field, which helps to protect the soil from wind and water erosion (Lal, 2001). This practice also enhances soil moisture retention, which is particularly important in semi-arid regions where water is scarce. By reducing the frequency and intensity of soil tillage, conservation tillage helps to maintain soil organic matter and improve the overall health of the soil. In the long term, this technique contributes to the sustainability of agricultural systems by improving soil fertility, reducing erosion, and conserving water (Derpsch et al., 2010).

4.4 Windbreaks and Shelterbelts

In semi-arid regions, where wind erosion can be a major problem, wind breaks and shelter belts are commonly used to protect soil from wind erosion. These are rows of trees or shrubs planted along the edges of fields to reduce wind speed and prevent the wind from blowing away topsoil. Windbreaks are particularly effective in flat or open landscapes where wind erosion is prevalent (Morgan et al., 2005).Windbreaks also provide additional benefits, such as reducing evaporation, improving microclimates for crops, and providing habitats for biodiversity. In many semi-arid regions, local communities have adopted this technique to protect their agricultural lands and reduce the impacts of desertification.

4.5 Water Harvesting for Soil Conservation

Water harvesting is a critical technique in semi-arid regions for both soil conservation and sustainable agriculture. By capturing and storing rainfall, water harvesting systems reduce soil erosion and improve soil moisture. Techniques such as stone bunds, check dams, and earthen bunds are used to slow down water flow during rainfall events, allowing more water to percolate into the soil and reduce surface runoff (Critchley & Siegert, 1991).In semi-arid areas like the Indian subcontinent and Sub-Saharan Africa, water harvesting techniques are widely used to enhance soil moisture, reduce erosion, and improve agricultural productivity during dry seasons (Agarwal & Narain, 1997). These systems, which combine indigenous knowledge with modern engineering, have proven to be effective in restoring degraded lands and improving water availability for crops.

4.6 Crop Diversification and Rotation

Crop diversification and crop rotation are essential strategies for promoting sustainable agriculture in semi-arid regions. These practices help to maintain soil fertility, reduce pest and disease pressure, and minimize the risks associated with monocropping (Altieri, 1999). By growing a variety of crops, farmers can improve soil health, increase biodiversity, and reduce their reliance on a single crop, which may be vulnerable to drought or pests. Crop rotation, where different crops are planted in a sequence on the same field, helps to improve soil structure, enhance nutrient cycling, and reduce the buildup of pests and diseases. Leguminous crops, such as beans and peas, are often included in crop rotations because they fix nitrogen in the soil, improving soil fertility for subsequent crops (Lal, 2001). In semi-arid regions, crop rotation also helps to conserve soil moisture and reduce the risk of soil degradation.

4.7 Drought-Resistant and Native Crops

In semi-arid regions, where water availability is a major constraint, the cultivation of drought-resistant crops and native species is essential for sustainable agriculture. Crops such as sorghum, millet, and drought-tolerant maize varieties are well adapted to semi-arid conditions and require less water than conventional crops (Mugalavai et al., 2008). These crops are able to thrive in areas with low rainfall, ensuring food security even during periods of drought. The use of native species is also an important aspect of sustainable agriculture in semi-arid regions. Native plants are often more resilient to local environmental conditions, including soil types, water availability, and pests. By cultivating native crops and integrating indigenous knowledge into agricultural practices, farmers can enhance the resilience of their agricultural systems (Altieri & Toledo, 2011).

4.8 Conservation Agriculture

Conservation agriculture combines several practices to promote sustainable land use and minimize soil disturbance. The core principles of conservation agriculture include minimal soil disturbance, permanent soil cover, and crop rotation (Derpsch et al., 2010). These practices help to improve soil structure, enhance water retention, and reduce soil erosion, making conservation agriculture an effective strategy for semi-arid regions. Conservation agriculture has been widely promoted in semi-arid regions of Africa and Latin America, where it has helped to improve crop yields, restore degraded soils, and increase resilience to climate variability (Kassam et al., 2009). By reducing the need for intensive tillage, conservation agriculture also helps to conserve soil organic matter and sequester carbon, contributing to climate change mitigation.

4.9 Integrated Pest Management (IPM)

Integrated Pest Management (IPM) is a sustainable approach to controlling pests and diseases in agriculture by combining biological, cultural, physical, and chemical tools in a way that minimizes risks to the environment and human health (Altieri, 1999). In semi-arid regions, where pests can cause significant crop losses, IPM strategies are essential for maintaining agricultural productivity. IPM techniques include the use of natural predators, crop rotation, intercropping, and the judicious use of pesticides. By reducing the reliance on chemical pesticides, IPM helps to preserve soil health, protect biodiversity, and reduce environmental contamination (Lal, 2001). IPM is particularly important in semi-arid regions, where fragile ecosystems can be easily disrupted by overuse of agrochemicals.

4.10 Policy Interventions for Agricultural Sustainability

Governments in semi-arid regions are increasingly implementing policies to promote sustainable agricultural practices. These policies often include subsidies for drought-resistant seeds, financial incentives for farmers to adopt conservation agriculture, and investments in irrigation infrastructure. In addition, international organizations like the FAO and World Bank are working to address the challenges of food security in semi-arid regions through programs that support sustainable agriculture and rural development.

5 Ecosystem Resilience and Biodiversity Conservation

Ecosystem services in semi-arid regions are vital for supporting biodiversity, regulating climate, providing food and water, and maintaining cultural and spiritual values. These services contribute not only to the well-being of local communities but also to the global environment. However, the degradation of semi-arid ecosystems due to unsustainable land use and climate change threatens the provision of these services. To sustain the critical ecosystem services provided by semi-arid regions, it is essential to implement conservation and restoration strategies, promote sustainable land management, and integrate traditional knowledge with modern practices (Altieri & Toledo, 2011; Falkenmark & Rockström, 2004).

5.1 Role of Ecosystem Services in Semi-Arid Regions

Ecosystems in semi-arid regions provide a range of services that are essential for human well-being, including water filtration, carbon sequestration, and the maintenance of biodiversity. However, these services are often under threat due to unsustainable land use practices and climate change (Honey, 2008; Lal, 2001; Lal, 2004a, 2004b; Millennium Ecosystem Assessment, 2005; Mortimore, 2009; Mugalavai et al., 2008). Protecting and restoring ecosystems in semi-arid regions is critical for maintaining these services and ensuring the long-term sustainability of the environment.

5.2 Conservation Strategies

Conservation strategies in semi-arid regions often focus on the restoration of degraded lands and the protection of biodiversity. Reforestation and afforestation are commonly used to restore vegetation cover and prevent further desertification. Additionally, the establishment of wildlife corridors helps protect endangered species by allowing them to migrate between different habitats. In some cases, governments and NGOs have worked to create protected areas and national parks to conserve the unique biodiversity of semi-arid regions. These efforts are often supported by international agreements, such as the Convention on Biological Diversity (Otte et al., 2012; Pimentel, 2006).

5.3 Community Involvement in Conservation

Community involvement is essential for the success of conservation efforts in semi-arid regions. Indigenous communities, in particular, have valuable knowledge about sustainable land management practices that have been passed down through generations. Involving these communities in decision-making processes and conservation projects can help ensure that traditional knowledge is incorporated into modern conservation strategies. Case studies from regions such as the Sahel in Africa and Andhra Pradesh in India demonstrate the success of community-led conservation initiatives. In these cases, local communities have played a key role in restoring degraded lands, conserving water resources, and protecting biodiversity.

5.4 Ecosystem-Based Adaptation (EbA)

Ecosystem-Based Adaptation (EbA) is an approach that uses biodiversity and ecosystem services to help people adapt to the adverse effects of climate change. In semi-arid regions, EbA strategies may include restoring degraded ecosystems, protecting water resources, and promoting sustainable land management practices. EbA is increasingly being integrated into national and international policies as a cost-effective and sustainable way to address the impacts of climate change.

6 Climate Change and Semi-Arid Regions

6.1 Impacts of Climate Change on Semi-Arid Climates

Climate change is exacerbating the challenges faced by semi-arid regions. Rising temperatures, changing precipitation patterns, and increasing frequency of extreme weather events, such as droughts and heat waves, are putting additional pressure on already fragile ecosystems. As a result, many semi-arid regions are experiencing longer and more severe droughts, reduced water availability, and increased risk of desertification. These changes are having a profound impact on agriculture, water resources, and biodiversity, threatening the livelihoods of millions of people.

6.2 Adaptive Measures

To cope with the impacts of climate change, semi-arid regions must adopt adaptive measures that increase resilience and reduce vulnerability. Climate-smart agriculture (CSA), which promotes practices that are both productive and resilient to climate change, is one such approach. CSA includes the use of drought-tolerant crops, efficient water management practices, and conservation agriculture techniques. In addition, early warning systems and climate forecasting tools are helping communities in semi-arid regions prepare for extreme weather events. By providing timely information on droughts and other climate-related risks, these systems allow communities to take proactive measures to mitigate the impact of such events.

6.3 Role of International Agreements and Frameworks

International agreements and frameworks, such as the Paris Agreement and the UN Sustainable Development Goals (SDGs), play a critical role in addressing the challenges of climate change in semi-arid regions. These agreements encourage countries to adopt policies and strategies that promote climate resilience and sustainable development. In many cases, international organizations are providing technical and financial support to help semi-arid regions implement climate adaptation measures. For example, the Global Environment Facility (GEF) and the Green Climate Fund (GCF) have provided funding for projects aimed at building resilience in semi-arid regions.

6.4 Future Climate Scenarios and Risk Management

Looking ahead, climate models predict that semi-arid regions will continue to experience rising temperatures and more extreme weather events. To manage the risks associated with these changes, semi-arid regions must prioritize disaster preparedness, resilient infrastructure, and sustainable land use practices. Investing in risk management and adaptation planning will be essential to ensure that communities in semi-arid regions can withstand the impacts of climate change and continue to thrive in the face of uncertainty.

7 Future Scenarios and Innovations

7.1 Technological Innovations

Technological innovations have the potential to significantly improve the resilience of semi-arid regions to environmental challenges. Artificial intelligence (AI), machine learning, and big data are being used to predict climate-related risks, optimize water use, and improve agricultural productivity. Innovations in water purification, desalination, and irrigation technologies are helping to address water scarcity in semi-arid regions. For example, solar-powered desalination plants and drip irrigation systems are providing reliable and sustainable solutions for water management in water-stressed areas.

7.2 Policy Shifts and Global Collaboration

As the challenges of semi-arid regions become more pronounced, global collaboration and policy shifts are needed to ensure sustainable development. International organizations, governments, and NGOs must work together to develop and implement policies that address the unique challenges of semi-arid regions. Global cooperation on issues such as climate change, water scarcity, and desertification will be critical to ensuring that semi-arid regions can adapt to changing environmental conditions. In particular, sustainable development policies that prioritize water conservation, soil health, and biodiversity protection will be essential for the long-term sustainability of these regions.

7.3 Sustainable Development and Resilience Planning

Sustainable development and resilience planning must be at the forefront of efforts to address the challenges of semi-arid regions. By promoting economic diversification, renewable energy, and sustainable agriculture, semi-arid regions can build resilience to environmental changes while ensuring the well-being of local communities. Investments in renewable energy, such as solar and wind power, can provide reliable energy sources while reducing the environmental footprint of semi-arid regions. Additionally, sustainable agricultural practices that conserve water and improve soil health will be essential for ensuring food security in the face of climate change.

8 Conclusion

The challenges of semi-arid regions require a multifaceted approach that integrates water conservation, soil management, biodiversity protection, and sustainable agriculture. Traditional knowledge, combined with modern technology, plays a critical role in addressing these challenges. Governments, NGOs, and international organizations are working together to develop and implement policies that promote the sustainable management of water and land resources in semi-arid regions. Technological innovations are also providing new solutions for water management, agriculture, and conservation.

As climate change continues to affect semi-arid regions, there is a growing need for continued innovation and adaptation. Climate-smart agriculture, technological advancements, and sustainable land management practices will be essential for building resilience in these regions. Governments and international organizations must continue to invest in climate adaptation strategies, disaster preparedness, and risk management to ensure that semi-arid regions can thrive in the face of environmental challenges. The challenges of semi-arid climates are global in nature, requiring coordinated action and collaboration at all levels. Addressing the impacts of climate change, water scarcity, and desertification in semi-arid regions is essential for achieving the UN Sustainable Development Goals and ensuring the long-term sustainability of the planet. Through global cooperation, innovative technologies, and sustainable development policies, semi-arid regions can build resilience and continue to support the millions of people who depend on them.

References

Agarwal, A., & Narain, S. (1997). *Dying wisdom: Rise*. Fall and Potential of India's Traditional Water Harvesting Systems.

Altieri, M. A. (1999). The ecological role of biodiversity in agroecosystems. *Agriculture, Ecosystems & Environment, 74*(1–3), 19–31. https://doi.org/10.1016/S0167-8809(99)00028-6

Altieri, M. A., & Toledo, V. M. (2011). The agroecological revolution in Latin America: Rescuing nature, ensuring food sovereignty and empowering peasants. *Journal of Peasant Studies, 38*(3), 587–612. https://doi.org/10.1080/03066150.2011.582947

Bouwer, H. (2002). Artificial recharge of groundwater: Hydrogeology and engineering. *Hydrogeology Journal, 10*(1), 121–142. https://doi.org/10.1007/s10040-001-0182-4

Chapin, F. S., et al. (2000). Consequences of changing biodiversity. *Nature, 405*(6783), 234–242. https://doi.org/10.1038/35012241

Chen, F., Wang, J., Jin, L., & Zhang, Y. (2019). Characteristics and climatic responses of cold-season precipitation in semi-arid Central Asia. *Climate Dynamics, 52*(1–2), 815–826. https://doi.org/10.1007/s00382-018-4166-6

Critchley, W., & Siegert, K. (1991). *Water harvesting: A manual for the design and construction of water harvesting schemes for plant production*. FAO.

Cruz-Angón, A., Greenberg, R., & Harvey, C. A. (2011). Vegetation structure and avian diversity in coffee agroforestry systems of Central America. *Journal of Applied Ecology, 38*(1), 53–63. https://doi.org/10.1046/j.1365-2664.2001.00582.x

Derpsch, R., Friedrich, T., Kassam, A., & Hongwen, L. (2010). Current status of adoption of no-till farming in the world and some of its main benefits. *International Journal of Agricultural and Biological Engineering, 3*(1), 1–25.

Eswaran, H., Lal, R., & Reich, P. F. (2001). Land degradation: An overview. In *Response to land degradation* (pp. 20–35). Science Publishers.

Falkenmark, M., & Rockström, J. (2004). *Balancing water for humans and nature: The new approach in Ecohydrology*. Earthscan.

FAO. (2011). *The state of the world's land and water resources for food and agriculture*. Food and Agriculture Organization of the United Nations.

FAO. (2019). *The state of the world's biodiversity for food and agriculture*. Food and Agriculture Organization of the United Nations.

Ghassemi, F., Jakeman, A. J., & Nix, H. A. (1995). *Salinization of land and water resources: Human causes, extent, management, and case studies*. UNSW Press.

Gleick, P. H. (2003). Water use. *Annual Review of Environment and Resources, 28*(1), 275–314. https://doi.org/10.1146/annurev.energy.28.040202.122849

Global Water Partnership. (2000). *Integrated water resources management*. TAC Background Papers No. 4. https://doi.org/10.1016/j.jhydrol.2010.07.022, 12.

Honey, M. (2008). *Ecotourism and sustainable development: Who owns paradise?* Island Press.

Kassam, A., Friedrich, T., Shaxson, F., & Pretty, J. (2009). The spread of conservation agriculture: Justification, sustainability and uptake. *International Journal of Agricultural Sustainability, 7*(4), 292–320. https://doi.org/10.3763/ijas.2009.0477

Keller, J., & Keller, A. (2003). *Affordable drip irrigation for small farms in developing countries*. International Water Management Institute.

Kumar, M. D., Singh, O. P., & Sharma, S. (2008). Groundwater management in India: Physical, institutional and policy alternatives. *International Journal of Water Resources Development, 24*(2), 285–308. https://doi.org/10.1080/07900620701723761

Lal, R. (2001). Soil degradation by erosion. *Land Degradation & Development, 12*(6), 519–539. https://doi.org/10.1002/ldr.472

Lal, R. (2004a). Soil carbon sequestration impacts on global climate change and food security. *Science, 304*(5677), 1623–1627. https://doi.org/10.1126/science.1097396

Lal, R. (2004b). Soil carbon sequestration to mitigate climate change. *Geoderma, 123*(1–2), 1–22. https://doi.org/10.1016/j.geoderma.2004.01.032

Millennium Ecosystem Assessment. (2005). *Ecosystems and human Well-being: Desertification synthesis*. World Resources Institute.

Molle, F. (2008). Nirvana concepts, narratives and policy models: Insights from the water sector. *Water Alternatives, 1*(1), 131–156.

Morgan, M. Granger, R.C., William C.C., Ann F., Henry D.J., Anthony C.J., Ann P.K., Jerry M., Roger B.S., and Thomas J.W. Learning from the US national assessment of climate change impacts. (2005): 9023–9032.

Mortimore, M. (2009). *Dryland opportunities: A new paradigm for people, ecosystems, and development*. IUCN.

Mugalavai, E. M., Kipkorir, E. C., Raes, D., & Rao, M. S. (2008). Analysis of rainfall onset, cessation, and length of growing season for western Kenya. *Agricultural and Forest Meteorology, 148*(6–7), 1123–1135. https://doi.org/10.1016/j.agrformet.2008.02.013

Myers, N., et al. (2000). Biodiversity hotspots for conservation priorities. *Nature, 403*(6772), 853–858. https://doi.org/10.1038/35002501

Nair, P. K. R. (1993). *An introduction to agroforestry*. Kluwer Academic Publishers.

Nicholson, S. E. (2013). The West African Sahel: A review of recent studies on the rainfall regime and its interannual variability. *ISRN Meteorology, 2013*, 453521. https://doi.org/10.1155/2013/453521

Otte, J., et al. (2012). *Livestock sector development for poverty reduction: An economic and policy perspective*. Food and Agriculture Organization of the United Nations.

Peel, M. C., Finlayson, B. L., & McMahon, T. A. (2007a). Updated world map of the Köppen-Geiger climate classification. *Hydrology and Earth System Sciences Discussions, 4*(2), 439–473. https://doi.org/10.5194/hessd-4-439-2007

Peel, M. C., Finlayson, B. L., & McMahon, T. A. (2007b). Updated world map of the Köppen-Geiger climate classification. *Hydrology and Earth System Sciences, 11*(5), 1633–1644.

Pfenninger, S., & Staffell, I. (2016). Long-term patterns of European PV output using 30 years of validated hourly reanalysis and satellite data. *Energy, 114*, 1251–1265. https://doi.org/10.1016/j.energy.2016.08.060

Pimentel, D. (2006). Soil erosion: A food and environmental threat. *Environment, Development and Sustainability, 8*(1), 119–137. https://doi.org/10.1007/s10668-005-1262-8

Postel, S., Polak, P., Gonzales, F., & Keller, J. (2001). Drip irrigation for small farmers: A new initiative to alleviate hunger and poverty. *Water International, 26*(1), 3–13. https://doi.org/10.1080/02508060108686882

Reij, C., Tappan, G., & Smale, M. (2009). *Agroenvironmental transformation in the Sahel: Another kind of green revolution*. IFPRI Discussion Paper.

SADC. (2012). *Southern African development community regional water strategy*. Southern African Development Community.

Schwinning, S., & Sala, O. E. (2004). Hierarchy of responses to resource pulses in arid and semi-arid ecosystems. *Oecologia, 141*(2), 211–220. https://doi.org/10.1007/s00442-004-1520-8

Thornton, P. K., et al. (2009). Climate change and livestock production: A review of threats and adaptation. *Livestock Science, 130*(1–3), 47–57. https://doi.org/10.1016/j.livsci.2010.02.015

Trenberth, K. E., Dai, A., van der Schrier, G., Jones, P. D., Barichivich, J., Briffa, K. R., & Sheffield, J. (2014). Global warming and changes in drought. *Nature Climate Change, 4*(1), 17–22. https://doi.org/10.1038/nclimate2067

UN. (2015). *Transforming our world: The 2030 agenda for sustainable development*. United Nations.

UNCCD (United Nations Convention to Combat Desertification). (2017). *The global land outlook* (first ed.). United Nations. https://www.unccd.int/global-land-outlook

UNEP. (2011). *The benefits of soil carbon: Managing soils for multiple economic, societal and environmental benefits*. United Nations Environment Programme.

Vermeulen, S. J., Campbell, B. M., & Ingram, J. S. I. (2012). Climate change and food systems. *Annual Review of Environment and Resources, 37*, 195–222. https://doi.org/10.1146/annurev-environ-020411-130608

Wheeler, T., & Von Braun, J. (2013). Climate change impacts on global food security. *Science*, 341(6145), 508–513.

Irrigation Water Quality for Sustainable Agriculture Development

Ankita Begam, Kalyan Sundar Das, and Susanta Dutta

Abstract Irrigation water is a crucial resource in sustainable agriculture development, directly affecting crop growth, soil health, and overall agricultural productivity. The contents and quality of irrigation water need to be carefully managed to ensure the long-term sustainability of agricultural practices. The soluble components consisting of anions such as sulphate, chloride, bicarbonate, carbonate, and nitrate, and cations like calcium, magnesium, sodium, and potassium determine the quality of water. Poor quality of irrigation water damages the plant in several aspects, viz., water extraction from soil, showing toxicity on crops, nutrient uptake mechanism, and inhibiting the production of proteins, cell division, and cell expansion. However, using poor-quality water of higher sodium content deflocculates the soil, which indirectly hampers crop root penetration. Salinity, boron, chloride hazard, and carbonates and bicarbonates concentrations are the indicators to standardize the quality of irrigation water. Anthropogenic managements like cropping (tolerant varieties sowing), fertilizer, irrigation management, and mulching help to minimize the effect and impurities present in poor-quality irrigation water. Good water used for irrigation purposes is a non-renewable resource, mostly depleted by bad agricultural practices (BAP). Therefore, it is advisable to use sustainable practices for conserving natural resources and go for conjunctive use of water (mixing of good plus bad water) for irrigation purposes.

Keywords Crop health · Irrigation water quality · Conjunctive use of water · Saline hazard · Poor-quality water

---

A. Begam · K. S. Das (✉)
ICAR-Agricultural Technology Application Research Institute Kolkata, Bhumi Vihar Complex, Kolkata, WB, India

S. Dutta
Horticulture College, BAU, Ranchi, Jharkhand, India

A. K. Singh et al. (eds.), *Sustainable Agriculture Management in Semi-Arid Climates*, https://doi.org/10.1007/978-3-031-94062-0_2

1 Introduction

Water, the most prevalent component of the earth, is necessary for all living things, including plants, animals, and humans. It covers more than 70% of the surface of the planet. The oceans contain about 97.16% of the total water on the earth, and it is salty. Only 2.60% of fresh water is available on the earth, which is used for navigation, energy production, industrial production, and other purposes. Increase in human population, rapid urbanization and industrialization, high economic growth, intensification of agricultural production, climate change, etc. are also putting lots of pressure, which has resulted in the significant degradation and threat to groundwater quality in the recent decades. The utility of aquifers as a source of water is really diminished and destroyed by these circumstances (Vaux, 2011; Houemenou et al., 2020). Many healthy soils become saline or sodic due to a number of factors, one of which is poor water quality.

Many salts dissolve in water because it is a universal solvent. Water's salinity is determined by the properties of the parent material of the soil it moves through. Salts that are dissolved in large amounts of water tend to accumulate in the top soil layers after rising to the surface through evaporation. These salts build up over seasons and cause the soil to become saline or sodic if there is not enough rainfall to allow them to evaporate. Consequently, ensuring the long-term viability of groundwater has emerged as a significant issue. The task at hand is to safeguard these resources for future generations while simultaneously addressing the demands of various economic sectors, especially agricultural systems (Elgettafi et al., 2013; Abdel-Fattah et al., 2020). Agriculture, predominantly situated in developed nations, represents the principal consumer of hydric resources on a global scale, with an approximate 60% share. In developing countries, groundwater supplies constitute about 90% of the available resources (Aliyu et al., 2017; Velasco-Munoz et al., 2018). Not all irrigation water is pure and may contain soluble salts. Crop cultivation in dry and semi-arid regions cannot be successful without additional irrigation. Thus, in order to use these resources for irrigation, it is imperative that their quality be assessed. In the majority of irrigation scenarios, salinity levels and soluble salt components are the main issues with water quality (Zaman et al., 2018; Malakar et al., 2019; Mirzavand et al., 2020). Because soil salinity has a negative effect on agricultural sustainability and productivity, it is a significant worldwide concern. Any climate can have salinity issues, which can be caused by both natural and man-made factors. Saline soils are generally found in semi-arid and dry regions with little rainfall, insufficient to sufficiently hydrate crops and draw up mineral salts from the root zone. Historical records illustrate that numerous civilizations have collapsed as a result of elevated saline levels in agricultural fields; Mesopotamia (now Iraq) is the most well-known example of this. Humans and salinity have been associated for centuries. Because it compromises the integrity of the soil's ability to self-regulate, soil salinity can arise naturally or as a result of abuse and poor management, which lowers soil quality and weakens the base of available resources. The operation of irrigation systems, crop yield, soil physical characteristics, and fertility requirements can all be adversely affected by high saline levels (Adimalla, 2018; Ramadan et al., 2019). There are more than 100 countries worldwide where soil salinity is increasing; not a single continent is totally free of salinity. Prospective climate change scenarios are expected to result in increased soil salinization because of rising sea levels and their effects on coastal areas, as well as temperature increases that will certainly cause increased evaporation and further salinization. Salinity and sodicity are also caused by a high subsurface water table, aridity, canal seepage, inadequate drainage, backwater flow, and seawater intrusion. Therefore, maintaining the health of the soil and ensuring the production of high-quality crops rely on improving the water quality.

2 Contents of Irrigation Water

Certain salts are always dissolved in irrigation water, regardless of their source—a river, canal, tank, or well. The principal soluble components consist of anions (negatively charged ions) such as sulphate, chloride, bicarbonate, carbonate, and nitrate, and cations (positively charged ions) like calcium, magnesium, sodium, and potassium. Minor quantities of lithium, silicon, bromine, iodine, copper, nickel, cobalt, boron, fluorine, titanium, vanadium, rubidium, barium, arsenic, caesium, manganese, lead, molybdenum, chromium, selenium, phosphates, and organic materials may also be found. If plants absorb enormous amounts of boron, selenium, molybdenum, or fluorine, it can be detrimental to animals that consume these plants or drink water containing these elements. The following elements are essential for determining the irrigation water's quality and suitability: calcium, magnesium, sodium, chloride, sulphate, bicarbonate, and boron. Bicarbonate ions can precipitate calcium and magnesium under specific circumstances, which raises the sodium hazard and increases the amount of sodium. From a chemical perspective, low salinity water has higher amounts of calcium, magnesium, and bicarbonate ions, whereas very saline waters are characterized by a higher concentration of chloride and sodium ions. Remarkably, certain groundwater sources contain significant levels of nitrates. When determining if water is suitable for irrigation, other factors, including the crop being cultivated, the type of soil, its drainage capacities, and the surrounding environment, all play an equally important role. Parameters including total salinity, sodium concentration relative to other cations, boron content, and bicarbonate concentration are frequently used in the evaluation of irrigation water quality.

3 Problems with Poor-Quality Water for Irrigation

3.1 Extraction of Water

Large concentrations of soluble salts can accumulate in irrigation water near the crop's roots, making it difficult for the crop to get enough water. Additionally, the growth of roots is inhibited, which makes it more challenging for the plant to absorb water. A common term for salinity stress in plants is physiological drought. Over a wide spectrum, the growth reduction imposed by rising salinity is linear. Plants impacted by salt have smaller, darker leaves than ordinary leaves. Succulent leaves result from higher chloride ion levels. Reduced yield is caused by decreased water absorption as well as other factors. In warmer climates than in colder ones, salt causes a greater yield drop.

3.2 Soil Permeability

Soil aggregates, which give soil its structure, are built and broken down by sodium ions found in irrigation water, which has higher sodium ions than calcium and magnesium ions. Reduced aeration, waterlogging, and poor-soil tilth can result from this. Sodium has a deflocculating action, which lowers soil permeability. In addition to other issues, decreased soil permeability can result in the development of an impermeable layer that can obstruct root penetration and lower water infiltration into and through the soil. A high salt level compared to calcium and magnesium can lead to issues such as crusting of seedbeds, waterlogging, decreased oxygen, and nutrition availability to the crops.

3.3 Toxicity Symptoms

When plants absorb excessive levels of boron, chloride, salt, sulphate, and bicarbonate, toxicity issues arise. Furthermore, hazardous to plants might be irrigation water containing high concentrations of salt and chloride ions. Both direct contact with the leaves and absorption by the roots are possible routes for these ions to pass through, with direct absorption causing greater damage. Salt poisoning symptoms include scorch, leaf burn, and dead tissue along the edges of leaves. As the soil solution's osmotic pressure rises, vegetative growth decreases. Growth is reduced even in the absence of any obvious harmful signs. As salinity rises, signs of salt damage become apparent. They have waxy blooms, strong cuticles, and deeply colored leaves that are blue-green. Barley, sorghum, and field beans get leaf burnt when exposed to excessive salinity.

3.4 Nutrient Uptake

Plant roots may be harmed by high soluble salt concentrations, which will hinder their ability to absorb nutrients and water. Greater sulphate concentration increases sodium uptake while decreasing calcium uptake. Plants exposed to excessive salt levels as a result of this action develop sodium toxicity. Excessive calcium and magnesium concentrations inhibit potassium and magnesium absorption, respectively.

3.5 Anatomical and Physiological Effects

Salinity inhibits the production of proteins, cell division, and cell expansion. It promotes swelling in the mitochondria and chloroplasts and alters the structure and integrity of plant membranes. Toxic concentrations of sodium and chloride cause structural changes in protein molecules. A high chloride level prevents xylem tissue from growing.

3.6 Soil Microorganisms

The pH range of 5.5–5.8 is preferred by most microorganisms, and variations in this range can hinder their growth. Numerous investigations have shown that salinity alters the structure of microbial communities and lowers microbial activity and biomass (Andronov et al., 2012). By raising the osmotic potential of soil water, soluble salts have the ability to dehydrate cells and destroy microorganisms and roots. Additionally, low osmotic potential may make it more difficult for microorganisms and roots to draw water out of the soil. More vulnerable to high salt concentrations than ammonium-producing bacteria are those that produce nitrate and nitrite. High concentrations of salt do not seem to harm Azotobacter greatly.

4 Quality of Irrigation Water

The amount and composition of soluble salts in the water determine its quality for use in agriculture, drinking by humans and animals, and other purposes. Therefore, the sustainable use of water for irrigated agriculture depends heavily on water quality, especially in regions where salinity development is expected to be a problem. When assessing water quality for irrigation, there are four fundamental criteria to consider:

- The entire amount of soluble salt (salinity hazard).
- The ratio of sodium(Na^+) to calcium(Ca^{2+}) and magnesium (Mg^{2+}) ions is known as the sodium adsorption ratio (also known as the sodium hazard).
- Residual sodium carbonates (RSC) are defined as the concentration of bicarbonate $\left(HCO_3^{2-}\right)$ and carbonate $\left(CO_3^{2-}\right)$ anions relative to Ca^{2+} and Mg^{2+} ions.
- Excessively high elevations can result in an ionic imbalance or plant toxicity.

- Since the soil acts as a buffer to the water's pH and most crops can withstand a wide pH range, the irrigation water's pH is not a suitable criterion for measuring water quality.

4.1 Salinity Hazard

A physiological drought condition may develop as a result of excess salt increasing the osmotic pressure of the soil solution. The plants will wilt even though it looks like there is plenty of moisture in the field's soil. This happens because soil has a high osmotic potential, which prevents plant roots from absorbing it. Wilting results from the inability of the plant to replace the water lost through transpiration from the plant shoot. Finding the true salt content and measuring irrigation water's electrical conductivity (EC), which is represented as milli Siemens per meter (mSm^{-1}), are two methods to quantify the total soluble salts (TSS) concentration (Table 1).

4.2 Sodium Hazard

To express the sodium hazard of irrigation water, utilize the 'sodium adsorption ratio (SAR)'. The main concern associated with a high sodium concentration is its negative effect on the physical properties of soil, leading to the destruction of the soil structure. Additionally, sodium contributes to the total saltiness of the soil and might pose a risk to delicate crops such as fruit trees. This indicates that both the SAR value and the proportion of sodium ions will simultaneously rise in comparison to each other. Consequently, the sodium threat in soil-water will escalate, surpassing the indicated SAR value. Thus, it is advisable to refrain from using water with an SAR value more than 26 if it will be the only irrigation source available for a considerable amount of time (Table 2).

Table 1 Determination of irrigation water quality based on the EC content

Kinds of water	EC (mSm^{-1})	Salt Concentration (gl^{-1})	Remarks
Low Salinity	0-25	< 0.16	Safely use
Medium Salinity	25-75	0.16-0.50	Use with moderate leaching
High Salinity	75-225	0.50-1.50	Not in use for Irrigation Purposes
Very High Salinity	225-500	1.50-3.0	

Table 2 Determination of irrigation water quality based on the SAR value

Kinds of water	SAR value	Remarks
S1: Low sodium level	0-10	No hazard
S2: Medium sodium level	10-18	Appreciable hazard, use with appropriate management
S3: High sodium level	18-26	Unsatisfactory for most of the crops
S4: Very High sodium level	>26	

Table 3 Determination of irrigation water quality based on the RSC value

Water quality	RSC value (mel^{-1})
Water can be used safely	<1.25
Water can be used with certain management	1.25–2.5
Inappropriate for irrigation purposes	>2.5

4.3 *Carbonates and Bicarbonates Concentration*

When the soil solution concentrates through evapotranspiration, calcium carbonate ($CaCO_3$), and magnesium carbonate ($MgCO_3$) will frequently precipitate from fluids with high carbonate $\left(CO_3^{2-}\right)$ and bicarbonate (HCO_3^-) contents. Consequently, this will elevate the sodium hazard in soil-water beyond the amount indicated by the RSC value (Table 3).

4.4 *Boron Hazard*

Despite being not essential in huge amounts, all plants require boron for proper development. Boron may cause damage if it exceeds an acceptable level that differs based on the kind of crop. The range of boron toxicity to deficiency is narrow for most crops. In order to ensure an adequate boron supply for the plant, it may be essential to have a minimum of 0.02 parts per million (ppm) of boron in the irrigation water. It is advisable to maintain boron levels in irrigation water below 0.3 ppm in order to prevent toxicity. The targeted crop type's boron tolerance will probably need to be assessed prior to using higher boron concentrations. While boron toxicity is generally not an issue, it can be a significant irrigation water quality factor. It is worth noting that plants raised in high-lime soils may be more resistant to boron concentrations than those grown in non-calcareous soils (Table 4).

Table 4 Determination of irrigation water quality based on boron concentration

Boron concentration (ppm)	Effect on crops
<0.5	Acceptable for irrigation of all crops
0.5–1.0	Acceptable for irrigation of most crops
1.0–2.0	Acceptable for semi-tolerant crops
2.0–4.0	Acceptable for tolerant crops only

4.5 Chloride Hazard

The most significant cause of crop toxicity is irrigation water containing chlorides. Every water contains the anion chloride (Cl); chlorides are soluble and easily absorbed by drainage waters. Although they can impede plant growth in high quantities and be extremely harmful to certain plant species, chlorides are essential for plant growth. Therefore, while evaluating the quality of water, the concentration of Cl must be assessed. Symptoms in vulnerable crops occur when Cl^- levels (0.3–1.0% dry weight basis) build up in the leaves. Cl toxicity, a highly common sign of chloride toxicity, begins at the leaf tips of plants, according to Ayers and Westcot (1985). As the toxic effect becomes more severe, the toxicity moves from the leaf tip back along the edges. A plant may experience early leaf drop or even complete defoliation in conjunction with excessive necrosis.

5 Significance of Poor-Quality Water for Mitigating the Impacts of Climate Change

Climate change is a process characterized by increased temperatures and the proportion of greenhouse gases (GHGs). It is responsible for sudden changes in drought and heavy precipitation, which are significant concerns in Indian agriculture (Dotaniya et al., 2016). The challenges encompass not only the management of scarce and depleting natural resources such as soil and water but also the mitigation of adverse effects on soil, crops, animals, and other related areas. Agriculture may act as a vital means of lowering greenhouse gas emissions and enhancing environmental quality (Kushwah et al., 2014). Farmers in dry or semi-arid locations are utilizing low-quality water for irrigating crops. The cultivation of vegetables in peri-urban regions, often using sewage or a combination of fresh water, is a widespread practice. These low-quality water resources are also utilized for the cultivation of forestry plants through the implementation of several agricultural engineering and agronomic interventions. Saline and alkaline water are utilized for cultivating a variety of agricultural crops and related activities in arid and semi-arid regions. The imperative to mitigate the impact of climate change necessitates the safe utilization of low-quality water in water-scarce regions for agricultural purposes.

6 Management Practices of Poor-Quality Irrigation Water

6.1 Choice of Crops/Tolerant Crops

Proper choice of crops and cropping sequences for saline irrigated soils is a crucial management practice aimed at reducing the impact of low-quality water on crop cultivation. This is because different crops have varying levels of tolerance to salinity, with some being highly sensitive while others are semi-tolerant or fully tolerant to specific levels of salinity. The choice of the initial crop will therefore be determined by the level of soil salinity once the fundamental reclamation procedures have been put into effect. Therefore, crops that exhibit tolerance and resilience to salinity are favored during the initial stages of land restoration. Adopting proper agricultural techniques and cultivating appropriate crops facilitate the removal of salts from the soil, while the crop cover inhibits the reoccurrence of salt buildup caused by evaporation. In the dry and semi-arid regions of India, where the soil is irrigated with saline water, crops such as cotton, sorghum, pearl millet, cluster bean, and moth bean can be cultivated during the *kharif* season. In the *rabi* season, crops like wheat, barley, mustard, safflower, and sugarcane can be cultivated, based on the water availability. A study conducted at CSSRI investigated the relationship between crop productivity and soil salinity under field circumstances. The study determined that the threshold tolerated values of soil salinity for wheat, barley, and mustard are 4, 7, and 6 dS/m, respectively. Upon resolving the salinity issue in the impacted region, the production of wheat, barley, and mustard can be augmented by 28%, 30%, and 13%, respectively. Additionally, it was documented that pearl millet and cotton have the ability to thrive on salty soils with electrical conductivity (ECe) levels of 6.5 and 7.5 dS/m, respectively, without suffering a substantial reduction in the yield of crops. The cropping sequence strategy involves cultivating cereal crops in one season and pulse crops in the next season, such as pearl millet-gram or mungbean-wheat. The optimal cropping sequences for saline soils include pearl millet-barley, pearl millet-wheat, pearl millet-mustard, sorghum-wheat/barley, sorghum-mustard, cluster bean-wheat/barley, and cotton-wheat/barley, which are more economically profitable. Mustard might be a suitable alternative to wheat in saline locations because of its lower water requirement in the cropping sequence. The growth of pulse crops and crop production were ultimately lowered due to the presence of different cations and anions in the saline water used for irrigation. Chloride ions exhibit more toxicity compared to sulphate ions due to the ability of sulphate ion-rich water to mitigate the detrimental effects of chloride (Dotaniya et al., 2019). When growing crops using salinized water is not only impractical but also expensive, this water can be utilized to cultivate trees, particularly on degraded soil. The recommended species include *A. indica*, *A. nilotica*, *A. tortilis*, *A. farnesiana*, *C. siamea*, *E. teretecornis*, *F. limonia*, *P. juliflora*, *P. cineraria*, *Pithecellobium dulce*, *S. persica*, *S. oleoides*, and others. Isabgol, Aloe, and Kalmeg are medicinal plants that have shown potential as alternative crops in saline irrigation situations. The growth phases of crops are not equally effective when exposed to poor-quality water, such as saline, alkaline, or water contaminated with heavy metals. Early developmental stages exhibit more sensitivity compared to later stages. The addition of chromium (Cr) at a concentration of 40 mg kg^{-1} soil resulted in a decrease in the germination rate of pigeon pea (Dotaniya et al., 2014). Consequently, it is not advisable to use low-quality water during the germination stage of crops (Table 5).

6.2 Method of Sowing

Modifying sowing methods can be done to get a more advantageous distribution of salt with respect to the location of seeds or growing roots. The germination of seeds is negatively affected by higher levels of salinity. Due to the tendency of salt to build on the ridges in furrow irrigation, using the furrow planting method may improve crop growth and yield under conditions that are saline. The optimal location for seed placement is either on the side or at the base of the ridge, but not on the top. Furthermore, it is recommended to plant the seeds in rows at the ideal depth using seed drills operated by either bullocks or tractors. It is advisable to use a seed-cum-fertilizer drill in order to achieve consistent and precise placement of seeds and fertilizers at the ideal depth, which promotes the early growth of seedlings. If a seed drill is unavailable, sowing should be conducted using either the kera or pora method.

6.3 Fertilizer Application

Crop nutrient uptake patterns were impacted by low-quality water, which also decreased soil fertility. Under saline irrigation, the response to supplied nitrogen is somewhat diminished. Therefore, although they do not significantly alter salinity

Table 5 Comparative salinity tolerance of crops

Sensitive	Moderately sensitive	Moderately tolerant	Tolerant
Apple	Cabbage	Safflower	Barley
Carrot	Cowpea	Soybean	Cotton
Lemon	Grape	Wheat	Date palm
Ladies finger	Groundnut		Sugar beet
Onion	Maize		
Orange	Potato		
Pulses	Rice		
	Sugarcane		
	Tomato		
	Radish		
	Sesbania		
	Sorghum		

tolerance, higher dosages of nitrogenous fertilizers are advised to make up for volatilization losses. Crop development was improved by adding soil organic carbon (SOC) to fields of crops by external application of farmyard manure (FYM) (Dotaniya, 2013). Through the incorporation of crop leftovers, organic matter was added, which improved the health of the soil and decreased salinity and trace metal toxicity in crops (Meena et al., 2022). When FYM was applied to low-quality irrigated soils, it improved the various forms of organic acids and regulated the rhizospheric microclimate, which improved plant nutrient uptake (Dotaniya and Meena, 2013; Dotaniya et al., 2013). There is no need to use phosphatic fertilizers in excess for sulphate-rich water; the prescribed dosage under ordinary circumstances can be followed. Soils that are irrigated with water containing high levels of chloride show a greater response to increased phosphate application. This happens because the chloride ions decrease the availability of soil phosphorus to plants, and the crop's demand for phosphoric fertilizers is consequently increased by around 50%. Not only that applying biofertilizers along with compost can decrease salinity and mitigate the toxicity of heavy metals in crop production systems that are irrigated with wastewater (Sarkar and Chourasia, 2017).

6.4 Soil Amendments

The impact of using low-quality water for agricultural cultivation affects groundwater, and the crop is influenced by the soil type and texture. Soils with high permeability are more likely to cause groundwater contamination in locations where sewage is used for irrigation. The presence of lower levels of organic matter in sandy or arid locations results in a reduced ability of the soil to bind metals, causing them to be washed down to the lower layers of the soil. The utilization of water with high sodium content, which is of low quality, resulted in a decrease in both soil fertility and soil productivity. Prolonged use of wastewater with high sodium concentration in crop cultivation leads to soil alkalinity. Gypsum application in a cultivated field mitigated the alkalinity effect, while the use of lime in a field watered with lower pH effluent enhanced the germination rate of the crop. If the soil is of high quality but the water is of low quality, it is recommended to apply gypsum to the water. The gypsum utilized for agricultural purposes should have a minimum purity level of 65%. Gypsum is a feasible option for achieving a favorable calcium to sodium (Ca : Na) or calcium to magnesium (Ca : Mg) ratio in irrigated water due to its low solubility (0.25–0.30%) and cost. The Ca : Na ratio or SAR can be improved by raising the concentration of calcium ions, reducing the Mg : Ca ratio, and causing excessive carbonate ions to precipitate.

$$2\mathrm{NaHCO_3} + \mathrm{CaSO_4} \rightarrow \mathrm{Ca}\left(\mathrm{HCO_3}\right)_2 + \mathrm{Na_2SO_4}$$

$$\mathrm{Ca}\left(\mathrm{HCO_3}\right)_2 \rightarrow \mathrm{CaCO_3} + \mathrm{H_2O} + \mathrm{CO_2}$$

6.5 Irrigation and Leaching Management

With each irrigation using saline water, a specific quantity of salts is added to the soil. Over time, these salts can build up in the root zone to harmful levels, leading to a decrease in crop yields if no leaching occurs. The leaching requirement refers to the proportion of irrigation water that needs to be flushed through the crop's root zone in order to maintain the soil salinity below a predetermined threshold. Leaching is the process of supplying an adequate amount of water to induce drainage from the zone where the roots are located. Salts are transported away from the root zone through the drainage water.

$$\mathrm{LF} = \frac{D_{\mathrm{d}}}{D_{\mathrm{i}}} = \frac{\mathrm{EC}_{\mathrm{i}}}{\mathrm{EC}_{\mathrm{d}}}$$

where LF = leaching fraction. EC_{i} = electrical conductivity of water entering the root zone (dS/m). EC_{d} = electrical conductivity of water draining from the root zone (dS/m). D_{d} = depth of water draining from the root zone (mm). D_{i} = depth of water entering the root zone (mm).

However, excessive salt buildup in the root zone can be avoided with appropriate irrigation and leaching techniques. When using high-quality water, the leaching needs might range from 15% over the water requirement to up to 70% above the water requirement when using relatively saline water and growing crops that are especially sensitive to salts. In order to leach the salts, heavy irrigation may be applied at the start of the season. Normal irrigation rates are then applied after that. An alternative method involves bundling the fields and collecting rainwater in a reservoir at the onset of the season. To achieve the needs for leaching, an additional 15–20% of water would need to be used for irrigation in arid locations. Depending on the irrigation technique, soils have different distributions of salts and water. A move toward drip and sprinkler systems, which allow for more precise control over the distribution of water and salt, has the potential to improve the efficiency of saline water use, particularly for crops of high value. Applying saline water through sprinklers before the appearance of plants helps maintain low levels of soluble salt in the soil where seeds are planted, which improves the establishment of the crop.

6.6 Mulching

Mulching is a suitable method to increase the effectiveness of irrigation water usage while also enhancing crop yields. Agricultural sustainability in dryland regions is threatened by the inadequate utilization of scarce water resources, as well as by drought or heat stress during crop cultivation. (Siddique et al., 2012). Climate change induces pronounced soil aridity, resulting in inadequate soil moisture for agricultural cultivation (Mohamed & Akladious 2014; Shalaby et al., 2022). Mulching is a widely used technique that involves the application of various materials to the field either before, during, or shortly after sowing. The purpose of mulching is to provide support and coverage to the soil surface. Examples of materials used for mulching include plastic, agricultural wastes, livestock manure, sand, and cement (El-Beltagi et al., 2022). Mulching reduces salt-related challenges by reducing water evaporation and erosion (Gan et al., 2013), increasing soil temperature, enhancing soil water retention capacity (Zohu et al., 2011), and controlling weed growth. Mulching enhances crop productivity, promotes plant development, and decreases water consumption. (Yu et al., 2018). Mulches facilitate the leaching of salts into lower layers, even in the presence of rainfall. Mulching can potentially address water constraints by implementing deficit irrigation (DI). DI subjects the crop to a certain level of water stress, however, it can nevertheless lead to substantial water conservation.

6.7 *Conjunctive Use of Water*

Water availability is an essential component for several activities, viz., agriculture, industries, and civilization. However, it is depleted day by day, although saline water availability is not in a deficit condition. So, scientists showed their interest in the conjunctive use of water, mixing of good water and bad water for irrigation purposes only. It can act as a sustainable option for using such action and save from any disruption in agriculture or increase the utilizable water.

7 Conclusion

The quality of irrigation water in agricultural production plays an important role in plant's proper growth, nutrition, production, and reproduction. Good-quality irrigation water can optimize the production potential of different crops. On the other hand, poor quality results in various problems like stunted growth, malnutrition, toxicity, low production, and poor grain quality in agricultural production, which indirectly causes human health hazards. Therefore, it is very essential to check the water quality before its use in agricultural production, for which some sustainable management practices like mulching, irrigation and leaching management, and conjunctive use of water can be used to improve irrigation water quality.

References

Abdel-Fattah, M. K., Abd-Elmabod, S. K., Aldosari, A. A., Elrys, A. S., & Mohamed, E. S. (2020). Multivariate analysis for assessing irrigation water quality: A case study of the Bahr Mouise Canal, Eastern Nile Delta. *Water, 12*(9), 2537.

Adimalla, N. (2018). Groundwater quality for drinking and irrigation purposes and potential health risks assessment: A case study from semiarid region of South India. *Exposure and Health Journal, 11*(2), 109–123.
Aliyu, T. H., Balogun, O. S., Namani, C. S., Olatinwo, L. K., & Aliyu, A. S. (2017). Assessment of the presence of metals and quality of water used for irrigation in Kwara State, Nigeria. *Pollution Journal, 3*(3), 461–470.
Andronov, E. E., Petrova, S. N., Pinaev, A. G., Pershina, E. V., Rakhimgalieva, S. Z., Akhmedenov, K. M., Gorobets, A. V., & Sergaliev, N. K. (2012). Analysis of the structure of microbial community in soils with different degrees of salinization using T-RFLP and real-time PCR techniques. *Eurasian Soil Science, 45*, 147–156.
Ayers, R. S., & Westcot, D. W. (1985). *Water quality for agriculture* (Vol. 29, p. 174). Food and agriculture organization of the United Nations.
Dotaniya, M. L. (2013). Impact of various crop residue management practices on nutrient uptake by rice-wheat cropping system. *Current Advances in Agricultural Sciences, 5*(2), 269–271.
Dotaniya, M. L., & Meena, V. D. (2013). Rhizosphere effect on nutrient availability in soil and its uptake by plants—A review. *Proceedings of the National Academy of Sciences, India Section B: Biological Sciences, 85*(1), 1–12.
Dotaniya, M. L., Meena, V. D., & Das, H. (2014). Chromium toxicity on seed germination, root elongation and coleoptile growth of pigeon pea (*Cajanus cajan*). *Legume Research, 37*(2), 227–229.
Dotaniya, M. L., Prasad, D., Meena, H. M., Jajoria, D. K., Narolia, G. P., Pingoliya, K. K., Meena, O. P., Kumar, K., Meena, B. P., Ram, A., Das, H., Chari, M. S., & Pal, S. (2013). Influence of phytosiderophore on iron and zinc uptake and rhizospheric microbial activity. *African Journal of Microbiology Research, 7*(51), 5781–5788.
Dotaniya, M. L., Rajendiran, S., Meena, B. P., Meena, A. L., Meena, B. L., Jat, R. L., & Saha, J. K. (2016). Elevated carbon dioxide (CO2) and temperature vis-a-vis carbon sequestration potential of global terrestrial ecosystem. In Bisht et al. (Eds.), *Conservation agriculture: An approach to combat climate change in Indian Himalaya* (pp. 225–256). Springer.
Dotaniya, M. L., Saha, J. K., Rajendiran, S., Coumar, M. V., Meena, V. D., Das, H., Kumar, A., & Patra, A. K. (2019). Reducing chromium uptake through application of calcium and sodium in spinach. *Environmental Monitoring and Assessment, 191*(12), 754.
El-Beltagi, H. S., Basit, A., Mohamed, H. I., Ali, I., Ullah, S., Kamel, E. A., Shalaby, T. A., Ramadan, K. M., Alkhateeb, A. A., & Ghazzawy, H. S. (2022). Mulching as a sustainable water and soil saving practice in agriculture: A review. *Agronomy, 12*(8), 1881.
Elgettafi, M., Elmandour, A., Himi, M., & Casas, A. (2013). The use of environmental markers to identify groundwater salinization sources in a Neogene basin, kert aquifer case, NE Morocco. *International Journal of Environmental Sciences and Technology, 10*, 719–728.
Gan, Y., Siddique, K. H., Turner, N. C., Li, X. G., Niu, J. Y., Yang, C., Liu, L., & Chai, Q. (2013). Ridge-Furrow mulching systems-an innovative technique for boosting crop productivity in semiarid rain-fed environments. *Advances in Agronomy, 118*, 429–476.
Houemenou, H., Tweed, S., Dobigny, G., Mama, D., Alassane, A., Slimer, R., Babic, M., Ruy, S., Chaigneau, A., Gauthier, P., Socochou, A., Dossou, H. J., Badou, S., & Leblanc, M. (2020). Degradation of groundwater quality in expanding cities in West Africa: a case study of the undergulated shallow aquifer in Cotonou. *Journal of Hydrology, 582*, 124438.
Kushwah, S. K., Dotaniya, M. L., Upadhyay, A. K., Rajendiran, S., Coumar, M. V., Kundu, S., & Subba Rao, A. (2014). Assessing carbon and nitrogen partition in kharif crops for their carbon sequestration potential. *National Academy Science Letters, 37*, 213–217.
Malakar, A., Snow, D., & Ray, C. (2019). Irrigation water quality-a contemporary perspective. *Water Journal, 11*(7), 1482.
Meena, M. K., Yadav, B. L., Dotaniya, M. L., & Meena, M. D. (2022). Can addition of organic manures mediated sodicity toxicity in mustard cultivation? *Communications in Soil Science and Plant Analysis, 53*(1), 77–88.

Mirzavand, M., Ghasemieh, H., Sadatinejad, S. J., & Bagheri, R. (2020). An overview on source, mechanism and investigation approaches in groundwater salinization studies. *International Journal of Environmental Sciences and Technology, 17*, 2463–2476.

Mohamed, H. I., & Akladious, S. A. (2014). Influence of garlic extract on enzymatic and non-enzymatic antioxidants in soybean plants (*Glycine max*) grown under drought stress. *Life Science Journal, 11*(3s), 46–58.

Ramadan, E. M., Fahmy, M. R., Nosair, A. M., & Badr, A. M. (2019). Using geographic information system (GIS) modeling in evaluation of canals water quality in Sharkia Governorate, East Nile Delta, Egypt. *Modeling Earth and Systems Environment, 5*, 1925–1939.

Sarkar, P., & Chourasia, R. (2017). Bioconversion of organic solid wastes into biofortified compost using a microbial consortium. *International Journal of Recycling of Organic Waste in Agriculture, 6*(4), 321–334.

Shalaby, T. A., Taha, N. A., Taher, D. I., Metwaly, M. M., El-Beltagi, H. S., Rezk, A. A., El-Ganainy, S. M., Shehata, W. F., El-Ramady, H. R., & Bayoumi, Y. A. (2022). Paclobutrazol improves the quality of tomato seedlings to be resistant to Alternaria solani Blight disease, Biochemical and histological perspectives. *Plants, 11*(3), 425.

Siddique, K. H., Johansen, C., Turner, N. C., Marie-Helene, J. M. H., Hashem, A., Sakar, D., Gan, Y., & Alghamdi, S. S. (2012). Innovations in agronomy for food legumes—A review. *Agronomy for Sustainable Development, 32*, 45–64.

Vaux, H. (2011). Groundwater under stress: the importance of management. *Environmental Earth Sciences Journal, 62*, 19–23.

Velasco-Munoz, J., Aznar-Sanchez, J., Belmonte-Urena, L., & Roman-Sanchez, M. (2018). Sustainability water use in agriculture: a review of worldwide research. *Sustainability, 10*(4), 1084.

Yu, Y. Y., Turner, N. C., Gong, Y. H., Li, F. M., Fang, C., Ge, L. J., & Ye, J. S. (2018). Benefits and limitations to straw- and plastic-film mulch on maize yield and water use efficiency, a meta-analysis across hydrothermal gradients. *European Journal of Agronomy, 99*, 138–147.

Zaman, M., Shabbir, A., Shahid, A., & Heng, L. (2018). *Irrigation water quality. Guideline for salinity assessment, mitigation and adaptation using nuclear and related techniques* (pp. 113–133). Springer Open.

Zhou, J. B., Wang, C. Y., Zhang, H., Dong, F., Zheng, X. F., Gale, W., & Li, S. X. (2011). Effect of water saving management practices and nitrogen fertilizer rate on crop yield and water use efficiency in a winter wheat-summer maize cropping system. *Field Crops Research, 122*(2), 157–163.

A Climate-Resilient Agriculture Technique for Semi-Arid Regions

Hirdayesh Anuragi, Rajesh Kumar Singhal, Ashok Yadav, Asha Ram, Naresh Kumar, Arun Kumar Handa, and Ayyanadar Arunachalam

Abstract Agroforestry is gaining global recognition as an effective, sustainable and climate-smart means of increasing farm productivity while also playing a significant role in conserving biodiversity, restoring ecosystems and providing plentiful of overlooked environmental services. Under the ongoing scenario of changing climate in the twenty-first century, where rainfall and temperature fluctuations severely impact the crop yields, food security and livelihood of farming communities, the role of agroforestry is becoming an increasingly essential and viable option for sustained production, climate change adaptation and mitigation. Developing and populated countries essentially need to adopt various climate-smart agroforestry (CSA) practices, such as alley cropping, agro-silvopastures, agro-horticulture, home gardening, agroforestry parklands, etc., to achieve triple goal of sustainable increase in the farm productivity, resilience and mitigating climate change for healthier food, environment and livelihood in the future. Interestingly, agroforestry alone has the potential to achieve at least nine out of the 17 SDGs through sustainable food production, ecosystem services and economic benefits. This chapter comprises different concepts of agroforestry to smartly deal with various challenges in agriculture under the scenario of global warming and climate change to sustainably meet the increasing food demand of a fast ever-growing global population.

Keywords Agroforestry · Climate change · Climate-smart technique · Adaptation · Mitigation

---

H. Anuragi (✉) · A. Yadav · A. Ram · N. Kumar · A. K. Handa · A. Arunachalam
ICAR-Central Agroforestry Research Institute, Jhansi, Uttar Pradesh, India
e-mail: hirdayesh.anuragi@icar.gov.in

R. K. Singhal
ICAR-Indian Grassland and Fodder Research Institute, Jhansi, Uttar Pradesh, India
e-mail: hirdayesh.anuragi@icar.gov.in

A. K. Singh et al. (eds.), *Sustainable Agriculture Management in Semi-Arid Climates*, https://doi.org/10.1007/978-3-031-94062-0_3

1 Introduction

Agriculture is a primary source of food, nutrition and livelihood security for the growing population, and thus acts as the backbone of the Indian economy. Meeting the rising demand of more food of higher quality by a fast ever-growing global human population, which is projected to cross 9 billion by next 30 years, is a serious challenge particularly when agriculture is already facing serious issues like high chemical inputs, erosion soil and soil quality deterioration, volatile weather, and extreme events like floods and drought, pests and disease resurgence and biodiversity loss (Arora, 2019; Malhi et al., 2021). Addressing this increasing food demand in a sustained way is a serious challenge before the scientific fraternity and policymakers. Moreover, despite having high-yielding varieties and other scientific interventions in agriculture, which have certainly boosted the food production potential of crops in the last few decades, the prevailing global climatic stress has set forth serious apprehensions in moving ahead with the same traditional farming strategies. Further, the agricultural sector is a significant contributor of emissions of greenhouse gas (GHGs) like carbon dioxide, nitrous oxide and methane, primarily through synthetic fertilizers, pesticides, irrigation systems and livestock farming, which immensely contribute towards global warming and climate change. This gentle balance between agricultural practices and the escalating climate challenges necessitates the adoption of innovative solutions that align with nature. Integrated sustainable development strategy like CSA offers a nature-based solution for sustained production and ecological distress adaptation and mitigation (Zerihun, 2020).

Agriculture is highly dependent on climatic parameters, and the changing climate has directly or indirectly affected agricultural farming practices and poses a threat to ensuring food and livelihood. Climatic factors like rising global temperatures and shifting weather patterns frequently result in reduced crop yields, which is projected to worsen in the future. Intergovernmental Panel on Climate Change (IPCC) reports clearly mentioned the temperature rise by 1.5 to 2 °C in the next 1–2 decades (IPCC, 2014; Masson-Delmotte et al., 2022). Without any adaptation measures, rainfed rice yields in India are anticipated to fall by 20% in 2050 and 47% in 2080, while irrigated rice yields are projected to fall by 3.5% in 2050 and 5% in 2080 scenarios, as per the latest future projection reports. Similarly, for wheat, climate change is expected to cut its yield by 19.3% in 2050 and 40% in 2080, with a significant regional and temporal inconsistency. Another important crop, Kharif maize, might also suffer yield losses by 18 and 23% in the 2050 and 2080 scenarios, respectively (Singh et al., 2020). In addition to the temperature fluctuation and water crisis, continuous diminishing of the soil physical and bio-chemical quality is another growing challenge that has already started affecting the crop yield and food quality. The possible attributes could be monocropping and high-input agricultural practices. Likewise, deterioration of nature, particularly a significant loss in the biological diversity and ecosystems, may bring the whole humanity to a stage of risk that, if not acted upon on a sustainable basis, would certainly affect the whole scenario of the farming system and the food chain. Likewise, degradation affects around 30% of the world's land, with 2.2 billion hectares of degraded land considered potentially vulnerable to restoration. Adopting a sustainable and climate-smart approach like agroforestry is now being recognized as a potential solution for sustained agricultural production, climate change adaptation and mitigation; therefore, farmers in different fields in various countries are increasingly opting for agroforestry adoption.

2 Agroforestry: A General Description

Agroforestry is basically an age-old practice of purposefully combining the perennial tree species with annual crops and livestock animals on the same piece of land for generating environmental, economic and social benefits to smallholder farmers. Integrating trees with crops and cattle, agroforestry systems can maintain production levels during both wetter and drier years, mitigate the harmful effects of wind and water flow, and create alternative income streams for farmers (Montagnini and Nair, 2004). This effective land use pattern effectively diversifies the production systems, increases farm profitability and enhances the overall sustainability and resilience of that particular ecosystem. Interestingly, agroforestry alone has the potential to meet the nine out of the total seventeen Sustainable Development Goals (SDGs) defined by the United Nations (van Noordwijk et al., 2018) as shown in the Fig 1. According to the CAFRI Jhansi report, the 28.74 million hectares (MHa) of land under agroforestry in Jhansi is a significant area.

Depending upon the nature of components being used in an agroforestry model, it can be referred to as the *agri-silvi-cultural system* that combines perennial trees with other components like annual or perennial crops on the same piece of land, in order to meet the grain and wood requirements; the *silvo-pastoral system* that associates the forage and livestock production along with the perennial trees or shrubs and the *agri-silvo-pastoral system* that combines perennial trees in the form of multifunctional hedgerows, woodlots or fodder trees with annual crops and livestock on the same land. Other systems of agroforestry include combinations of perennial trees/shrubs providing habitat or fodder, with fish ponds or bees in the same habitat. Similarly, agroforestry, depending upon the arrangement of different components, can be classified into spatial (Mixed, Strip, Boundary) and temporal (Sequential, Simultaneous, Overlapping) for harnessing the maximum possible productivity from a piece of land. Likewise, agroforestry can be classified based on the function, i.e. productive (Food, Fodder, Fuel, Fibre, Fertilizer) and protective (Windbreaks, Shelterbelts, Soil conservation, Soil restoration, Shade) in order to obtain a particular function. A brief flowchart is depicted in Fig. 2 to understand different aspects of agroforestry.

As agroforestry involves perennial tree components, the system has the advantage of adapting the fluctuation in climatic events such as flooding, drought, and high winds, soil degradation rates and water scarcity, and hence the system as a whole provides the basis for sustainability in long term which is otherwise not

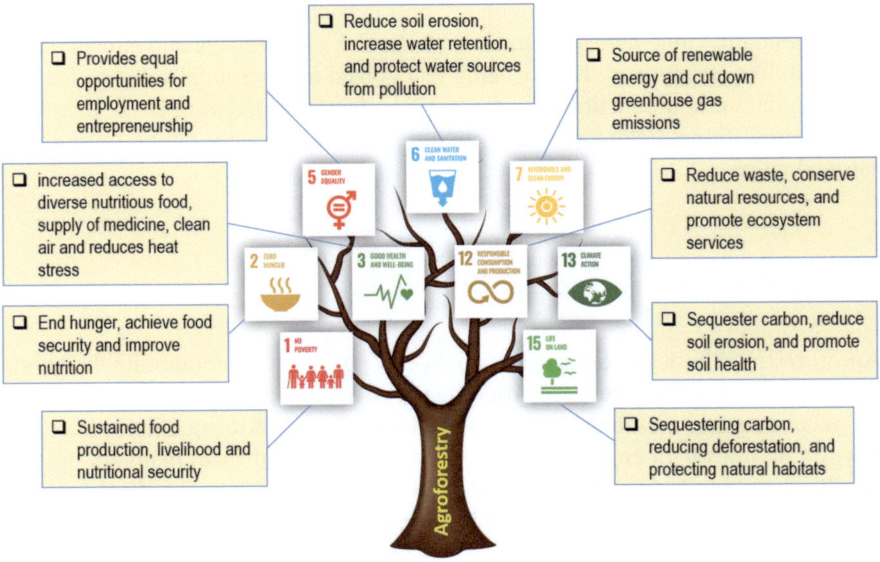

Fig. 1 Nine out of 17 SDGs can be accomplished through agroforestry

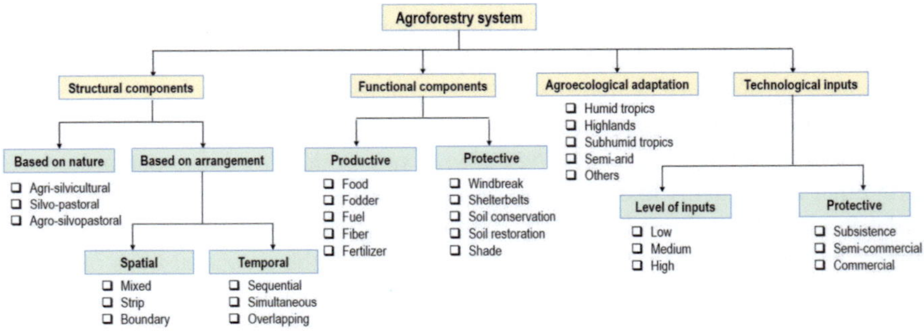

Fig. 2 Broader classifications of agroforestry systems

possible in agriculture alone. This not only offers stability in productivity but also assures the food, nutritional, livelihood and environmental security. Further, the global warming owing to high GHG emissions has enhanced the ambient temperature which directly alters the pest and disease resurgence and weed distributions and subsequently the crop yield potentials. The agroforestry system influences the pest occurrence and abundance through enhanced top-down regulation by natural enemies as well as bottom-up factors such as microclimate, soil nutrients and moisture content management. Further, the high-input agriculture deteriorates the soil biological and physico-chemical properties and the ecosystem, which significantly reduces the life of the system and the quality of the produce. However, agroforestry promotes ecological balance while reducing the usage of chemical inputs. Perennial trees create a diversified ecosystem that suppresses weeds, decreases soil erosion

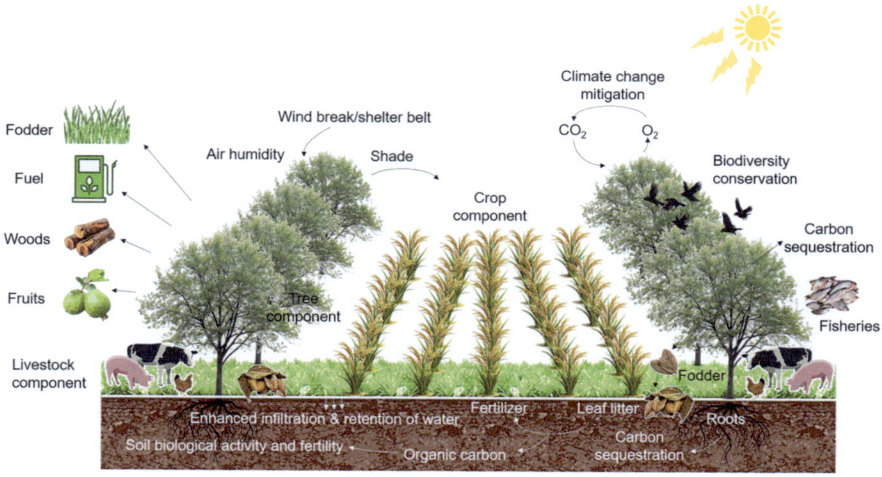

Fig. 3 A holistic representation of potential services of an agroforestry system

and offers crop shade that can greatly reduce the need for chemical pesticides, fertilizers and herbicides, resulting in a healthier and more resilient agroecosystem. Agroforestry also increases biodiversity, soil health and ecosystem services, resulting in a more sustainable and environmentally friendly farming system (Fig. 3).

3 Is Agroforestry Really a Climate-Smart Tool

Agriculture practices are mostly dependent on the climatic factors, and changing climate is drastically affecting the crop yield potentials and deteriorating the whole ecosystem through rising temperatures, altered precipitation patterns and extreme weather events, leading to habitat loss, changes in species distributions and decreased biodiversity. Additionally, climate change can exacerbate factors such as pollution, habitat fragmentation and over-exploitation of natural resources, further impacting ecosystem health. These changes have profound implications for the balance of ecosystems and the services they provide to both wildlife and humans. On the contrary, agroforestry has been considered climate smart for several reasons, such as it helps to sequester carbon, mitigates climate change and creates more resilient farming systems by providing shade and windbreaks. Additionally, it supports biodiversity, improves soil health and often leads to more sustainable land management. Further, the ecosystem rehabilitation potential of agroforestry also offers an added advantage to restore the degraded lands and bring them back under cultivation. These combined factors make agroforestry a nature-friendly and climate-resilient agricultural practice. Therefore, under all sorts of climatic adversities, an agroforestry system is always expected to provide higher productivity and climate resilience as compared to an agricultural system. This not only protects the yield losses during odd times but also assures income and livelihood to the farmers

(Fig. 3). A comparative analysis of different aspects of agriculture and agroforestry is briefed in the Table 1.

4 Advantages of an Agroforestry System

4.1 High and Sustainable Productivity

Monocropping or intensive cropping using high-yielding hybrid varieties in agriculture, although produces high crop yield diminishes the whole ecosystem, including soil biological and physico-chemical properties because of the high incorporation of chemical and synthetic fertilizers into the soil. After a few years, the overall productivity of the system goes significantly down. Agroforestry involving various components like trees and livestock not only restores the degraded ecosystem but also enhances the crop yield by improving the soil moisture content, organic matter, microbial activities, nutritional quality and finally soil fertility (Muchane et al., 2020; Bishaw et al., 2022). Interestingly, agroforestry reduces the usage of synthetic chemical fertilizers and creates a healthy ecosystem for high crop production and overall productivity on a sustainable basis.

Table 1 A brief and holistic comparison of agriculture vs the agroforestry system

S. no.	Characteristics	Agriculture	Agroforestry
1.	Climate smart	No	Yes
2.	Mono or single cropping	Yes	No
3.	Tree components	No	Yes
4.	High input	Yes	No
5.	Perennial trees + Crops + Livestock	No	Yes
6.	Climate change adaptation	No	Yes
7.	Chemical usage	Yes	No or least
8.	Insect pest resurgence	Yes	No
9.	Ecosystem restoration	No	Yes
10.	Biodiversity conservation	No	Yes
11.	Recycling and mobilization of resources	No	Yes
12.	Year-round production	No	Yes
13.	Application to wastelands	No	Yes
14.	Improvement in soil health	No	Yes
15.	Improvement in water and air quality	No	Yes
16.	Sustainable food and nutritional security	No	Yes
17.	Sustainable income and livelihood security	No	Yes
18.	Long term CO_2 sequestration	No	Yes
19.	Wind break and shelterbelts	No	Yes
20.	Number of SDGs	12 SDGs	09 SDGs

4.2 Adaptation to Climatic Adversities

The events of climatic adversities like drought, flood, high and low temperature are becoming more severe and frequent owing to global warming and climate change, which directly affect the crop growth and yield potentials. The tree component of agroforestry provides resilience to climate change, restores degraded ecosystems and provides long-term sustainability (Quandt et al., 2023). Financial losses incurred through crop failures can be easily compensated with the tree and livestock components.

4.3 Climate Change Mitigation

The process of carbon sequestration by perennial tree species in agroforestry systems involves capturing atmospheric CO_2 and storing it for extended periods, potentially over a century (Goswami et al., 2014; Abbas et al., 2017). This practice plays a significant role in reducing ambient temperatures by mitigating the levels of CO2 in the atmosphere. By incorporating perennial trees into agroforestry systems, not only is carbon dioxide absorbed effectively, but the environmental impact can be long lasting. This sustainable approach helps combat climate change by removing and sequestering carbon from the atmosphere, thereby contributing to the mitigation of global warming trends (Nair, 2012; Raj et al., 2018).

4.4 Ecological Restoration

Ecosystem restoration by agroforestry involves designing and managing farming systems to mimic natural ecosystems, improving biodiversity, soil health and water cycles. Agroforestry practices help reverse degradation, sequester carbon, enhance ecosystem services and increase overall yields, reduce environmental impacts and contribute to ecosystem restoration, ultimately leading to more resilient and productive ecosystems.

4.5 Biodiversity Conservation

Agroforestry practices play a crucial role in conserving biological diversity by providing a habitat for various plant and animal species within agricultural landscapes (Udawatta et al., 2021). By integrating trees, shrubs or other woody plants alongside crops or livestock, agroforestry systems create diverse ecosystems that support a wide range of flora and fauna. This enhanced biodiversity not only promotes ecological resilience but also contributes to increased soil health, water retention and overall ecosystem productivity. Agroforestry's multifunctional approach fosters a balanced and sustainable environment, benefiting both agricultural production and conservation efforts.

4.6 *Sustainable Food, Financial and Livelihood Security*

Agroforestry provides a diversified and resilient food production system. Trees and shrubs can yield fruits, nuts, timber and other products alongside traditional crops, ensuring a more reliable and varied food supply. The integration of trees can create additional revenue streams for farmers through the sale of timber, fruits, nuts and other tree products. This diversification of income sources can help protect against market fluctuations and increase financial stability (Singh et al., 2023). Agroforestry systems often enhance soil fertility, water retention and overall farm productivity. This can lead to increased yields and improved crop resilience, ultimately supporting the long-term livelihood security of farmers and rural communities.

5 Different Services of Agroforestry

Agroforestry practices on crop or pasture lands with livestock can integrate various ecological, economic and social functions to achieve multiple benefits (Duffy et al., 2021). Tree components in the agroforestry interventions play a very crucial role in the supply of a variety of services like regulating services, ecosystem services and provisioning services, which assures the win-win situation for the farmers, the environments, consumers, etc.

5.1 *Provisioning Services*

Provisioning services are the concrete benefits that humans receive from agroforestry ecosystems, such as fodder (fodder grasses or tree-based fodder), fuelwood (woody biomass from trees used as a source of bioenergy, cooking, heating, generating electricity, etc.) and fruits (as obtained from horticulture trees). India is deficient in green fodder production by 35.6% (Singh et al., 2022), and agroforestry involving pastures and trees producing green leaf fodder certainly provides additional opportunity to enhance the availability of green fodder throughout the year. Likewise, agroforestry for a sustainable source of energy supports environmental stewardship by reducing reliance on fossil fuels and lowering greenhouse gas emissions. Furthermore, fuelwood production helps to improve the overall livelihoods of small-scale farmers and forest dwellers, hence promoting rural development and poverty reduction. Horticulture tree-based agroforestry systems not only enhance fruit production but also ensure nutritional and livelihood security among the farmers. These services are vital to rural lives and food security. Agroforestry can provide a consistent supply of fresh produce, enhancing nutrition and income in local populations. Furthermore, it promotes soil health, lowers erosion and improves biodiversity, making it an essential component of sustainable agriculture.

5.2 *Regulating Ecosystem Services*

Agroforestry's regulating services relate to the ecosystem benefits that result from incorporating trees into agricultural settings. These services can be pollination, pest control, and climate change adaptation and mitigation. Tree component creates a microclimate that promotes biodiversity, habitats for beneficial insects, birds, microbes, etc., and enhances the overall productivity and resilience of the system. Tall perennial trees give shade that reduces the soil temperature and the moisture loss. Tree rows act as a shelter belt or wind break for effective control of high winds in coastal regions, conserve soil moisture and create a more hospitable habitat for improving the health of an ecosystem. They also augment the soil organic carbon matter, microbial activities, soil structure and fertility, positively enhancing the capacity of the soil to support plant growth and reducing erosion. This, in turn, also minimizes the need for artificial fertilizers and pesticides, which benefits both the environment and human health. Additionally, agroforestry can combat climate change by sequestering carbon and lowering greenhouse gas emissions (FAO, 2018). Agroforestry, by controlling these ecosystem services, can improve agricultural productivity, food security and environmental sustainability.

5.3 *Cultural Services*

One important consideration is cultural services, which refer to the non-market benefits that agroforestry systems provide to society. These services include ecotourism, recreation and spiritual or religious acts. Agroforestry settings frequently maintain traditional cultural heritage, hence promoting biodiversity and cultural variety. Farmers can retain or restore traditional land use practices by incorporating trees into agricultural ecosystems, hence preserving cultural identity and community cohesion. Agroforestry systems can also facilitate teaching, research and capacity building, ultimately contributing to long-term cultural development.

6 Challenges Impede the Adoption of Agroforestry Practices

Agroforestry adoption raises several practical challenges in the farmer's field. Lack of quality knowledge or expertise in technical and agronomic problems, as well as the knowledge diffusion required to implement and sustain agroforestry systems. Furthermore, the perceived socioeconomic issue related to the market is another challenge that hinders its quicker adoption. The amount of effort or time necessary to construct or maintain agroforestry systems is much higher, which might be a barrier to adoption by farmers. Few farmers may also have difficulty making the first economic investment required to establish an agroforestry system. Furthermore, a lack of technical help required to implement or sustain agroforestry systems could be a contributing factor to low agroforestry adoption.

7 Conclusion

Climate change is clearly apparent and poses a threat to the sustainability and livelihood of the communities. In the face of such challenges, farmers devise adaptation tactics such as climate-smart agroforestry practices for sustainable production and climate change adaptation and mitigation. Trees are less vulnerable to climate shock than crops, therefore, they play an important role in providing fodder, food and lumber products during droughts. Agroforestry plays a crucial role in carbon sequestration, soil preservation and promoting biodiversity. Through the agroforestry adoption, farmers can not only improve their livelihoods but also diminish poverty levels. Furthermore, these sustainable practices contribute to ensuring food security and fostering resilience against the impacts of climate change, thereby paving the way for a more sustainable future. It can be effective or ideal under certain circumstances, while being less beneficial or unsuitable in others. Its acceptance and feasibility vary among different groups, with certain groups being more adept at and successful in implementing agroforestry practices. Agroforestry is not a universal solution for land use, but certainly can act as a potential nature-based solution to deal with major agricultural and climate-related issues in present and future scenarios. Promoting the adoption of better agroforestry practices through policy support, capacity building programmes, research and community engagement, partnerships, monitoring and scaling up successful models would undoubtedly increase agroforestry's potential to address climate change and food insecurity, fostering sustainable agricultural development.

References

Abbas, F., Hammad, H. M., Fahad, S., Cerdà, A., Rizwan, M., Farhad, W., Ehsan, S., & Bakhat, H. F. (2017). Agroforestry: A sustainable environmental practice for carbon sequestration under the climate change scenarios—A review. *Environmental Science and Pollution Research, 24*, 11177–11191. https://doi.org/10.1007/s11356-017-8687-0

Arora, N. K. (2019). Impact of climate change on agriculture production and its sustainable solutions. *Environmental Sustainability, 2*(2), 95–96.

Bishaw, B., Soolanayakanahally, R., Karki, U., & Hagan, E. (2022). Agroforestry for sustainable production and resilient landscapes. *Agroforestry Systems, 96*(3), 447–451.

FAO. (2018). *World food and agriculture: Staistical pocket book*. FAO.

Duffy, C., Toth, G.G., Hagan, R.P., McKeown, P.C., Rahman, S.A., Widyaningsih, Y., & Spillane, C. (2021). Agroforestry contributions to smallholder farmer food security in Indonesia. *Agroforestry Systems*, 95(6), 1109-1124.

Goswami, S., Verma, K. S., & Kaushal, R. (2014). Biomass and carbon sequestration in different agroforestry systems of a Western Himalayan watershed. *Biological Agriculture Horticulture, 30*(2), 88–96.

IPCC. (2014). *Climate change 2014: Impacts, adaptation, and vulnerability. Contribution of Working Group II to the Fifth Assessment Report of the Intergovernmental Panel on Climate Change* (32 p). Geneva, Switzerland.

Malhi, G. S., Kaur, M., & Kaushik, P. (2021). Impact of climate change on agriculture and its mitigation strategies: A review. *Sustainability, 13*(3), 1318.

Masson-Delmotte, V., Zhai, P., Pörtner, H. O., Roberts, D., Skea, J., & Shukla, P. R. (2022). *Global Warming of 1.5 C: IPCC special report on impacts of global warming of 1.5 C above pre-industrial levels in context of strengthening response to climate change, sustainable development, and efforts to eradicate poverty*. Cambridge University Press.

Montagnini, F., Nair, P. Carbon sequestration: An underexploited environmental benefit of agroforestry systems. *AgroforestrySystems* 61, 281–295 (2004).

Muchane, M. N., Sileshi, G. W., Gripenberg, S., Jonsson, M., Pumariño, L., & Barrios, E. (2020). Agroforestry boosts soil health in the humid and sub-humid tropics: A meta-analysis. *Agriculture, Ecosystems & Environment, 295*, 106899.

Nair, P. K. R. (2012). Climate change mitigation and adaptation: A low hanging fruit of agroforestry. In P. K. R. Nair & D. Garrity (Eds.), *Agroforestry – The future of global land use* (pp. 31–67). Springer.

Quandt, A., Neufeldt, H., & Gorman, K. (2023). Climate change adaptation through agroforestry: Opportunities and gaps. *Current Opinion in Environmental Sustainability, 60*, 101244.

Raj, A., Jhariya, M. K., & Bargali, S. S. (2018). Climate smart agriculture and carbon sequestration. In Pandey et al. (Eds.), *Climate change and agroforestry* (pp. 1–19). New India Publishing Agency.

Singh, D. N., Bohra, J. S., Tyagi, V., Singh, T., Banjara, T. R., & Gupta, G. (2022). A review of India's fodder production status and opportunities. *Grass and Forage Science, 77*(1), 1–10.

Singh, N. P., Anand, B., Srivastava, S. K., Rao, K. V., Bal, S. K., & Prabhakar, M. (2020). Assessing the impacts of climate change on crop yields in different agro-climatic zones of India. *Journal of Atmospheric Science Research, 3*(4), 16–27.

Singh, P., Choudhary, B. B., Dwivedi, R. P., Arunachalam, A., Kumar, S., & Dev, I. (2023). Agroforestry improves food security and reduces income variability in semi-arid tropics of central India. *Agroforestry Systems, 97*(4), 509–518.

Udawatta, R. P., Rankoth, L. M., & Jose, S. (2021). Agroforestry for biodiversity conservation. In *Agroforestry and ecosystem services* (pp. 245–274). Springer.

van Noordwijk, M., Duguma, L. A., Dewi, S., Leimona, B., Catacutan, D. C., Lusiana, B., et al. (2018). SDG synergy between agriculture and forestry in the food, energy, water and income nexus: Reinventing agroforestry? *Current Opinion in Environmental Sustainability, 34*, 33–42.

Zerihun, M. (2020). Institutional analysis of adoption of agroforestry practices in the Eastern Cape Province of South Africa. *South African Journal of Environmental Education, 36*, 37–55.

Agroforestry: Climate-Smart Agriculture Technique for Semi-Arid Climate

Alok Kumar Singh, Saurbh Shukla, Anwesha Dey, Abhinav Singh, and Girish Tantuway

Abstract Agroforestry is a multifaceted land use system that integrates trees, crops, and/or livestock on the same unit of land, emerges as a promising solution to deal with the problems caused by climate change in semi-arid climates. With its multiple benefits, agroforestry stands as a cornerstone for bolstering agricultural resilience and fostering environmental sustainability in regions vulnerable to extreme weather events and climate variability. This chapter delineates the paramount significance of agroforestry as a climate-smart agriculture technique tailored to the unique needs of semi-arid landscapes. It seeks to explain agroforestry's critical role in reducing the effects of climate change while boosting agricultural output and livelihoods by offering a thorough review of its concepts and uses. The chapter highlights the inherent adaptability and resilience of agroforestry systems to climatic pressures through an analysis of diverse systems and their characteristics. This positions agroforestry systems as effective solutions for resolving local agricultural and environmental concerns in semi-arid regions. It also explores the many advantages of climate-smart agroforestry, which include biodiversity preservation, soil moisture retention, carbon sequestration, and economic diversification. The chapter emphasizes the critical role that agroforestry plays in promoting sustainable agricultural practices and boosting the resilience of semi-arid communities by highlighting the synergistic relationship between agroforestry and climate change mitigation/adaptation. In elucidating implementation strategies, the chapter navigates through critical considerations such as site and species selection, integration with agricultural crops and livestock, and management practices geared toward optimiz-

A. K. Singh · A. Singh · G. Tantuway
Rajiv Gandhi South Campus, Banaras Hindu University, Barkachha, Mirzapur, Uttar Pradesh, India

S. Shukla
Department of Silviculture and Agroforestry, Dr. Yashwant Singh Parmar University of Horticulture and Forestry, Solan, Himachal Pradesh, India

A. Dey (✉)
Department of Agricultural Economics, Institute of Agricultural Sciences, Banaras Hindu University, Varanasi, Uttar Pradesh, India

A. K. Singh et al. (eds.), *Sustainable Agriculture Management in Semi-Arid Climates*, https://doi.org/10.1007/978-3-031-94062-0_4

ing productivity and sustainability. Additionally, it underscores the indispensable role of policy and institutional support in fostering the widespread adoption of agroforestry, thereby facilitating its integration into mainstream agricultural practices in semi-arid regions. Despite its myriad benefits, the chapter acknowledges the challenges associated with agroforestry adoption in semi-arid climates, ranging from logistical barriers to research gaps. However, it also identifies scaling-up opportunities and outlines future directions for advancing agroforestry as a preeminent climate-smart agriculture technique in semi-arid regions. Through a comprehensive synthesis of existing knowledge and innovative perspectives, this chapter endeavors to contribute to the advancement of agroforestry as a cornerstone strategy for climate resilience and sustainability in semi-arid climates.

Keywords Agroforestry · Carbon sequestration · Climate change · Semi-arid · Smart agriculture

1 Introduction

The supply of food and energy for human sustenance is one of the prime challenges that we should seriously consider in this era of an overgrowing population around the world (WHO, 2018), which is likely to reach 9.7 billion by 2050. Global warming and climate change issues substantially limit the sustainability of many agricultural ecosystems during the twentieth century, which has serious implications on global food security (Shayanmehr et al., 2020). The global agricultural production is likely to fall by 16% because of global climate change (Perdinan, 2010). The global report of the state of food security and nutrition says that one in ten people currently face challenges of hunger and food shortage. The rainfed agriculture constitutes 79% of global crop land area, which supports a large section of resource-poor farmers in arid and semi-arid tropics (Wani et al., 2009; FAO, 2021). Climate-smart agriculture refers to a comprehensive strategy that tackles the problems caused by climate change and can improve the food and agricultural production system, with a primary emphasis on the resilience, sustainability, and health of the ecosystem.

Agroforestry, defined as the intentional integration of trees and shrubs into crop and livestock systems, represents a dynamic and sustainable land management approach that synergizes agricultural and forestry techniques. In semi-arid regions with scarce resources and severe environmental stressors, this approach is especially important as it not only diversifies production but also improves ecosystem services. In semi-arid areas, where droughts occur more frequently and intensely, soil degradation and biodiversity loss make addressing climate change crucial. These areas have particular difficulties that jeopardize livelihoods and agricultural productivity. Agroforestry offers a climate-resilient approach that can lessen these negative effects and advance sustainable development by increasing soil fertility, optimizing water use, and offering microclimatic advantages. The purpose of this chapter is to examine the fundamentals and practical uses of agroforestry as a climate-smart farming method that works particularly well in semi-arid regions. It will examine the ecological, economic, and social benefits of agroforestry practices, supported by empirical evidence and case studies. The scope of the chapter includes an in-depth analysis of various agroforestry systems, their role in climate change adaptation and mitigation, and their contribution to sustainable land management and food security in semi-arid regions.

2 Agroforestry Systems in Semi-Arid Climates

2.1 Overview of Semi-Arid Climates and Their Agricultural Challenges

Over 40% of the world's land surface is made up of drylands, which also house about 2.5 billion people, 50% of cattle, and 45% of the world's food output (Gaur & Squires, 2018). Semi-arid regions are subsets of dryland regions where mean annual precipitation is in the range of 200–700 mm (Gallart et al., 2002) with an annual variability of 20–50% (Barakat, 2009; IFAD, 2000) and aridity index ranging between 0.20 and 0.50 (Lal, 2004; Higgins & Kassam, 1981). These regions constitute 15.2% of the global land area and provide a livelihood to about 1.1 billion people (Scholes, 2020). Precipitation plays an important role in the growth of vegetation in arid and semi-arid regions (Wang et al., 2023), but when annual potential evapotranspiration exceeds annual precipitation, frequent drought stresses occur. Due to their low yearly rainfall, substantial evaporative losses, and significant temporal and geographical variability in vegetation dynamics and water availability, semi-arid regions mostly experience drought (Snyder & Tartowski, 2006). Reduced agricultural output is the result of drought stress, which affects the rate of photosynthesis, leaf area, the mobilization of growth reserves, crop development period, and causes advancement in phenological stages (Shah & Paulsen, 2003; Dietz et al., 2021). Developing countries are more prone to the consequences of droughts because they are more susceptible to the effects of climate change (Kheiri et al., 2023). According to IPCC (2018), in the twenty-first century, rapid population growth and drastically changed economic development would increase greenhouse gas concentrations, which will strain water supplies and alter the amount of precipitation that falls each year. The low precipitation, rising temperature, low vegetation cover triggered by the steep slopes, and high intensity storms make these ecosystems more vulnerable to wind and water erosion (Kašanin-Grubin et al., 2018) and are sensitive to global climate change (Ding, 2007). The agriculture sector in semi-arid areas is mostly rainfall dependent but contributes substantially to crop production (Karimi et al., 2018). For example, Ghamghami and Beiranvand (2022) estimated that in Iran, approximately 60% of total cultivated land is rainfed and supports over 30% of the total production of agricultural crops. However, crops that receive irrigation are heavily dependent on groundwater resources, which lead to overexploitation of water resources, and climate change shocks further affect the agricultural activities (Radmehr et al., 2021). The primary agricultural land use in semi-arid regions is pastoralism, which involves cattle, sheep, goats, and camels. This is because crop production under drought stress conditions is more prone to crop failure (Scholes, 2020).

2.2 *Classification and Characteristics of Agroforestry Systems Suitable for Semi-Arid Regions*

Agroforestry is the practice of integrating trees and shrubs into systems that support livestock and agriculture, and has many positive ecological, economic, and social effects. Semi-arid regions are especially affected by environmental issues such as soil erosion, water scarcity, and climate variability. This chapter explores the categorization and traits of agroforestry systems that are most suited for semi-arid regions, emphasizing their special qualities and advantages.

2.2.1 Agroforestry Parklands

As a historic and very successful agroforestry system, parklands are best suited for semi-arid areas. The intentional planting or preservation of sporadic trees within agricultural landscapes characterizes these systems, resulting in a mosaic of crops and trees that provides a multitude of social, economic, and ecological advantages. These systems are often a result of traditional land management practices where trees are either naturally regenerated or selectively planted and maintained by farmers. The parklands are dominated by a few favored tree species, viz. *Faidherbia albida, Prosopis cineraria, Parkia biglobosa, Vitellaria paradoxa, Prosopis africana, Cordyla pinnata,* etc., along with staple cereal crops (millets and maize), together with legumes (Félix et al., 2018; Bayala et al., 2014). Although the tree density varies by location, <20 trees are usually maintained per acre. Besides trees, shrubs are also an important component of agroforestry parklands, which are repeatedly coppiced during land preparation for agricultural crops. Successful examples from West Africa, Sahelian Africa, and East Africa demonstrate the potential of parklands to sustain agriculture and improve livelihoods in semi-arid regions.

2.2.2 Windbreaks and Shelterbelts

In semi-arid areas, windbreak agroforestry, the deliberate planting of trees and shrubs to slow wind and shield agricultural landscapes, is a vital tactic. As a natural defense against the extreme weather, windbreaks have several positive ecological and economic effects that support sustainable agriculture and improved living standards. Windbreaks are primarily used to lessen the negative effects of strong winds, which can destroy crops, induce soil erosion, and lower soil moisture levels. Windbreaks lower the crop water consumption due to reduction in wind velocity, air temperature, and higher humidity inside the system (Perry et al., 2009; Alemu, 2016). On the leeward side of the windbreaks, the wind speed decreases to 0%, 50%, and 70% of the wind speed at a distance of five times, 15–25 times, and 20–30 times the windbreak height, respectively (Vasilyev, 1980). Windbreaks aid in preventing topsoil loss, which is essential for preserving soil fertility and agricultural output, by slowing down the wind (Cleugh, 1998). This is especially crucial in semi-arid areas where desertification and soil deterioration are major problems. The trees and shrubs used as windbreaks can produce additional income in addition to protecting crops and increasing yields. Shelter from trees protects the crops from direct mechanical damage, reduces temperature, soil evaporation, and evapotranspiration losses, thus creating a microclimate conducive to increased crop yield as compared to open lands. Additionally noteworthy are the social and cultural advantages of windbreak agroforestry. The construction of windbreaks is frequently a community-based endeavor in many semi-arid areas, encouraging collaboration and group action among farmers. In addition to fostering social cohesiveness, this cooperative approach makes it easier to share resources and information, which results in more efficient and sustainable land management techniques. Additionally, the visual value of the landscape can be increased by the presence of trees and green areas, which will improve the standard of living in rural settlements.

2.2.3 Alley Cropping

In an agroforestry system hedgerows of trees or shrubs create the "alleys" between which agricultural crops are cultivated. In order to lessen competition, trees are trimmed during the rainy season, and the trimmings are then used as mulch to enrich the soil (Singh et al., 1989). The alley cropping systems, particularly suitable for humid regions with narrow hedgerows (4m), pose significant competition to crops in semi-arid regions where water is a limiting constraint. The benefit of alley cropping in semi-arid regions is that tree pruning can be used as a fodder source during the lean period for livestock, as hedgerow pruning does not consistently improve the crop production (Singh et al., 1989). The development of management techniques to reduce crop–hedgerow competition and a greater variety of woody or perennial grass species for the hedgerows will dictate the future role of alley cropping in the semi-arid tropics. There is a greater chance of success if alley cropping is limited to areas with guaranteed annual rainfall of more than 750 mm on soils with adequate water-holding ability.

Some prevalent agroforestry systems in different parts of the semi-arid regions have been shown in Table 1.

Table 1 Agroforestry systems followed in the semi-arid tropics

Serial no.	Region	Agroforestry system	Reference
1	Northern Africa	Olive (*Oleaeuropea* L.) based agroforestry system	
2	Northern Africa	Cork oak (*Quercussuber* L.) based Silvipastoral system	(Campos et al., 2009)
3	Morocco	*Arganaspina*-based Silvipastoral system	(Chebli et al., 2012)
4	Morocco	*Artiplex nummularia* and barley-based alley cropping system	(Chebli et al., 2012)
5	Western India	*Prosopis cineraria* (L.), *Acacia leucocephala* (Roxb.), *Acacia nilotica* (L.), Willd, scattered in agricultural crops (*Pennisetum typhoides*, *Arachis hypogea*, and *Cymopsis tetragonloba,* etc.	(Yadav et al., 2011)
6	Kenya	*Grevillea robusta*—Maize-based agroforestry system	(Lott et al., 2009)
7	Central India	*Dendrocalamus strictus* + sesame-chickpea based agroforestry system	(Dev et al., 2020)
8	Central Punjab, Pakistan	*Populus deltoides, Dalbergia sissoo, Azadirachta indica, Vechillia nilotica, Zizyphus mauritiana, Eucalyptus camaldulensis* and *Citrus reticulata* with *Triticum aestivum, Zea mays, Saccharum officinarum*	(Yasin et al., 2023)

2.3 *Case Studies Highlighting Successful Agroforestry Practices in Semi-Arid Climates*

Agroforestry as a climate-smart agriculture technique has shown in its potential in semi-arid arid countries like Niger in West Africa where it gives employment to more than 80% of the population (Traore et al., 2024). The water resource limitation in semi-arid regions requires the development of water harvesting techniques within the farming system to utilize the excess rain water during extreme events that will eventually lead to increased resilience. Palsaniya et al. (2023), in a study carried out in Central Indian semi-arid landscapes, suggested that the inclusion of rainfed crops in an agroforestry system (silvipasture and agrisilviculture) with small ruminants and a farm pond for rainwater harvesting increases the land productivity, profitability, and resilience of the agroecosystem. The harvested rainwater supports the crops having low water requirements in dry winter, which in usual cases is not possible in these regions. The perennial woody component, being hardy in nature, is least affected by weather vagaries and performs multiple roles, the inclusion of animals increases the profitability of the system. A modeling approach was adopted by Reyes et al. (2021), employing IPCC scenarios (history (1951–1990), present (1991–2030), and future (2031–2070)) to compare the growth of winter durum wheat (single cropped) with the walnut-based alley cropping system, it was discovered that the stresses, viz., nitrogen, drought, and heat increased two times from past, and agroforestry contributes to reduction of 20–35% of these stresses with medium-sized trees on farmlands.

3 Climate-Smart Benefits of Agroforestry

3.1 Carbon Sequestration and Climate Change Mitigation

The expansion of industrialization since the beginning of the industrial era has led to an increase in the concentration of CO_2 and other persistent gases, which results in global warming and become a threat to global biodiversity. The consequence of global warming is the rising earth temperature which resulted in a well-known phenomenon called climate change. The global warming and resulting unpredictable climatic variability are primarily due to the rising CO_2 levels whose present concentration in earth atmosphere has reached 426 ppm as of April 2024 (https://www.co2.earth/). In semi-arid regions, the dependence on rainfed agriculture, unpredictable climate and weather events, and urban and industrial growth intensify the impact of climate change. According to Abbas et al. (2017), agroforestry and natural forests are two types of ecosystems that have been developed to store and absorb large amounts of carbon dioxide, which makes them essential for mitigating climate change. Forests are considered to be the significantly largest active carbon pool that store 60% of total terrestrial carbon (Wilson & Daf, 2003) but the degradation of forest and deforestation activities releases significant amount of greenhouse gas emission (20%) (Sukhdevb et al., 2015) and contributes toward climate change. The greenhouse gas sequestration potential of agriculture ranges from approximately 1500–4300 Mt. CO_2e per year and soil carbon restoration has the largest potential to mitigate (Smith & Wollenberg, 2013). Thus, there is need to develop some sustainable system like agroforestry which could be a safe sequester (long term) of carbon and in the meantime meet food security goals. Agroforestry is the third greatest carbon sink in Africa, behind forests and long-term fallows, and yields three to four times as much biomass as treeless agricultural land uses (IPCC, 2000). Three strategies—reducing greenhouse gas emissions, blocking emissions, and storing carbon in biomass and soils—allow agroforestry to mitigate the effects of weather exchange (Bogale & Bekele, 2023). Agroforestry systems result in lower emission of greenhouse gases through lower equipment use in tree areas, higher nutrient uptake by woody perennials, and decreased need for lower nutrient application (Bogale & Bekele, 2023). In agroforestry systems, trees have the ability to accumulate carbon for a long term in their biomass as well as in soil through litter decomposition (Nair et al., 2010). A conservative approach to the estimation of mitigation potential of agroforestry on global scale shows it can sequester of 9.4 Gt CO_2 eq year^{-1} (Chapman et al., 2020). The soil is the greatest sink of CO_2 accumulation in the agroforestry system, which accounts for 77–92% of carbon stock, followed by trees and herbaceous vegetation, which constitute only 7–22% and 1%, respectively (Bogale &

Bekele, 2023). The crop production is the result of various biotic and abiotic factors acting independently or the interaction among them and the way it responds to management practices (Moreira et al., 2018). Trees in agroforestry systems regulate the microclimate by providing shade and reducing diurnal variations in temperature. In semi-arid regions, high temperature and intense light can stress the crops and livestock and affect their physiological activities. The shade reduces the ambient temperature and lowers transpiration water loss, promoting water use efficiency (Partelli et al., 2014). The cooling effect creates more favorable microenvironmental conditions for crop and livestock, enhancing their growth and productivity, thus preventing global warming without lowering agricultural output (Moreira et al., 2018).

3.2 Soil Moisture Retention and Water Management

In semi-arid areas, moisture limitation is the prime reason for crop failure. The infiltration following rainfall is the major source for soil water, but higher temperatures in these regions led to the loss of a significant amount of rainfall water due to high temperature when rainfall intensity is low or by percolation to deeper soil horizons when rainfall intensity is high. Low stomatal conductance under moisture-stressed conditions reduces the internal CO_2 availability, which leads to lesser photosynthetic activity and light use efficiency (Zhou et al., 2016). Intercropping trees in an agricultural system can optimize the usage of below ground resource pool where woody perennials can capture nutrients and moisture from the deeper layer of soil that cannot be used by the crops either temporally or spatially. The root system of component species occupies different niches (soil depth and width) in soil horizons in agroforestry (Zhang et al., 2014), and higher productivity under agroforestry systems depends upon the complementarity among trees and arable crops. In an agroforestry system, the water requirement of trees and arable crops varies temporally and spatially, which results in complementarity between woody and herbaceous components. The shading caused by the companion trees can reduce the soil evaporation and help to conserve water, as demonstrated by Wallace et al. (1999) in the *Grevelliarobusta*-based agroforestry system in Kenya. The enhanced tree cover creates microclimatic conditions in agroforestry entirely different from land use involving monocropping of agricultural crops. For example, Lazarev (2006) found that shelterbelts could increase the precipitation by 22% and reduce the specific water uptake by 20%. Trees can enhance groundwater recharge by increasing soil infiltrability and preferential flow. This is because water flows through the vadose zone more quickly and is accessible in smaller amounts for plant root uptake and evaporation (Bargués Tobella et al., 2014). In semi-arid regions, the presence of trees in agricultural lands increases the organic content through higher root and leaf litter inputs, which increase the organic matter and soil faunal activity due to favorable microclimatic conditions beneath the trees (De Bruyn & Conacher, 1990; Bargués Tobella et al., 2014). The rational use of low-lying and upland areas can reduce the risk of crop failure in semi-arid areas, as low-lying areas are less prone to drought, but vulnerable to flooding (Murungweni et al., 2016).

3.3 *Biodiversity Conservation and Ecological Resilience*

Agroforestry, the integration of trees and shrubs into agricultural landscapes, offers significant potential to enhance biodiversity and ecological resilience, particularly in semi-arid regions. This land-use system, which combines woody perennials with crops and/or livestock, creates a more complex habitat structure compared to conventional agriculture. This complexity supports a wider variety of species, thereby enhancing biodiversity. Agroforestry systems provide multiple ecosystem services, including habitat for wildlife, improved soil fertility, and increased carbon sequestration, which collectively contribute to ecological resilience. In semi-arid regions, agroforestry can play a critical role in mitigating the harsh effects of climate variability. These regions often face challenges such as soil degradation, water scarcity, and reduced agricultural productivity due to extreme weather events. By incorporating trees and shrubs, agroforestry systems help stabilize soils, reduce erosion, and enhance water retention. Trees act as windbreaks, reducing the impact of strong winds and preventing soil erosion. Their roots help in binding the soil, while their canopy provides shade, reducing soil temperature and evaporation rates. This creates a microclimate that can sustain agricultural productivity even under adverse conditions (Pantera et al., 2021).

The biodiversity benefits of agroforestry are particularly evident in the increased presence of beneficial insects, birds, and other wildlife. For instance, studies have shown that agroforestry systems support a higher diversity of insect species, including pollinators and natural predators of crop pests. This biological control reduces the need for chemical pesticides, promoting a more sustainable and environmentally friendly agricultural practice (Akesse-Ransford et al., 2021). To further improve biodiversity, agroforestry systems' diverse plant structure offers birds and other wildlife food supplies and nesting places. Additionally, agroforestry aids in carbon sequestration, a crucial function given the state of climate change. Greenhouse gas concentrations are lowered by the absorption and storage of carbon dioxide from the atmosphere by trees and bushes. By boosting the amount of organic matter in the soil, this not only slows down climate change but also enhances its health. For agriculture to be sustainable in semi-arid areas, there must be an increase in soil organic matter because this improves soil structure, increases nutrient availability, and increases water-holding capacity (Pantera et al., 2021). Agroforestry systems also foster economic and social resilience. By producing timber, fruits, nuts, and other tree products, they give farmers a variety of revenue streams. This diversification reduces farmers' dependence on a single crop, thereby decreasing their vulnerability to market and climate shocks. Additionally, the environmental benefits of agroforestry, such as improved soil fertility and water availability, can lead to higher and more stable agricultural yields over time (Jose, 2009). Despite these benefits, the adoption of agroforestry in semi-arid regions faces several challenges, including a lack of awareness, insufficient technical knowledge, and policy barriers. Education and training for farmers, as well as supportive policies and incentives, are crucial for the widespread adoption of agroforestry practices. Integrating agroforestry into national and regional agricultural strategies can help unlock its full potential to enhance biodiversity and ecological resilience (Mosquera-Losada et al., 2020).

3.4 Economic Diversification and Livelihood Enhancement

Climate-smart agroforestry is essential for improving livelihoods and diversifying the economy, especially in semi-arid areas where conventional farming methods frequently face environmental challenges. Agroforestry systems produce many layers of production that can generate a range of products, such as fruits, nuts, lumber, fodder, and medicinal plants, by integrating trees and shrubs with crops and cattle. Farmers' economic risk is decreased by this diversification because they are not dependent on just one kind of crop or commodity. Additionally, the addition of organic matter and nutrient cycling that trees provide to the soil enhances soil fertility, which over time may result in greater and more consistent agricultural yields. Agroforestry techniques also improve resistance to shocks associated with climate change by producing shade and lowering evapotranspiration, which aids in preserving soil moisture levels. This resilience is critical in semi-arid climates where water scarcity can severely impact agricultural productivity. The diversified income streams from agroforestry systems can significantly improve household income, allowing farmers to invest in education, health, and other essential services, thereby enhancing overall quality of life.

Furthermore, agroforestry systems can create employment opportunities both on-farm and in associated value chains, such as processing and marketing of agroforestry products. This contributes to rural economic development and reduces migration pressures by providing viable livelihoods in rural areas. Overall, climate-smart agroforestry not only bolsters economic stability and enhances livelihoods but also promotes sustainable land management practices that are crucial for long-term ecological health and productivity.

4 Implementation Strategies for Agroforestry in Semi-Arid Climates

4.1 Site Selection and Species Selection Considerations

First and foremost, choosing a site is very crucial. In semi-arid environments, soil fertility and structure can vary greatly, thus, it is crucial to consider the kind and quality of the soil. Testing the soil thoroughly to determine its pH, organic matter content, and nutrient levels is crucial. Soils with good water-holding capacity are preferable, as they help mitigate the frequent droughts typical of semi-arid climates. Water availability is another crucial factor. Semi-arid regions often face irregular rainfall, so identifying reliable water sources, whether through natural rainfall, groundwater, or irrigation, is vital. Implementing water conservation techniques such as rainwater harvesting and contour farming can enhance water management and retention. The land use history also plays a significant role in site selection. Once a suitable site is selected, species selection becomes the next focal point. Species must be carefully chosen for their adaptability to semi-arid conditions. The choice of species is greatly determined by the suitability to the local context and should also conform to the acceptance by the farmers of the concerned region (Tadesse et al., 2021). Drought tolerance is a primary criterion, with species having deep root systems that can access groundwater being particularly valuable. Several essential features are present in woody plant species that, especially in semi-arid areas, have the greatest potential to boost water and nutrient availability for primary production in agroforestry systems. The ability to produce stem flow is one of these characteristics, along with deep rooting systems, tall, closed canopies with evergreen phenology, unappealing foliage and fruits, long life spans, slow growth rates, high internal nutrient cycling efficiency, nitrogen fixation skills, and mycorrhizal relationships. Furthermore, leaf characteristics have a big impact on the condition of soil organic matter (Breman & Kessler, 1997).

4.2 Integration of Trees with Agricultural Crops and Livestock

In semi-arid areas, integrating trees with livestock and agricultural crops necessitates a deliberate, multifaceted approach. In order to choose appropriate tree and crop species that can survive semi-arid environments, the first stage entails careful site selection and assessment. This includes in-depth soil and water investigation to determine fertility, pH, texture, and water availability, as well as assessments of climate and terrain. It is important to choose drought-resistant and multipurpose tree species like Acacia, Prosopis, and *Emblica officinalis*. It is also important to choose crops and livestock that work well with these trees and can withstand drought. Agroforestry systems need to be carefully planned in terms of both space and time. Alley cropping, silvopasture, and windbreaks are examples of spatial configurations that must maximize interactions between trees, crops, and livestock while maintaining sufficient spacing to reduce resource competition. Temporal arrangements should plan for staggered planting and harvesting to maintain continuous productivity. Soil and water conservation practices, including mulching, cover crops, and water harvesting techniques like contour bunds and check dams, are essential to retain soil moisture and reduce erosion. Integrating livestock involves developing silvopasture systems where livestock graze within tree plantations, selecting suitable forage species, and managing grazing intensity to prevent overgrazing and soil compaction. Planting fodder trees and shrubs provides reliable feed sources during dry seasons. Regular management and maintenance, such as pruning, thinning, and implementing integrated pest management (IPM) practices, are necessary to optimize light penetration, reduce competition, and control pests and diseases.

Building capacity and involving the community are essential to sustainability and success of agroforestry operations. Adoption and ownership of agroforestry systems are encouraged by holding farmer training programs and incorporating neighborhood communities in planning and decision-making processes. Establishing performance indicators to evaluate growth, productivity, soil health, water use efficiency, and economic returns is the first step in monitoring and evaluation. Based on these data, management procedures are adjusted and improved as necessary. Promoting agroforestry requires encouraging government policies and providing grants, subsidies, and technical assistance, among other things. The economic feasibility of agroforestry systems is further enhanced by facilitating market access for agroforestry products through value chain development, certification programs, and cooperative marketing. In semi-arid regions, agroforestry may greatly increase production, resilience, and sustainability by putting this all-encompassing plan into practice.

4.3 Management Practices for Optimizing Agroforestry Productivity and Sustainability

Optimizing agroforestry productivity and sustainability in semi-arid climates require meticulous management practices that balance the needs of trees, crops, and livestock while enhancing the overall ecosystem. Trees must be regularly pruned and thinned in order to minimize competition for nutrients and sunlight, which will allow crops to grow alongside trees. Important biomass is also produced by this method, which can be applied as mulch to increase soil fertility and moisture retention. To manage diseases and pests without significantly depending on chemical pesticides, it is imperative to put integrated pest management (IPM) techniques into practice. IPM includes biological controls, resistant varieties, and cultural practices like crop rotation and intercropping, which can disrupt pest cycles and reduce infestations.

Soil and water conservation practices are vital for maintaining soil health and optimizing water use. Contour plowing, terracing, and check dam construction are some of the techniques that improve water infiltration and inhibit soil erosion. A more robust agroecosystem is promoted by mulching and planting cover crops, which shield the soil's surface, lower evaporation, and increase the amount of organic matter in the soil. In semi-arid areas, effective water management is essential. To enhance water use efficiency and guarantee that plants receive enough moisture during dry seasons, drip irrigation and rainwater collection devices must be used.

Careful control of grazing intensity is necessary when integrating livestock into agroforestry systems through silvopasture in order to avoid overgrazing and soil compaction. While preserving the fertility and structure of the soil, rotational grazing and the construction of fodder banks with drought-resistant forage species can offer a sustainable source of food. Continuous monitoring and evaluation of agroforestry systems are essential to track growth, productivity, and environmental impact. This involves using performance indicators such as soil health parameters, crop yields, and biodiversity assessments to make data-driven adjustments to management practices. Engaging local communities and providing ongoing education and training are critical for the long-term success of agroforestry. Farmers need to be well versed in agroforestry techniques, understand the benefits, and be equipped to implement best practices effectively. Establishing farmer cooperatives and knowledge-sharing networks can facilitate the exchange of information and support collective action toward sustainable agroforestry. Overall, these management practices, when integrated and executed effectively, can enhance agroforestry productivity and sustainability, leading to more resilient agricultural landscapes and improved livelihoods in semi-arid climates.

4.4 *Policy and Institutional Support for Promoting Agroforestry Adoption*

Effective promotion of agroforestry adoption in semi-arid climates necessitates robust policy frameworks and strong institutional support. Governments play a crucial role in creating enabling environments through policies that recognize the benefits of agroforestry and incentivize its adoption. The creation and execution of national agroforestry policies, which provide precise guidelines and goals for incorporating trees into agricultural landscapes, is one of the main policy initiatives. The many uses of agroforestry, such as improving food security, boosting climate change resilience, and supplying ecosystem services, should be highlighted in these strategies. Financial incentives and subsidies are essential instruments for persuading farmers to use agroforestry techniques. Governments can provide tax benefits for investments in sustainable land management, as well as direct subsidies for planting trees and financial support for setting up agroforestry systems. Furthermore, giving farmers access to microcredit and low-interest loans might assist them in covering the upfront expenditures associated with making the switch to agroforestry, which can be a major obstacle, particularly for smallholders.

Institutional support is equally vital in promoting agroforestry. It is necessary to improve agricultural extension services in order to give farmers access to training, technical support, and capacity building. Extension personnel should be knowledgeable about agroforestry techniques and able to provide farmers with practical assistance so they may successfully establish and maintain agroforestry systems. Creating model farms and demonstration plots can operate as real-world examples and educational hubs for farmers to see effective agroforestry techniques in action. Universities and research centers are essential to the advancement of agroforestry because they carry out studies on the best combinations of trees, crops, and livestock, create drought-tolerant tree species, and discover novel methods for conserving water and soil. Collaborating research facilities and extension organizations ensure that farmers can reap the benefits of cutting-edge scientific breakthroughs and technology advances. Moreover, future generations of farmers and agricultural professionals can become more aware of agroforestry by including it in the agricultural education curriculum.

Market development is another crucial aspect of promoting agroforestry adoption. Market development and expansion for agroforestry goods, including fruits, nuts, wood, and non-timber forest products, should be the focus of governments and other institutions. The implementation of certification programs for agroforestry goods produced sustainably can improve market accessibility and yield higher prices for producers. Furthermore, small-scale farmers can pool their resources, save transaction costs, and strengthen their negotiating position in the market by supporting cooperative marketing agreements. Strong legal frameworks are also essential for securing land tenure and property rights, which are fundamental for encouraging long-term investments in agroforestry. Clear land tenure policies that recognize and protect the rights of farmers and communities to use and benefit from trees on their land can remove significant barriers to agroforestry adoption. Policies should also address issues related to land use conflicts and provide mechanisms for resolving disputes.

5 Conclusion

Agroforestry stands out as a transformative, climate-smart agricultural technique, particularly well suited to the unique challenges of semi-arid climates. By integrating trees with crops and livestock, agroforestry systems offer a multitude of benefits, including enhanced soil fertility, improved water use efficiency, increased biodiversity, and diversified income streams for farmers. Together, these advantages help people to be more resilient to climate unpredictability and extreme weather events, which are occurring more frequently and with greater intensity in semi-arid locations. We have studied the tenets and methods of agroforestry in this chapter, highlighting its potential to reduce the negative effects of climate change while advancing sustainable land use and financial stability. We have covered in depth the strategic approaches—such as species selection, site selection, temporal and spatial arrangements, conservation strategies for soil and water, and livestock integration—that are required for successful implementation. We have also emphasized the significance of capacity building, community engagement, and ongoing monitoring and assessment in order to guarantee the long-term viability and flexibility of agroforestry systems. Agroforestry adoption on a large scale is greatly aided by institutional and policy assistance. Supportive policies, financial incentives, robust extension services, research advancements, and market development are essential components that can significantly enhance the viability and scalability of agroforestry practices. Secure land tenure and property rights further encourage farmers to invest in and maintain agroforestry systems, providing long-term benefits to both the environment and rural livelihoods. Agroforestry presents a viable and sustainable solution to the pressing challenges faced by agriculture in semi-arid climates. We can create a more resilient, fruitful, and sustainable agricultural landscape that promotes food security, economic growth, and environmental preservation by implementing and expanding agroforestry practices. In addition to meeting farmers' immediate needs, this all-encompassing strategy supports larger initiatives to fend off climate change and advance sustainable development for coming generations.

References

Abbas, F., Hammad, H. M., Fahad, S., Cerdà, A., Rizwan, M., Farhad, W., & Bakhat, H. F. (2017). Agroforestry: A sustainable environmental practice for carbon sequestration under the climate change scenarios—A review. *Environmental Science and Pollution Research, 24*, 11177–11191.

Akesse-Ransford, G., Owusu, E. O., Kyerematen, R., & Adu-Acheampong, S. (2021). Arthropod diversity of cocoa farms under two management systems in the eastern and central regions of Ghana. *Agroforestry Systems, 95*(5), 791–803.

Alemu, M. M. (2016). Ecological benefits of trees as windbreaks and shelterbelts. *International Journal of Ecosystems, 6*, 10–13.

Barakat, H. N. (2009). Arid lands: Challenges and hopes. In *Earth system: History and natural variability-Vol. III*. EOLSS Publishers/UNESCO.

Bargués Tobella, A., Reese, H., Almaw, A., Bayala, J., Malmer, A., Laudon, H., & Ilstedt, U. (2014). The effect of trees on preferential flow and soil infiltrability in an agroforestry parkland in semiarid Burkina Faso. *Water Resources Research, 50*(4), 3342–3354.

Bayala, J., Sanou, J., Teklehaimanot, Z., Kalinganire, A., & Ouédraogo, S. J. (2014). Parklands for buffering climate risk and sustaining agricultural production in the Sahel of West Africa. *Current Opinion in Environmental Sustainability, 6*, 28–34.

Bogale, G. A., & Bekele, S. E. (2023). Sustainability of agroforestry practices and their resilience to climate change adaptation and mitigation in sub-saharanafrica: A review. *Ekológia (Bratislava), 42*(2), 179–192.

Breman, H., & Kessler, J. J. (1997). The potential benefits of agroforestry in the Sahel and other semi-arid regions. *European Journal of Agronomy, 7*(1–3), 25–33.

Campos, P., Daly-Hassen, H., Ovando, P., Chebil, A., & Oviedo, J. L. (2009). Economics of multiple use cork oak woodland: Two case studies of agroforestry systems. In A. Rigueiro Rodriguez, J. McAdam, & M. R. Mosquera-Losada (Eds.), *Agrofrestry in Europe: Current status and future prospects*. Springer.

Chapman, M., Walker, W. S., Cook-Patton, S. C., Ellis, P. W., Farina, M., Griscom, B. W., & Baccini, A. (2020). Large climate mitigation potential from adding trees to agricultural lands. *Global Change Biology, 26*(8), 4357–4365.

Chebli, Y., Mrabet, R., & Chentouf, M. (2012). Alley cropping as a durable alternative for pasture land development in the drought prone region of Eastern Mrocco. In Z. Acar, A. López-Francos, & C. Porqueddu (Eds.), *New approaches for grassland research in a context of climate and socio-eco nomic changes*. CIHEAM.

Cleugh, H. A. (1998). Effects of windbreaks on airflow, microclimates and crop yields. *Agroforestry Systems, 41*, 55–84.

De Bruyn, L. L., & Conacher, A. J. (1990). The role of termites and ants in soil modification-a review. *Soil Research, 28*(1), 55–93.

Dev, I., Ram, A., Ahlawat, S. P., Palsaniya, D. R., Singh, R., Dhyani, S. K., Kumar, N., Tewari, R. K., Singh, M., Babanna, S. K., Newaj, R., Dwivedi, R. P., Kumar, R. V., Yadav, R. S., Chand, L., Kumar, D., & Prasad, J. (2020). Bamboo-based agroforestry system (Dendrocalamusstrictus+ sesame–chickpea) for enhancing productivity in semi-arid tropics of central India. *Agroforestry Systems, 94*, 1725–1739.

Dietz, K. J., Zörb, C., & Geilfus, C. M. (2021). Drought and crop yield. *Plant Biology, 23*(6), 881–893.

Ding, X. J. (2007). *Reconstruction of temperature variations since 1500ad in the arid-semiarid central-eastern Asia using tree-ring chronology-based om EOF iteration scheme*. Master Dissertation, Lanzhou University.

FAO. (2021). *FAO statistical year book 2021.*

Félix, G. F., Scholberg, J. M., Clermont-Dauphin, C., Cournac, L., & Tittonell, P. (2018). Enhancing agroecosystem productivity with woody perennials in semi-arid West Africa. A meta-analysis. *Agronomy for Sustainable Development, 38*(6), 57.

Gallart, F., Solé, A., Puigdefábregas, J., & Lázaro, R. (2002). Badland systems in the Mediterranean. In *Dryland Rivers: Hydrology and geomorphology of semi-arid channels* (pp. 299–326). Wiley.

Gaur, M. K., & Squires, V. R. (2018). *Climate variability impacts on land use and livelihoods in drylands* (p. 348). Springer International Publishing.

Ghamghami, M., & Beiranvand, J. P. (2022). Rainfed crop yield response to climate change in Iran. *Regional Environmental Change, 22*, 1–17.

Higgins, G. M., & Kassam, A. H. (1981). The FAO agro-ecological zone approach to determination of land potential. *Pédologie, 31*(2), 147–168.

International Fund for Agriculture Development (IFAD). (2000). *The rangelands of Arid and Semiarid areas*. Retrieved from https://www.ifad.org/en/

IPCC. (2000). *Land use, land use change and forestry*. UNEP and WMO.

IPCC. (2018). Summary for policymakers. In: V. Masson-Delmotte, et al. (Eds.), *Global Warming of 1.5°C. An IPCC Special Report on the impacts of global warming of 1.5°C above pre-industrial levels and related global greenhouse gas emission pathways, in the context of strengthening the global response to the threat of climate change, sustainable development, and efforts to eradicate poverty* (p. 32). World Meteorological Organization.

Jose, S. (2009). Agroforestry for ecosystem services and environmental benefits: An overview. *Agroforestry Systems, 76*, 1–10.

Karimi, V., Karami, E., & Keshavarz, M. (2018). Climate change and agriculture: Impacts and adaptive responses in Iran. *Journal of Integrative Agriculture, 17*(1), 1–15.

Kašanin-Grubin, M., Vergari, F., Troiani, F., & Della Seta, M. (2018). The role of lithology: Parent material controls on badland development. In *Badlands dynamics in a context of global change* (pp. 61–109). Elsevier.

Kheiri, M., Kambouzia, J., Sayahnia, R., Soufizadeh, S., Damghani, A. M., & Azadi, H. (2023). Environmental and socioeconomic assessment of agroforestry implementation in Iran. *Journal for Nature Conservation, 72*, 126358.

Lal, R. (2004). Carbon sequestration in dryland ecosystems. *Environmental Management, 33*, 528–544.

Lazarev, M. M. (2006). Transformation of the annual water budget of soils under shelterbelts. *Eurasian Soil Science, 39*, 1318–1322.

Lott, J. E., Ong, C. K., & Black, C. R. (2009). Understorey microclimate and crop performance in a Grevillea robusta-based agroforestry system in semi-arid Kenya. *Agricultural and Forest Meteorology, 149*(6–7), 1140–1151.

Moreira, S. L., Pires, C. V., Marcatti, G. E., Santos, R. H., Imbuzeiro, H. M., & Fernandes, R. B. (2018). Intercropping of coffee with the palm tree, macauba, can mitigate climate change effects. *Agricultural and Forest Meteorology, 256*, 379–390.

Mosquera-Losada, M. R., Santiago-Freijanes, J. J., Rigueiro-Rodríguez, A., Rodríguez-Rigueiro, F. J., Martínez, D. A., Pantera, A., & Ferreiro-Domínguez, N. (2020). The importance of agroforestry systems in supporting biodiversity conservation and agricultural production: A European perspective. In *Reconciling agricultural production with biodiversity conservation* (pp. 243–258). Burleigh Dodds Science Publishing.

Murungweni, C., Van Wijk, M. T., Smaling, E. M. A., & Giller, K. E. (2016). Climate-smart crop production in semi-arid areas through increased knowledge of varieties, environment and management factors. *Nutrient Cycling in Agroecosystems, 105*, 183–197.

Nair, P. K. R., Nair, V. D., Mohan Kumar, B., & Showalter, J. M. (2010). Carbon sequestration in agroforestry systems. In L. Donald (Ed.), *Sparks, advances in agronomy* (Vol. 108, pp. 237–307). Academic Press.

Palsaniya, D. R., Kumar, S., Das, M. M., Rai, S. K., Kumar, T. K., Kumar, S., et al. (2023). Rain water harvesting, agroforestry and goat based intensification for livelihood resilience in drought prone rainfed smallholder farming system: A case for semi-arid tropics. *Agroforestry Systems, 97*(8), 1405–1419.

Pantera, A., Mosquera-Losada, M. R., Herzog, F., & Den Herder, M. (2021). Agroforestry and the environment. *Agroforestry Systems, 95*(5), 767–774.

Partelli, F. L., Araújo, A. V., Vieira, H. D., Dias, J. R. M., Menezes, L. F. T. D., & Ramalho, J. C. (2014). Microclimate and development of 'Conilon' coffee intercropped with rubber trees. *Pesquisa Agropecuária Brasileira, 49*, 872–881.

Perdinan, P. (2010). A rationale for international cooperation in implementing adaptation strategies to climate change in the face of global inequality. *Journal of Indonesia Focus, 1*(1).

Perry, C. H., Woodall, C. W., Liknes, G. C., & Schoeneberger, M. M. (2009). Filling the gap: Improving estimates of working tree resources in agricultural landscapes. *Agroforestry Systems, 75*, 91–101.

Radmehr, R., Ghorbani, M., & Ziaei, A. N. (2021). Quantifying and managing the water-energy-food nexus in dry regions food insecurity: New methods and evidence. *Agricultural Water Management, 245*, 106588.

Reyes, F., Gosme, M., Wolz, K. J., Lecomte, I., & Dupraz, C. (2021). Alley cropping mitigates the impacts of climate change on a wheat crop in a Mediterranean environment: A biophysical model-based assessment. *Agriculture, 11*(4), 356.

Scholes, R. J. (2020). The future of semi-arid regions: A weak fabric unravels. *Climate, 8*(3), 43.

Shah, N. H., & Paulsen, G. M. (2003). Interaction of drought and high temperature on photosynthesis and grain-filling of wheat. *Plant and Soil, 257*, 219–226.

Shayanmehr, S., Rastegari Henneberry, S., Sabouhi Sabouni, M., & Shahnoushi Foroushani, N. (2020). Climate change and sustainability of crop yield in dry regions food insecurity. *Sustainability, 12*(23), 9890.

Singh, R. P., Saharan, N., & Ong, C. K. (1989). Above and below ground interactions in alley-cropping in semi-arid India. *Agroforestry Systems, 9*, 259–274.

Smith, P., & Wollenberg, E. (2013). Achieving mitigation through synergies with adaptation. In *Climate change mitigation and agriculture* (pp. 50–57). Routledge.

Snyder, K. A., & Tartowski, S. L. (2006). Multi-scale temporal variation in water availability: Implications for vegetation dynamics in arid and semi-arid ecosystems. *Journal of Arid Environments, 65*(2), 219–234.

Sukhdevb, P., Prabhua, R., Kumara, P., Bassic, A., Patwa Shaha, W., Entersa, T., Labbatea, G., & Greenwalta, J. (2015). *REDD+ and a green economy: Opportunities for a mutually supportive relationship*. UN-REDD Programme Policy Brief, Issue 01.

Tadesse, S., Gebretsadik, W., Muthuri, C., Derero, A., Hadgu, K., Said, H., & Dilla, A. (2021). Crop productivity and tree growth in intercropped agroforestry systems in semi-arid and sub-humid regions of Ethiopia. *Agroforestry Systems, 95*(3), 487–498.

Traore, B., Illiasso, A. D., Soulé, M., Arzika, M. Z., Abdoussalam, I., Moussa, A. A., et al. (2024). Low-cost small-scale irrigation for developing an agroforestry system in the semi-arid zone of

Niger: Case of the apple of the Sahel (*Ziziphus mauritiana* L.). *Frontiers in Sustainable Food Systems, 8*, 1282235.
Vasilyev, M. E. (1980). *Melioration through forests and harvest*. Kainar, Almaty, Kazakhstan.
Wallace, J. S., Jackson, N. A., & Ong, C. K. (1999). Modelling soil evaporation in an agroforestry system in Kenya. *Agricultural and Forest Meteorology, 94*(3–4), 189–202.
Wang, C., Wang, H., & Yuan, S. (2023). Precipitation governing vegetation patterns in an arid or semi-arid environment. *Journal of Mathematical Biology, 87*(1), 22.
Wani, S. P., Rockstrom, J., Oweis, T. (2009). *Rainfed agriculture: Unlocking the potential*. CAB International. Comprehensive assessment of water management in agriculture series 7.
Wilson, B. R. S. A., & Daf, J. T. (2003). *Australia's state of the forests report*. Department of Agriculture, Fisheries and Forestry, Govt. of Australia.
World Health Organization. (2018). *The state of food security and nutrition in the world 2018: Building climate resilience for food security and nutrition*. Food & Agriculture Organization.
Yadav, R. S., Yadav, B. L., Chhipa, B. R., Dhyani, S. K., & Ram, M. (2011). Soil biological properties under different tree based traditional agroforestry systems in a semi-arid region of Rajasthan, India. *Agroforestry Systems, 81*, 195–202.
Yasin, G., Nawaz, M. F., Zubair, M., Azhar, M. F., Mohsin Gilani, M., Ashraf, M. N., Qin, A., & Ur Rahman, S. (2023). Role of traditional agroforestry systems in climate change mitigation through carbon sequestration: An investigation from the semi-arid region of Pakistan. *Land, 12*(2), 513.
Zhang, D., Zhang, L., Liu, J., Han, S., Wang, Q., Evers, J., Liu, J., van der Werf, W., & Li, L. (2014). Plant density affects light interception and yield in cotton grown as companion crop in young jujube plantations. *Field Crops Research, 169*, 132–139.
Zhou, Z., Andersen, M. N., & Plauborg, F. (2016). Radiation interception and radiation use efficiency of potato affected by different N fertigation and irrigation regimes. *European Journal of Agronomy, 81*, 129–137.

Multifaceted Applications of Biochar for Sustainable Agriculture

Ruchi Rajpoot, Chitranshi Shrivastava, Priya Sinha, Sudipta Dhara, Pradeep Ranwa, Shubham A. Durgude, and Shiv Vendra Singh

Abstract Biochar, a solid carbon material formed through the thermochemical conversion of biomass, has become a viable instrument for carbon sequestration, soil improvement, and sustainable farming. This chapter critically overviews biochar production techniques, highlighting methods such as pyrolysis, gasification, and hydrothermal carbonization, and considers how variations in feedstock and process conditions influence the physicochemical properties of the resulting biochar. Its high porosity, extensive surface area, and variable carbon composition are fundamental to its effectiveness as a soil amendment. Empirical evidence presented in the chapter demonstrates that biochar addition can drastically alter soil structure by reducing bulk density, improving soil aggregation, and increasing water retention capacities. It also enhances chemical properties, such as cation exchange capacity, pH stabilization, and nutrient retention, which in turn support enhanced nutrient availability for plant growth. Biochar has been shown to neutralize soil acidity effectively and to promote the growth of beneficial microbial communities, fostering an environment that boosts overall soil health. Furthermore, the chapter explores biochar's role in environmental remediation through its ability to immobilize heavy metals and adsorb pollutants; a promising strategy for managing contaminated soils. The development of biochar-based slow-release fertilizers, alongside its integration with precision agriculture methodologies, points to its future potential in reducing chemical fertilizer dependency and mitigating environmental impacts. To fully harness these benefits, the chapter recommends systematic future research to standardize production conditions and evaluate long-term agronomic impacts, thereby establishing biochar as a critical component in advanced agroecological management and sustainable land use planning.

R. Rajpoot · C. Shrivastava · P. Sinha · S. Dhara · P. Ranwa · S. V. Singh (✉)
College of Agriculture, Rani Lakshmi Bai Central Agriculture University, Jhansi, Uttar Pradesh, India

S. A. Durgude
College of Agriculture, Rani Lakshmi Bai Central Agriculture University, Jhansi, Uttar Pradesh, India

Boorlaug Institute for South Asia-CIMMYT, Bihar, India

A. K. Singh et al. (eds.), *Sustainable Agriculture Management in Semi-Arid Climates*, https://doi.org/10.1007/978-3-031-94062-0_5

Keywords Biochar · Sustainable agriculture · Pyrolysis · Soil amendment · Climate change mitigation · Soil fertility

1 Introduction

Over time, net-cultivated areas have decreased and the world's population has increased. Even though crop output has increased significantly over the previous century, 14.3% of people worldwide suffer from malnutrition, and 8.9% of people still go hungry (Tian et al., 2021). Concern over sustainable agricultural waste management is growing along with food security. Some progress has been achieved in the realm of agriculture to address the developing issues, and numerous novel interventions that align with the sustainability goal are in the works (Barman et al., 2024a, b). Recent developments like integrated farming systems, precision agriculture, integrated nutrient management, pest control for cultivation methods, and the use of microbes like Pusa decomposer to break down waste in an environmentally benign manner have opened the road for more organic interventions. Fertilizers and other carbon-based inputs are among the new technologies that contribute to increased agronomic efficiency. The field of biochar is one of the newer ones. Biomass is the collective term for organic compounds derived from living matter or combinations of organic and inorganic materials from those sources. In addition to living things like plants and animals, biomass also consists of waste wood, sludge, and animal feces (Fang et al., 2024). Biochar is the solid material created during the thermochemical breakdown of biomass (Labanya et al., 2024). Proven due to its huge surface area, ion exchange capabilities, and porous structure. Biochar is often produced in clay charcoal furnaces. Biochar improves the soil's capacity to withstand rainfall during drought and fortifies the soil's biological and physical mineralogy (Singh et al., 2019a). Biochar can be used for composting, wastewater treatment, soil modification, carbon sequestration, and energy storage. Pesticide mobility and reactivity in soils and sediments can be influenced by biochar. Biochar can alter the chemical and physical characteristics of natural organic materials through its sorption and binding sites for pesticide chemicals. Biochar is a porous solid with a large surface area that is a viable alternative for wastewater treatment because of its adsorption and porosity qualities. Consequently, it is acknowledged for consistently boosting wastewater pollutant removal (Amalina et al., 2023). Climate change has increased the necessity of reducing CO_2 emissions into the atmosphere. The carbon cycle depends heavily on soil, which directly impacts global warming (Barman et al., 2024a, b). Biochar helps sequester carbon in soil because of its aromatic content, which makes it only marginally resistant to microbial degradation (Brown et al., 2020).

2 Production of Biochar

This section summarizes several different carbonization processes to produce biochar along with their characteristics.

2.1 *Pyrolysis*

Pyrolysis is the thermal breakdown of organic compounds at temperatures between 300 and 900 °C without the presence of oxygen (Singh et al., 2020; Chaturvedi et al., 2021). The biomass's constituent cellulose, hemicellulose, and lignin go through separate reaction pathways during thermal decomposition, such as cross-linking, depolymerization, and fragmentation, at their own temperature, resulting in solid, liquid, and gaseous products. The gaseous combination comprising CO, CO_2, H_2, and hydrocarbons is known as syngas, while the solid and liquid products are called char and bio-oil, respectively. The properties of the raw biomass materials and the modified pyrolysis methods determine the yields of the pyrolysis products.

The reaction temperature, heating rate, and residence time are the variables that affect the pyrolysis operations' end products. According to factors like temperature, pressure, residence time, and heating rate, pyrolysis is divided into three categories: slow, intermediate, and fast pyrolysis (Ahmed et al., 2020). Slow pyrolysis has a slower heating rate (0.1–1 °C/min) and a longer heating duration (up to many hours or days) than quick pyrolysis. Because of its high solid yield, slow pyrolysis is used extensively to create biochar. The mass fraction range of the pyrolysis solid grows as the temperature rises. However, more liquid and gas, such as bio-oil, CO, H2, and syngas, are produced by rapid pyrolysis at a higher temperature and quicker heating rate (Ahmed et al., 2020; Kumar & Panwar, 2024). Although intermediate pyrolysis has circumstances that fall between slow and fast pyrolysis, it is rarely used to produce biochar because it had a better balance between the yields of solid (biochar) and liquid (bio-oil). Even though traditional biochar has many great qualities for its intended use, it is necessary to modify it physically or chemically to enhance its qualities for use in specific application domains. For example, oxidant oxidation, compositing, sulphuration, nitrogenation, physical activation, acid treatment, alkaline treatment, and so on could all be used to modify biochar. Nitro-generation, physical activation modification, and acid treatment of biochar can improve polarity and alkalinity, increase porosity and surface area, and eliminate contaminants, respectively. When used as a catalyst, adsorbent, soil amendment, or electrochemical material, modified biochar with superior qualities will be more advantageous (Kazawadi et al., 2021). Future surprises will come from the advantages of using modified biochar in applications.

2.2 *Gasification*

Gasification is a thermochemical partial oxidation process that uses gasification agents (air, steam, oxygen, CO_2, or gas mixture) to transform carbonaceous materials—such as biomass, coals, and plastics—into gaseous products. Gaseous products (H_2, CO, CO_2, N_2, etc.), liquid products (tar and oil), and solid products (char and ash) are all produced during a gasification process. The production of biochar is

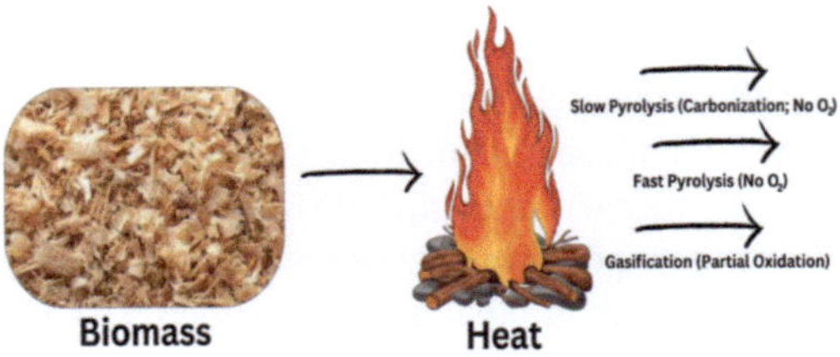

Product Distribution (wt%)		
Solid	Liquid	Gas
35	30	35
10	70	20
10	5	85

Fig. 1 Pyrolysis and gasification product yields (Mohan et al., 2014)

only about 5–10% of the raw biomass material mass due to the fact that gasification is intended to produce gaseous products; this is less than that of fast pyrolysis (15–20%), which contrasts the yields of products from gasification, quick pyrolysis, and slow pyrolysis. Although the gasification mechanism can be broken down into multiple phases (drying–pyrolysis–oxidation–combustion–gasification), as seen below, it is difficult to distinguish between them in terms of temperature and pressure (Fig. 1).

2.3 Hydrothermal Carbonization

When the biomass has low moisture content, dry procedures like pyrolysis and gasification can produce large product yields with little energy loss. However, the majority of biomass materials have high moisture contents, so in order to achieve high product yields and lower process energy, a separate drying step is needed. It is expected that hydrothermal procedures will be able to address this drawback of dry processes. In this article, hydrochar is referred to as "biochar," even though the char from a hydrothermal process is frequently named "hydrochar" to differentiate it from the biochar from dry procedures. In a hydrothermal process, water and biomass are combined in a closed reactor, and after a predetermined amount of time, the temperature is increased for stabilization. In order to keep water liquid over 100 °C, the pressure is also increased. Biochar, bio-oil, and gaseous products including CO, CO_2, CH_4, and $H_2$are the primary results of a hydrothermal process below 250 °C, at 250–400 °C, and above 400 °C, respectively, depending on the temperature under saturated pressure. Hydrothermal carbonization (HTC), hydrothermal liquefaction (HTL), and hydrothermal gasification (HTG) are the names given to the hydrothermal processes in each temperature range, respectively. Reaction temperature, pressure, residence duration, and water biomass ratio are said to be the primary factors influencing the properties of the products (Sabio et al., 2016), and the char generated by the HTC method has a greater C content than the char generated by dry techniques.

2.4 Other Thermochemical Technologies

Besides pyrolysis, gasification, and hydrothermal carbonization, discussed in the foregoing, flash carbonization (Nunoura et al., 2006) and torrefaction are other thermochemical biomass conversion processes.

3 Properties of Biochar

The physicochemical properties of biochar are highly dependent on the feedstock and production conditions. Key properties include:

- Porosity and Surface Area: Biochar typically exhibits a highly porous structure, which contributes to its large surface area. These characteristics enhance its ability to adsorb nutrients, pollutants, and water (Downie et al., 2009).
- Chemical Composition: Biochar is predominantly composed of carbon, along with varying amounts of hydrogen, oxygen, and mineral ash. The carbon content typically ranges between 60% and 90% depending on the pyrolysis conditions (Spokas et al., 2012). High-temperature biochars often contain more aromatic carbon, which is more stable and contributes to long-term carbon sequestration (Lehmann & Joseph, 2009).
- pH and Cation Exchange Capacity (CEC): Biochars often have a neutral to alkaline pH, which can help in ameliorating acidic soils. The CEC of biochar, which determines its ability to retain nutrients, increases over time as the material ages and undergoes oxidation (Cheng et al., 2008).
- Thermal Stability: Biochar's thermal stability is primarily attributed to its high aromatic carbon content, making it resistant to microbial decomposition. This stability underpins its utility as a long-term carbon sink (Woolf et al., 2010).
- Nutrient Content: Biochar contains essential nutrients such as nitrogen, phosphorus, and potassium, though these are often present in forms that require microbial activity for release (Kammann et al., 2011).

4 Agricultural Applications of Biochar

4.1 Soil Amendment and Carbon Sequestration

According to Bagrecha et al. (2024), in addition to being a key strategy for enhancing soil nutrient and water retention and supporting soil biological functioning, biochar is considered a crucial tool for addressing climate change by sequestering CO_2 in the soil. Biochar is a powerful soil amendment, thanks to its stable carbon content, high porosity, and large surface area. These qualities help improve soil health in many ways, making it a valuable tool for sustainable agriculture (Das et al., 2021; Zhang et al., 2021). Biochar has been found to improve soil's physical, chemical, and biological properties (Zhang et al., 2021), especially over a long-term application (Joseph et al., 2021)—therefore enhancing agricultural productivity (Diatta et al., 2020). In addition, biochar is essential for sequestering carbon, which helps combat climate change. Its stability lowers greenhouse gas emissions by allowing carbon to stay trapped in the soil for long periods of time. Furthermore, soils modified with biochar can withstand erosion and compaction better, fostering agricultural resilience over the long run. A key element in the shift to sustainable farming methods, the use of biochar in contemporary agriculture represents the meeting point of increased productivity and environmental stewardship. Its many advantages highlight how it could transform soil management in response to changing agricultural needs (Singh et al., 2023).

4.2 *Biochar Compost*

A straightforward technique for converting biodegradable materials into a humus-like substance that may be applied to soil or utilized as bio-manure is composting. However, when it comes to nitrogen loss and greenhouse gas emissions, manure fertilizers have some serious disadvantages (Figueiredo et al., 2024). Because of its special qualities, including chemical recalcitrance, porosity, adsorption capacity, and large surface area, biochar increases soil nutrient supply, microbial growth, soil organic matter, moisture retention, and plant growth while lowering fertilizer use, greenhouse gas emissions, nutrient loss, degradation, pollutant migration, and pollutant bioavailability (Wu et al., 2017). Because of its beneficial effects on each other's qualities, biochar has drawn more attention as a composting and soil remediation supplement. Using biochar as a compost additive has several advantages, such as raising compost temperature, reducing soil bulk density, and improving microbial activity through aeration.

4.3 *Slow-Release Fertilizers*

A crucial factor in increasing agricultural output and ensuring food security is the efficient application of fertilizers (Kharbach & Chfadi, 2021). The nutrients in synthetic fertilizers are highly soluble and swiftly released into the soil. Nonetheless, the nutritional requirements of crops are associated with their vegetative stage, which is sometimes low at first (Beig et al., 2022). A disparity between the availability of synthetic fertilizer and the nutrient needs of crops is the most frequent result of this (Meyer et al., 2018). According to Beig et al., crops that receive synthetic fertilizers only consume roughly 40–60% nitrogen (N), 15–20% phosphorous (P), and 50–60% potassium (K). The environment loses some of the nutrients that the plants do not use (Remya et al., 2021). Therefore, a barrier to sustainable agricultural output is the low effectiveness of synthetic fertilizers (Zulfiqar et al., 2019). Next, an eco-friendly and sustainable method of fertilizing the soil needs to be created right away (Beig et al., 2022). Using smart fertilizers, which use a controlled or gradual release mechanism to make nutrients available based on crop demands, is one such alternative (Roy et al., 2022; Venugopalan et al., 2022). According to Calabi-Floody et al. (2018), one such option is the use of smart fertilizers, which release nutrients gradually or under control in response to crop needs.

4.4 Plant Growth Promoter

It has been shown that biochar made from a range of materials, such as agricultural and forestry wastes, as well as animal, human, and agricultural wastes, can successfully boost plant production or growth. Plant growth can be facilitated by a variety of mechanisms, such as improved soil physicochemical properties, changes in soil microbial populations (Zheng et al., 2018a, b), increased nutrient availability and uptake (Xi et al., 2020), or modifications to gene expression patterns that may impact plant growth. Utilizing rice-derived biochar has shown promising results in terms of boosting plant growth and productivity. Rice straw that had been pyrolyzed at 400 °C and applied at a rate of 2.5% (w/w) enhanced the availability of P and K and showed a positive association between the increase in ryegrass biomass and accessible K. According to Xi et al. (2020), applying 2% (w/w) rice biochar increased the quantity of N and K available in the soil and produced higher lettuce plants with longer roots, stronger leaves and stems, and more leaf area.

4.5 Pollutant Immobilization

Because of industrialization, water contamination has become a major global problem. The quality of the water sources around the world has been declining in recent years, and the aquatic environment has been worse. Therefore, it is imperative to regulate and reduce water pollution. Among the main methods for treating wastewater are the physicochemical approach, membrane technology, and microbiological approach. Because of its many uses and superior elimination efficiency, the microbial approach has attracted the most attention among these (Pandey et al., 2024a, b). The goal of the microbial wastewater treatment method is to cleanse wastewater by breaking down organic materials in wastewater through the metabolic activity of microorganisms (Pandey et al., 2024a, b). The microbial approach does, however, still have several drawbacks, including the loss or decrease of microbial activity and the challenge of recovery and separation following adsorption. Microbial immobilization technology (MIT) is regarded as a promising and successful technique. MIT wastewater treatment has become a global research center because of its ability to sustain high microbial activity and shield cells from environmental harm.

4.6 Heavy Metal Immobilization

Since biochar's surface area (SA), microporosity, surface functional groups, pH, and cation exchange capacity are all better than those of raw feedstock, applying it to polluted soil is thought to be a potential technique for HM immobilization (Singh et al., 2023). A range of biochar made from diverse feedstocks (such as manure, crop residue, and sewage sludge) and under varied production circumstances (such as gasification, hydrothermal carbonization, slow pyrolysis, and quick pyrolysis) (Li et al., 2022) have been used to immobilize HMs (and metalloids) such as, Cd, Cu, Pb, Cr, Ni, Co, and Zn in soils. The type of biochar (e.g., production conditions and physicochemical properties), soil characteristics (e.g., pH, organic matter content, and electrical conductivity (EC)), and HM characteristics (e.g., valency and ionic radius) all affect the HM immobilization efficiency in biochar-amended soils. Numerous adsorption/immobilization processes and elements, including cation exchange, electrostatic interaction, precipitation, and complexation by surface functional groups, have been found to affect immobilization efficiency (Pramanick et al., 2025).

5 Impact of Biochar

This section highlights the effects of biochar on soil health, its physical and chemical properties, nutritional status, and biological activities within the soil rhizosphere and subsequently, on crop growth.

5.1 Soil Health

In many parts of the world, low soil fertility is a major problem (FAO, 2019). Soil is deemed fertile if it can provide the water and nutrients required to sustain plant growth while being devoid of dangerous materials that could impede it (Voltr, 2012). The fertility of the soil is determined by its physical, chemical, and biological properties. Low soil fertility is a widespread issue in several regions of the world (FAO, 2019). For example, soil in arid and semi-arid areas usually produces insufficient amounts of nutrients and has a low capacity to hold water for the majority of crops. Rainforest areas find it difficult to sustain agricultural productivity because of high rainfall, a restricted ability to bind cations, and a quick loss of essential nutrients from the topsoil. Furthermore, the high temperature and frequent breakdown caused notable rates of soil organic matter (SOM) mineralization (Bruun et al., 2015).

Therefore, proper SOM levels and nutrient life cycles are essential to a soil management system's effectiveness. Soil health could be significantly improved by using biochar. It has also been looked at if applying biochar could increase the productivity of damaged pastureland. According to (Brtnicky et al., 2021) and Joseph et al., 2021), Among its many potential applications, biochar is likely to: (1) improve soil properties; (2) increase soil protection and water retention; (3) prevent soil degradation and losses; (4) increase soil nutrient content and sequestration; (5) lessen the impact of potentially toxic substances; (6) promote the well-being of organisms in the soil environment; (7) enhance plant growth and biomass production and quality; and (8) increase crop yields and agricultural sector profits.

5.1.1 Physical Properties

- **Bulk Density and Soil Compaction**
  The density and weight of the soil have a direct and substantial influence on its qualities. Another important factor affecting root development is the physical stress of soil compaction, which normally occurs at a depth of around 30 cm in the soil. When the bulk density exceeds 1.7 g cm^{-3}, insufficient oxygenation and stamina hinder root growth. The reduced soil pore diameter inhibits root growth because roots cannot grow via soil pores smaller than the root cap diameter (Xiang et al., 2017). Compacted soils with little oxygen diffusion and little porosity can lead to anoxic conditions and slow plant growth. The physical properties of deteriorated and nutrient-poor soils can be improved and these effects can be mitigated by using biochar. Applying biochar has been shown to decrease soil compaction by over 10% (Peake et al., 2014).

- **Porosity and Water-Holding Capacity**
  Several field tests have demonstrated that applying biochar can improve soil porosity and water retention. Biochar's internal porosity, or pore size of <10 μm, determines its ability to hold water in the soil. Because biochar reduces hydraulic conductivity, traps water in pores, and replenishes water lost owing to gravity, it promotes water retention. It has been shown that as biochar inputs rise, soil water availability decreases due to its hydrophobic properties. Generally, biochar generated at temperatures below 450 °C exhibits hydrophobicity. Furthermore, it has been suggested that biochar can be used as a binder to increase the strength of soil aggregates, porosity, and water absorption (Jien et al., 2021) (Fig. 2).

5.1.2 Soil Chemical Properties

The use of biochar has shown significant potential to enhance the chemical properties of soil. The application of biochar and other organic amendments has been shown to improve agricultural productivity by increasing cation exchange capacity and nitrogen content in soil.

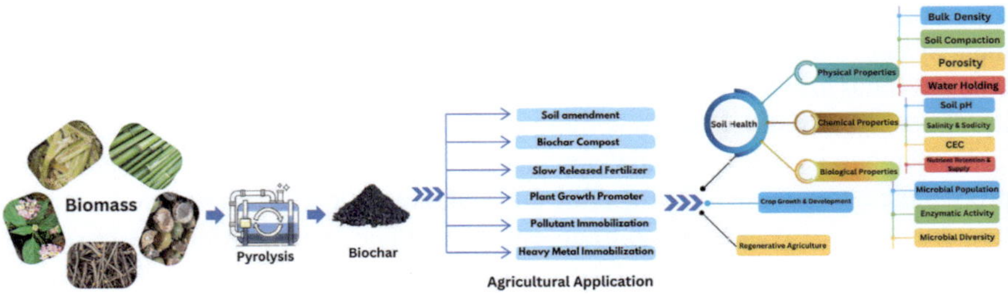

Fig. 2 Agricultural applications and potential impacts of biochar

- **Soil pH**
  Biochar can be utilized to counteract soil acidification because of its high alkalinity, potent buffering capabilities, and functional groups. It also raises the pH of the soil and the availability of plant nutrients. Furthermore, it liberates cations from charcoal, including potassium, magnesium, calcium, and sodium (Steiner et al., 2007).

- **Salinity and Sodicity**
  Because of saline irrigation and artificial fertilizers, certain soils have high levels of endemic salinity. Turf features include increased pH, ESP, and EC. By means of processes like dispersion, clay swelling, and hardness, these soils lessen agglomeration. Additionally, dryness and osmotic stress brought on by high soil salinity can slow down the soil's microbial and metabolic activity (Singh, 2023a, b). Because they have less organic matter, saline soils are less structurally stable. Therefore, reducing salt stress and improving plant growth in soils can be achieved by adding organic additions like biochar. Adding acidic (low pH) charcoal to these soils can help minimize salt and fertility issues (Hagner et al., 2016). By a variety of methods, biochar reduces soil salinity and sodicity. First of all, it improves soil structure, which lessens the effects of elevated sodium levels and encourages greater water infiltration. Second, the cation exchange ability of biochar aids in ion balance in the soil, reducing plant stress from salt. Furthermore, it promotes the growth of salt-tolerant organisms by facilitating microbial activity, which enhances soil health and lowers salinity.

- **Cation Exchange Capacity**
  When newly added biochar comes into contact with oxygen and soil water, it conducts surface oxidation processes that raise the CEC and net negative charge. The amount of soil organic matter (SOM) and surface negative mineral charge enhances the cation exchange capacity (CEC). The high reactivity of the biochar surface is caused by the presence of several reactive functional groups (COOH, OH, CO, C-O, N, siloxane), some of which are pH-dependent high CEC soils have the ability to hold onto plant nutrients and cations on the surface of clay, humus, and charcoal, preventing nutrients from leaking out or being absorbed by plants. The addition of alkaline or acidic chemicals had little to no effect on the

pH of the soil because of its high CEC concentration, which gave it a high buffering capacity (Table 1).

- **Nutrient Retention & Supply**
  Especially in tropical areas, the monsoon season causes nutrient loss (food waste imported from outside), accelerates soil acidification, reduces crop yields, and raises fertilizer needs. Biochar is a useful addition to soil to stop nitrogen loss (Luo et al., 2020). The gradual release of nutrients from more organic matter, improved organic matter stabilization, improved nutrient utilization, and retention of cations and CEC are some long-term advantages of adding biochar (Sohi et al., 2010).

5.1.3 Soil Biological Properties

Soil contains a variety of microbial populations that are always shifting in response to land management practices, climate, and soil characteristics. Plant production, soil nutrient cycling, and organic carbon content are all strongly correlated with the diversity and activity of soil microbes. The kind of soil, the amount and quality of biochar, and the activity of soil microbes all affect how adding biochar affects the soil. By giving soil bacteria carbon and nutrients to flourish, biochar promotes their activity. Furthermore, it offers an environment that is conducive to growth (Lu et al., 2020). Furthermore, biochar decreased the pH shift microhabitat on charcoal pellets and improved the soil's capacity to absorb acid changes. Biochar has improved beneficial bacteria that are essential to agriculture and soil enzymes that have the potential to transform soil microbial ecology. The quantity of microorganisms in the soil has a direct bearing on soil ecology. By giving soil microbial communities substrates and homes, biochar can boost microbial diversity. Furthermore, soil microorganisms are susceptible to modifications in the soil environment. Biochar can interact with soil particles to affect microbial populations by generating microenvironments, supplying unstable organic molecules for bacterial development, and activating nutrients (Conte et al., 2014).

5.2 Crop Growth and Productivity

Studies on the application of soil biochar have examined its potential impacts on plant growth and development, with varying results depending on the type of soil, the type of plant, and the particular characteristics of the biochar. Biochar increases soil aeration by increasing soil porosity, which facilitates improved root penetration and, consequently, plant growth by facilitating the roots' more effective access to resources. Surface area, another characteristic of biochar, is essential for enhancing plant growth. Its large surface area and abundance of functional groups let it absorb more nitrogen, phosphorous, and potassium, which lowers leaching and increases

Table 1 Effect of biochar on physical properties, chemical properties, biological, and GHG emission

S. no.	Type of biochar	Application rate/dose	Experimental details	Effect	Reference
Physical properties					
1.	Corn feedstock	0, 11.3, and 10 ton biochar ha^{-1}	Incubation study	1. Biochar applicationincreased Ag. St. and θ_V, decreased bulk density(increased porosity). No effect on soil water repellence. 2. Biochar helped drainage capacity in poorly drained soils.	Yu (2022)
2.	Wheat straw biochar	0, 5, 10, 15, 25,and 40 ton biochar ha^{-1}	Field experiment	1. PAP increased at 10 and 40 ton ha^{-1}, PAW at 40 ton ha^{-1} increase in soil EC and BS; additive effect on SOC. 2. Increased crop yield of maize and mustard in years 2 and 3	Smit (2023)
3.	Corn	4 ton ha^{-1}	–	Reduced bulk density 21.68%	Saffari et al. (2021)
4.	Birch	9 ton ha^{-1}	–	Increased porosity 1.9%	Karhu et al. (2011)
Chemical properties					
1.	Corn cob	10 ton ha^{-1}	–	pH increased from 5.9 to 6.2	Amoakwah et al. (2017)
2.	Woody shrub	4 ton ha^{-1}	–	CEC (cmol ha^{-1} kg^{-1}) increased from 11.19 to 45.66	Pandit et al. (2017)
3.	Wheat straw	0, 5, 10, 15, 25, and 40 t ha^{-1}	Field experiment	Increase in soil EC and BS	Smit (2023)
4.	Hard wood	30 t ha^{-1}	Sandy loam	Increased total nitrogen content from 0.18 to 0.21	Adekiya et al. (2020)

Biological					
1.	Corn stover (550 °C)	1%	–	*Actinobacteria, Chloroflexbacteria, Patellobacteria, Saccharimonadales, Arthrobacter*, and *Bacillus* abundance enhanced in biochar-added soils.	Ma et al. (2023)
2.	Pine and Spruce (550 °C)	3 ton ha^{-1}	–	Microbial biomass carbon was highest in biochar added with manure and fertilizer	Jiang et al. (2023)
3.	Peanut shells and wheat straw	10% (v/v)	Subtropical landfill cover soil	Increased Acidobacteria population from 4.86% to 5.91%	Lu et al. (2020)
4.	Rice husk Biochar(300 °C)	10 ton ha^{-1}	Sandy soil	The soil microbial biomass of carbon, nitrogen, and phosphorus exhibited a notable increase	Singh et al. (2018)
Contamination					
1.	Millet stalks (600 °C)	Application of biochar (5% w/w)	Pot experiment on Pakchoi (*Brassica chinensis L.*) soils to the impact of H_3PO_4 • modified and unmodified biochar on Cd, OTC, and antibiotic resistance genes distribution	Cd levels were significantly reduced in shoot and root of pakchoi by 14.41% and 20.19%, respectively	Qin et al. (2024)
2.	Wood chips	Application of biochar (0.3%, 0.6, 1% w/w)	Pot experiment in corn crop under brown soil for 51 days • Assess the alkaline reaction, metal sorption, complexation capacity, and macro- and microelements content	Increase in the content of available form of Pb and decrease in Zn was observed after biochar addition.	Rolka et al. (2024)

(continued)

Table 1 (continued)

S. no.	Type of biochar	Application rate/dose	Experimental details	Effect	Reference
3.	Wheat straw (600 °C)	Application of biochar (2% w/w)	Pot experiment • Phosphate modified wheat straw biochar	Biochar reduced the bioavailability by 13.86%–72.37% in soil.	Yu et al. (2024)
4.	Co-pyrolyzed Wheat straw, rice husk, pig manure, and oyster shell.	Application of biochar (2% w/w)	Pot experiment with Chinese cabbage • Compare the effects of biochar with limestone on soil Cd and Pb immobilization at two harvest periods	Co-pyrolyzed biochar soil application reduced the concentration of Cd and Pb by 3.0–38.0% and 42.7–75.6%	Lian et al. (2023)
GHG emission					
1.	Rice straw (500 °C)	7.5	Submerged paddy soil with a pH value of 6.31.	Conventional fertilization (CF), CF with straw, and optimal fertilization 1 15% less fertilizer all resulted in substantial increases in cumulative CH_4 emissions of11.80% and 2.35%, respectively, while CF_1 Biochar (B) significantly reduced cumulative CH_4 emissions by 27.80% and 28.46%, respectively. Optimal fertilization (OF) boosted the ability of biochar to reduce N_2O emissions, with CF1B and OF 1 B obtaining the greatest N_2Oreductions of 30.56% and 32.21%, respectively	Li et al. (2023)

2.	Activated rice hull biochar (850 °C) pellets with40% of N (ARHBP40%),and activated palm biochar pellets with40% of N (APBP-40%)	ARHBP-40%, 36 kg ha^{-1} based on total nitrogen requirement, APBP-40%, 36 kg ha^{-1} basedon total nitrogenrequirement.	Clay loam soil, biochar applied to rice crop.	During rice cultivation, N_2O emissions occurred in the following order: Control >APBP 40% >ARHBP 40%.	Shin et al. (2021)
3.	Wheat straw (450 °C)	Wheat straw (St): 30 t ha^{-1}, biochar (Bc): 20 t ha^{-1}, and wheat straw plus its biochar (StBc): 30 t ha^{-1} of wheat straw and 20 t ha^{-1} of biochar application.	Black Chernozem, silty clay loam texture.	–	Duan et al. (2020)
4.	Wheat straw (450 °C)	N fertilizer: 0 and 300 kg ha^{-1}. Biochar: 0, 20 (B_1), and 40 (B_2) t ha^{-1} (N_0B_0, N_0B_1, N_0B_2, N_1B_0, N_1B_1, N_1B_2)	Littoral clay salt soil. Biochar was applied to paddy	In comparison to the N_0B_0 value of the treatment, biochar amendment raised the total N_2O emissions from the N_0B_1 and N_0B_2 treatments by 23.4% and 30.5%, respectively. Regardless of the addition of N fertilizer, biochar amendment raised the cumulative CH_4 emissions by an average of 14.4%and 14.6% in the B_1 and B_2 treatments, respectively, in comparison to values from sites without biochar treatments.	Sun et al. (2020)

plant availability of these nutrients. Applications of biochar may have varying effects on different plant species. The application of biochar may have little to no effect on some plants, depending on their particular nutritional and environmental needs. Research on maize showed that when the rate of biochar application was between 2 and 4 t/ha, plant root traits, leaf area, plant growth, morphological, and yield-related parameters improved, while when the rate exceeded 6 t/ha, plant height, number of grains per cob, grains, and biological yield decreased. According to some other research, applying biochar to maize can raise its chlorophyll content up to a point, but once it reaches 12 t/ha, negative consequences begin to manifest. Research on how biochar affects the physiological and biochemical traits of plants is still lacking, though. Adding biochar to acidic soils can raise their pH because it is frequently alkaline. This improves nutrient availability and makes the environment more favorable for most plants. In acidic soils, many nutrients become less available to plants, though biochar helps offset this effect. Only when the optimality is understood and the application is properly customized can the addition of biochar to the soil be extremely beneficial in promoting plant growth (Singh et al., 2019a, b) (Table 2).

5.3 GHG Mitigation

Biochar, generated by thermochemical conversion of biomass, has a multifactorial strategy to reduce greenhouse gas (GHG) emissions. Its utility does not only depend on long-term carbon sequestration but also on the capacity to change the physico-chemical properties of soil and nitrogen cycle dynamics, in turn regulating GHG fluxes.

- **Carbon Sequestration and Stability**
- Biochar contains large amounts of stable aromatic carbon forms that are immune to microbial decomposition. Integrated in the ground, biochar behaves as a perpetual carbon storage entity, binding well away the otherwise oxidizable carbon into CO_2 through processes of natural degradation. Carbon mineralization rates in biochar are quite low in comparison to untreated biomass, assuring that sequestration can hold for over a century. It is primarily its production temperature, residence time, and nature of the raw materials that condition biochar to such stability. High-temperature pyrolysis generally leads to more carbonized and stable biochars and hence optimizes their sequestration potential.
- **Influence on Soil Microbial Activity and Nitrogen Cycling**
- Addition of biochar alters major soil properties including pH, porosity, and nutrient retention. These alterations impact the structure of the soil microbial community and their activity, which plays a crucial role in nitrogen cycling processes. A properly structured biochar-amended soil is likely to have enhanced aeration and better drainage, hence lowering conditions favorable to denitrification—the dominant mechanism for nitrous oxide (N_2O) production. In addition, through

Table 2 Effect of biochar on plant growth and development

S. no.	Type of biochar	Experimental details	Effect	Reference
1.	Wheat straw (550 °C)	Cotton plants (@2%, w/w)	Salt stress inhibited growth and yield but improved water use efficiency	Hou et al. (2023)
2.	Peanut Shell (300 °C)	Sesbania (@1.5–10% w/w)	Improved soil health, enhanced nutrient availability, elevated bacterial activities, and abundances related to nutrient transformations	Zheng et al., (2018a, b)
3.	Morus alba	*Brassica napus* L	Increase water use efficiency	Ullah et al. (2023)
4.	Pruning waste	Bread wheat	Fe and Zn uptake rates increases	Vahedi et al. (2022)
5.	Mushroom spent	Cauliflower	Positive correlation with growth, yield, and biochemical response of cauliflower	Širić et al. (2022)
6.	Wheat straw	Paddy field	Decrease the nitrogen, phosphorus concentration, and organic pollutants, inhibit the diffusion of nonpoint source pollutants, and reduce the risk of water pollution caused by straw returning	Liu et al. (2020)
7.	Wheat straw	Wheat	Improved spike length (6.52%), thousand-grain weight (6.42%), grains per spike (3.07%), biological (9.43%) and economic yield (~14%); increased WUE, chlorophyll content	Haider et al. (2020)
8.	Plant bur (700 °C)	Batch experiment of sulfamethazine sorption studies was conducted; sulfamethazine was used at 50 mg kg^{-1}	Steam-treated biochar enhanced diffusion of herbicide	Rajapaksha et al. (2014)
9.	Olive mill waste	Soil column leaching and dissipation experiment was taken up • Metalaxyl was applied at 2 kg ha^{-1}	Sorption of fungicide by biochar	Gámiz et al. (2016)
10	Peanut Shell (350 °C)	Sesbania (*Sesbania grandiflora* L. at1–10% w/w)	Pot experiment, BC-compost is superior to BC, with higher rates of depressed growth	Luo et al. (2017)

the augmentation of cation exchange capacity (CEC), biochar enhances fertilizer use efficiency by keeping ammonium (NH_4^+) in the soil. This action prevents the loss of nitrogen by volatilization or leaching, which are typical precursors to N_2O emissions. The modulation of microbial enzymatic processes in the presence of biochar can also inhibit the emission of N_2O by affecting denitrifiers and nitrifiers, hence aiding in GHG reduction.

- **Reduction in Methane Emissions**
- An anaerobic soil conditions, e.g., rice paddies or waterlogged fields, methane (CH_4) production is a major source of emissions. Biochar addition can have a major impact on the redox state of such soils. By enhancing soil aeration and decreasing anaerobic microsites, biochar suppresses methanogenic activity. Furthermore, biochar surfaces can adsorb organic substrates that are normally utilized by methanogens, thus directly inhibiting the biochemical pathways toward CH_4 formation.
- **Life Cycle Considerations**
- An adequate assessment of GHG reduction by biochar needs to account for the whole life cycle—ranging from biomass collection and conversion to the production of biochar and its application to the soil. While the thermochemical conversion process does release CO_2 and trace quantities of other GHGs, these emissions are typically compensated by the long-term carbon sequestration and the ancillary soil enhancements leading to reduced N_2O and CH_4 emissions. Life cycle assessments (LCAs) have shown that with optimization, the net GHG balance of biochar systems is highly positive. Parameter sensitivity studies indicate that feedstock choice, pyrolysis temperature, and application rate are important in deciding the overall climate mitigation potential.

6 Future Prospects

Biochar has emerged as a versatile and sustainable solution for addressing critical challenges in agriculture and environmental management. The application of biochar presents an environmentally and economically viable alternative in farming. Its affordability, contribution to reducing greenhouse gas emissions, enhancement of soil fertility, carbon sequestration, and role in soil and water remediation creates vast opportunities for research and innovation. However, to fully harness its benefits, targeted research and strategic implementation are essential. The following future directions outline key areas for advancing biochar research and application.

- Perform in-depth research on carbon-rich agricultural inputs (biochar, manure, and compost) by examining their interactions in various agroecosystems and soil types to develop effective ecological strategies. Evaluate economic viability to provide sustainable application.
- Develop nano-biochar further to utilize its increased surface area for better absorption and chemical interactions (Nepal et al., 2023). Standardize production procedures to provide uniform quality across applications.
- Examine short and long-term impacts of biochar on soil properties such as aggregation, pH, and organic carbon content (Singh et al., 2024). Formulate models to forecast performance over various time scales and environments.
- Tailor biochar applications according to soil type, since effects differ with soil type. Design decision support tools to assist farmers in choosing the right biochar types for their conditions.

- Improve biochar manufacturing by testing varied biomass feedstocks and manufacturing methods that affect the properties of biochar. Implement quality control parameters for reproducible performance.
- Investigate biochar's adsorption potential as a carbon-free approach to the cleanup of soil and water pollution. Create targeted formulations for site-specific environmental cleanup.
- Promote fertilizer-activated biochar advancement to increase nutrient use efficiency toward more sustainable agriculture. Determine the economic advantage of enhanced nutrient retention in multiple cropping systems.
- Combine biochar with precision farming technologies for selective applications that enhance soil health and crop yields (Khan et al., 2024). Create sensing technologies to track performance and inform applications.
- Estimate the role of biochar in curbing greenhouse gas emissions and improving carbon sequestration (Nepal et al., 2023). Establish standardized approaches to facilitate carbon credit schemes.
- Install demonstration plots illustrating the environmental and economic sustainability of biochar. Establish guidelines to promote quick uptake by farmers and land managers.

7 Conclusion

Based on the integrated assessment of biochar's multidimensional effects, it is clear that biochar possesses its revolutionary role in sustainable agriculture. Its addition enhances soil physical qualities by increasing aggregate stability, water-holding capacity, and porosity, forming a more stable soil structure. Concurrently, biochar alters soil chemical qualities by optimizing pH, enhancing cation exchange capacity, and raising organic carbon levels, thus enhancing the availability of nutrients and enhancing the overall fertility of the soil. Aside from these agronomic advantages, biochar is also a long-term carbon sink that reduces greenhouse gas emissions. Such a role is supported by its stable aromatic carbon structures that are resistant to fast decomposition, equating to good carbon sequestration. Its porosity not only enhances water retention but also provides microhabitats that can harbor complex microbial communities, hence enabling critical biochemical processes and nutrient cycling. Emerging innovations like nano-enhancement and interfacing with precision agriculture further highlight the scientific potential of biochar. Such developments are focused on enhancing its reactivity and making it application-targeted, paving the way for even higher levels of environmental and productivity improvement. Overall, the results promote widespread use of biochar as a foundation for taking sustainable, resilient, and science-based agriculture forward, which is instrumental in addressing impending environmental issues.

References

Adekiya, A. O., Agbede, T. M., Olayanju, A., Ejue, W. S., Adekanye, T. A., Adenusi, T. T., et al. (2020). Effect of biochar on soil properties, soil loss, and cocoyam yield on a tropical sandy loam Alfisol. *The Scientific World Journal, 2020*, 9391630. https://doi.org/10.1155/2020/9391630

Ahmed, A., Abu, M. S., Sukri, R. S., Hussain, M., Farooq, A., Moogi, S., & Park, Y. (2020). Sawdust pyrolysis from the furniture industry in an auger pyrolysis reactor system for biochar and bio-oil production. *Energy Conversion and Management, 226*, 113502.

Amalina, F., Krishnan, S., Zularisam, A. W., & Nasrullah, M. (2023). Recent advancement and applications of biochar technology as a multifunctional component towards a sustainable environment. *Environmental Development, 46*, 100819.

Amoakwah, E., Frimpong, K. A., Okae-Anti, D., & Arthur, E. (2017). Soil water retention, air flow, and pore structure characteristics after corn cob biochar application to a tropical sandy loam. *Geoderma, 307*, 189–197.

Bagrecha, S., Nithinkumar, K., Saikia, N., Meena, R. S., & Singh, S. V. (2024). Microbial dynamics and carbon stability under biochar-amended soils. In *Biochar production for green economy* (pp. 285–309). Academic Press, Elsevier.

Barman, A., Bera, A., Saha, P., Dey, S., Sen, S., Meena, R. S., Singh, S. V., & Singh, A. K. (2024a). Biochar enhances carbon stability and regulates greenhouse gas flux under crop production systems. In *Biochar production for green economy* (pp. 359–387). Academic Press, Elsevier. https://doi.org/10.1201/9781003376446-8

Barman, A., Roy, S., Saha, P., Dey, S., Patel, S., Meena, D. K., Bera, A., Singh, S. V., Mandal, S., Nath, S., & Bagrecha, S. (2024b). Sustainable management and diversification of problematic wastes: Prospects and challenges. In *Biochar production for green economy* (pp. 57–79). Academic Press, Elsevier. https://doi.org/10.1016/B978-0-443-15506-2.00012-2

Beig, B., Niazi, M. B. K., Sher, F., Jahan, Z., Malik, U. S., Khan, M. D., et al. (2022). Nanotechnology-based controlled release of sustanable fertilizers: A review. *Environmental Chemistry Letters, 20*, 2709–2726.

Brown, A. E., Adams, J. M. M., Grasham, O. R., Camargo-Valero, M. A., & Ross, A. B. (2020). An assessment of different integration strategies of hydrothermal carbonisation and anaerobic digestion of water hyacinth. *Energies, 13*(22), 5983.

Brtnicky, M., Datta, R., Holatko, J., Bielska, L., Gusiatin, Z. M., Kucerik, J., Hammerschmiedt, T., Danish, S., Radziemska, M., Mravcova, L., et al. (2021). A critical review of the possible adverse effects of biochar in the soil environment. *Science of the Total Environment, 796*, 148756.

Bruun, T. B., Elberling, B., Neergaard, A., & Magid, J. (2015). Organic carbon dynamics in different soil types after conversion of forest to agriculture. *Land Degradation & Development, 26*(3), 272–283.

Calabi-Floody, M., Medina, J., Rumpel, C., Condron, L. M., Hernandez, M., Dumont, M., et al. (2018). Smart fertilizers as a strategy for sustainable agriculture. In D. L. Sparks (Ed.), *Advances in agronomy* (Vol. 147, pp. 119–157). Academic Press.

Chaturvedi, M., Verma, S., Tiwari, R., Singh, S., & Kumar, R. (2021). Characterization, bioenergy value, and thermal stability of biochars derived from diverse agriculture and forestry lignocellulosic wastes. *Biomass Conversion and Biorefinery*, 13, 879–892.

Cheng, C. H., Lehmann, J., Thies, J. E., Burton, S. D., & Engelhard, M. H. (2008). Oxidation of black carbon by biotic and abiotic processes. *Organic Geochemistry, 39*(11), 1479–1491.

Conte, P., Hanke, U. M., Marsala, V., Cimo, G., Alonzo, G., & Glaser, B. (2014). Mechanisms of water interaction with pore systems of hydrochar and pyrochar from poplar forestry waste. *Journal of Agricultural and Food Chemistry, 62*(21), 4917–4923.

Das, S., Mohanty, S., Sahu, G., Rana, M., & Pilli, K. (2021). Biochar: A sustainable approach for improving soil health and environment. In A. Vieira & S. Carlos Rodrigues (Eds.), *Soil erosion - Current challenges and future perspectives in a changing world*. IntechOpen.

Diatta, A. A., Fike, J. H., Battaglia, M. L., Galbraith, J. M., & Baig, M. B. (2020). Effects of biochar on soil fertility and crop productivity in arid regions: A review. *Arabian Journal of Geosciences, 13*, 595.

Downie, A., Crosky, A., & Munroe, P. (2009). Physical properties of biochar. In J. Lehmann & S. Joseph (Eds.), *Biochar for environmental management*. Earthscan.
Duan, M., Wu, F., Jia, Z., Wang, S., Cai, Y., & Chang, S. X. (2020). Wheat straw and its biochar differently affect soil properties and field-based greenhouse gas emission in a Chernozemic soil. *Biology and Fertility of Soils, 56*(7), 1023–1036.
Fang, Y., Singh, B., & Singh, B. P. (2024). Biochar for future and futuristic biochar. *Science of the Total Environment, 713*, 136633.
Figueiredo, C. C., Melo, L. C. A., Silva, C. A., Fachini, J., de Carneiro, J. S. S., de Morais, E. G., Ndoung, O. C. N., Singh, S. V., & Nandipamu, T. M. K. (2024). Nutrient enriched and co-composted biochar: System productivity and environmental sustainability. In *Biochar production for green economy* (pp. 311–331). Academic Press, Elsevier. https://doi.org/10.1016/B978-0-443-15506-2.00010-9
Food and Agriculture Organization (FAO). (2019). *World fertilizer trends and outlook to 2022.*
Gámiz, B., López-Cabeza, R., Facenda, G., Velarde, P., Hermosín, M. C., Cox, L., & Celis, R. (2016). Effect of synthetic clay and biochar addition on dissipation and enantioselectivity of tebuconazole and metalaxyl in an agricultural soil: Laboratory and field experiments. *Agriculture, Ecosystems & Environment, 230*, 32–41.
Hagner, M., Kemppainen, R., Jauhiainen, L., Tiilikkala, K., & Setälä, H. (2016). The effects of birch (*Betula* spp.) biochar and pyrolysis temperature on soil properties and plant growth. *Soil & Tillage Research, 163*, 224–234.
Haider, I., Raza, M. A. S., Iqbal, R., Aslam, M. U., Habib-ur-Rahman, M., Raja, S., Khan, M. T., Aslam, M. M., Waqas, M., & Ahmad, S. (2020). Potential effects of biochar application on mitigating the drought stress implications on wheat (*Triticum aestivum* L.) under various growth stages. *Journal of Saudi Chemical Society, 24*(12), 974–981.
Hou, J., Zhang, J., Liu, X., Ma, Y., Wei, Z., Wan, H., & Liu, F. (2023). Effect of biochar addition and reduced irrigation regimes on growth, physiology, and water use efficiency of cotton plants under salt stress. *Industrial Crops and Products, 198*, 116702.
Jiang, R. W., Mechler, M. A., & Oelbermann, M. (2023). Exploring the effects of one-time biochar application with low dosage on soil health in temperate climates. *Soil Security, 12*, 100101.
Jien, S.-H., Kuo, Y.-L., Liao, C.-S., Wu, Y.-T., Wu, A. D., Tsang, D. C. W., & Ok, Y. S. (2021). Effects of field-scale in situ biochar incorporation on soil environment in a tropical highly weathered soil. *Environmental Pollution, 272*, 116009.
Joseph, S., Cowie, A. L., Van Zwieten, L., Bolan, N., Budai, A., Buss, W., Cayuela, M. L., Graber, E. R., Ippolito, J. A., Kuzyakov, Y., et al. (2021). How biochar works, and when it doesn't: A review of mechanisms controlling soil and plant responses to biochar. *GCB Bioenergy, 13*, 1731–1764.
Kammann, C., Ratering, S., Eckhard, C., & Müller, C. (2011). Biochar and hydrochar effects on greenhouse gas (CH_4, N_2O) fluxes from soils. *Journal of Environmental Quality, 41*(4), 1052–1066.
Karhu, K., Mattila, T., Bergström, I., & Regina, K. (2011). Biochar addition to agricultural soil increased CH_4 uptake and water holding capacity—Results from a short-term pilot field study. *Agriculture, Ecosystems & Environment, 140*(1–2), 309–313.
Kazawadi, D., Ntalikwa, J., & Kombe, G. (2021). A review of intermediate pyrolysis as a technology of biomass conversion for coproduction of biooil and adsorption biochar. *Journal of Renewable Energy, 2021*, 5533780.
Khan, S., Irshad, S., Mehmood, K., Hasnain, Z., Nwaz, M., Rais, A., & Gul. (2024). *Biochar production and characteristics, its impacts on soil health, crop production, and yield enhancement: A review* (Vol. 13, p. 166). Plants.
Kharbach, M., & Chfadi, T. (2021). General trends in fertilizer use in the world. *Arabian Journal of Geosciences, 14*, 2577.
Kumar, K. V., & Panwar, N. L. (2024). Pyrolysis technologies for biochar production in waste management: A review. *Clean Energy, 8*(4), 61–78.
Labanya, R., Sahu, P., Mandal, S., Singh, S. V., & Meena, R. S. (2024). Crop waste conversion into biochar: An overview. In *Biochar production for green economy* (pp. 1–23). Academic Press, Elsevier. https://doi.org/10.1016/B978-0-443-15506-2.00027-4

Lehmann, J., & Joseph, S. (2009). *Biochar for environmental management: Science and technology*. Earthscan.

Li, D., He, H., Zhou, G., He, Q., & Yang, S. (2023). Rice yield and greenhouse gas emissions due to biochar and straw application under optimal reduced N fertilizers in a double season rice cropping system. *Agronomy, 13*(4), 1023.

Li, J., Li, L., Suvarna, M., Pan, L., Tabatabaei, M., Ok, Y. S., & Wang, X. (2022). Wet wastes to bioenergy and biochar: A critical review with future perspectives. *Science of the Total Environment, 817*, 152921.

Lian, W., Shi, W., Tian, S., Gong, X., Yu, Q., Lu, H., Liu, Z., Zheng, J., Wang, Y., Bian, R., Li, L., & Pan, G. (2023). Preparation and application of biochar from co-pyrolysis of different feedstocks for immobilization of heavy metals in contaminated soil. *Waste Management, 163*, 12–21.

Liu, Y., Li, J., Jiao, X., Li, H., An, Y., & Liu, K. (2020). Effects of straw returning combined with biochar on water quality under flooded conditions. *Water, 12*(6), 1633.

Lu, H., Yan, M., Wong, M. H., Mo, W. Y., Wang, Y., Chen, X. W., & Wang, J.-J. (2020). Effects of biochar on soil microbial community and functional genes of a landfill cover three years after ecological restoration. *Science of the Total Environment, 717*, 137133.

Luo, C., Yang, J., Chen, W., & Han, F. (2020). Effect of biochar on soil properties on the Loess Plateau: Results from field experiments. *Geoderma, 369*, 114323.

Luo, X., Liu, G., Xia, Y., Chen, L., Jiang, Z., Zheng, H., & Wang, Z. (2017). Use of biochar-compost to improve properties and productivity of the degraded coastal soil in the Yellow River Delta, China. *Journal of Soils and Sediments, 17*, 780–789.

Ma, B., Shao, S., Ai, L., Chen, S., & Zhang, L. (2023). Influences of biochar with selenite on bacterial community in soil and Cd in peanut. *Ecotoxicology and Environmental Safety, 255*, 114742.

Meyer, G., Frossard, E., Mäder, P., Nanzer, S., Randall, D. G., Udert, K. M., et al. (2018). Water-soluble phosphate fertilizers for crops grown in calcareous soils – An outdated paradigm for recycled phosphorus fertilizers? *Plant and Soil, 424*, 367–388.

Mohan, D., Sarswat, A., Ok, Y. S., & Pittman, C. U., Jr. (2014). Organic and inorganic contaminants removal from water with biochar, a renewable, low-cost, and sustainable adsorbent–A critical review. *Bioresource Technology, 160*, 191–202.

Nepal, J., Ahmad, W., Munsif, F., Khan, A., & Zou, Z. (2023). Advances and prospects of biochar in improving soil fertility, biochemical quality, and environmental applications. *Frontiers in Environmental Science, 11*. https://doi.org/10.3389/fenvs.2023.1114752

Nunoura, T., Wade, S. R., Bourke, J. P., & Antal, M. J., Jr. (2006). Development of a continuous high-yield biomass charcoal reactor. *Industrial & Engineering Chemistry Research, 45*(2), 585–590.

Pandey, D., Savio, N., Naudiyal, N., Srivastava, R. K., Nayak, P., Cabañas, B., Moreno, A., & Singh, S. V. (2024a). Bioremediation of organic pollutants soil and water through biochar for a healthy ecosystem. In *Biochar production for green economy* (pp. 479–505). Academic Press, Elsevier. https://doi.org/10.1016/B978-0-443-15506-2.00007-9

Pandey, D., Singh, S. V., Savio, N., Bhutto, J. K., Srivastava, R. K., Yadav, K. K., Sharma, R., Nandipamu, T. K. M., & Sarkar, B. (2024b). Biochar application in constructed wetlands for wastewater treatment: A critical review. *Journal of Water Process Engineering, 69*, 106713. https://doi.org/10.1016/j.jwpe.2024.106713

Pandit, N. R., Mulder, J., Hale, S. E., Schmidt, H. P., & Cornelissen, G. (2017). Biochar from "Kon Tiki" flame curtain and other kilns: Effects of nutrient enrichment and kiln type on crop yield and soil chemistry. *PLoS One, 12*(4), e0176378.

Peake, L. R., Reid, B. J., & Tang, X. (2014). Quantifying the influence of biochar on the physical and hydrological properties of dissimilar soils. *Geoderma, 235–236*, 182–190.

Pramanick, B., Singh, S. V., Daw, S., & Nandipamu, T. M. K. (2025). Biochar and mitigation of heavy metal stress in plants. In *Biochar in mitigating abiotic stress in plants* (pp. 271–280). Academic Press. https://doi.org/10.1016/B978-0-443-24137-6.00015-X

Qin, J., Li, J., Pei, H., Li, Q., Cheng, D., et al. (2024). Effective remediation and phytotoxicity assessment of oxytetracycline and Cd co-contaminated soil using biochar. *Environmental Technology & Innovation, 32*, 103649. https://doi.org/10.1016/j.eti.2024.103649

Rajapaksha, A. U., Vithanage, M., Zhang, M., Ahmad, M., Mohan, D., Chang, S. X., & Ok, Y. S. (2014). Pyrolysis condition affected sulfamethazine sorption by tea waste biochars. *Bioresource Technology, 166*, 303–308.

Remya, V. R., George, J. S., & Thomas, S. (2021). Polymer formulations for controlled release of fertilizers. In F. B. Lewu, T. Volova, S. Thomas, & K. R. Rakhimol (Eds.), *Controlled release fertilizers for sustainable agriculture* (pp. 183–194). Academic Press.

Rolka, E., Wyszkowski, M., Załnowski, A. C., & Skorwider-Namiotko, A. (2024). Role of woody biomass ash material in immobilization of cadmium, lead, and zinc in soil. *Materials, 17*(10), 2206.

Roy, A., Chaturvedi, S., Singh, S. V., Govindaraju, K., Dhyani, V. C., & Pyne, S. (2022). Preparation and evaluation of two enriched biochar-based fertilizers for nutrient release kinetics and agronomic effectiveness in direct-seeded rice. *Biomass Conversion and Biorefinery., 14*, 1–12. https://doi.org/10.1007/s13399-022-02488-z

Sabio, E., Álvarez-Murillo, A., Román, S., & Ledesma, B. (2016). Conversion of tomato-peel waste into solid fuel by hydrothermal carbonization: Influence of the processing variables. *Waste Management, 47*, 122–132.

Saffari, N., Hajabbasi, M. A., Shirani, H., Mosaddeghi, M. R., & Owens, G. (2021). Influence of corn residue biochar on water retention and penetration resistance in a calcareous sandy loam soil. *Geoderma, 383*, 114744.

Shin, J., Park, D., Hong, S., Jeong, C., Kim, H., & Chung, W. (2021). Influence of activated biochar pellet fertilizer application on greenhouse gas emissions and carbon sequestration in rice (*Oryza sativa* L.) production. *Environmental Pollution, 285*, 117457.

Singh, C., Tiwari, S., Gupta, V. K., & Singh, J. S. (2018). The effect of rice husk biochar on soil nutrient status, microbial biomass, and paddy productivity of nutrient-poor agricultural soils. *Catena, 171*, 485–493. https://doi.org/10.1016/j.catena.2018.07.042

Singh, S. V. (2023a). Distinct behavior of biochar modulating biogeochemistry of salt-affected and acidic soil: A review. *Journal of Soil Science and plant Nutrition, 23*(3), 2981–2997. https://doi.org/10.1007/s42729-023-01370-9

Singh, S. V. (2023b). Occurrence of emerging contaminants in soils and impacts on rhizosphere. In *Emerging contaminants: Sustainable agriculture and the environment* (pp. 95–111). Elsevier. https://doi.org/10.1016/B978-0-443-18985-2.00014-6

Singh, S. V., Chaturvedi, S., & Datta, D. (2019a). Biochar: An eco-friendly residue management approach. *Indian Farming, 69*(08), 27–29. https://doi.org/10.1016/j.rineng.2024.102433

Singh, S. V., Chaturvedi, S., Dhyani, V. C., & Datta, D. (2019b). Biochar: An eco-friendly residue management approach. *Indian Farming, 69*(8) Retrieved from https://epubs.icar.org.in/index.php/IndFarm/article/view/94471

Singh, S. V., Chaturvedi, S., Dhyani, V. C., & Govindaraju, K. (2020). Pyrolysis temperature influences the characteristics of rice straw and husk biochar and sorption/desorption behavior of their biourea composite. *Bioresource Technology, 314*, 123674. https://doi.org/10.1016/j.biortech.2020.123674

Singh, S. V., Chaturvedi, S., & Nayak, P. (2023). Carbon offset potential of biochar based straw management under rice – Wheat system along Indo-Gangetic plains of India. *Science of the Total Environment, 897*, 165176. https://doi.org/10.1016/j.scitotenv.2023.165176

Singh, S. V., Mandal, S., Meena, R. S., Chaturvedi, S., & Govindaraju, K. (2024). *Biochar production for green economy: Agricultural and environmental perspectives*. Academic Press, Elsevier.

Širić, I., Eid, E. M., Taher, M. A., El-Morsy, M. H. E., Osman, H. E. M., Kumar, P., Adelodun, B., Abou Fayssal, S., Mioč, B., Andabaka, Ž., Goala, M., Kumari, S., Bachheti, A., Choi, K. S., & Kumar, V. (2022). Combined use of spent mushroom substrate biochar and PGPR improves growth, yield, and biochemical response of cauliflower (*Brassica oleracea* var. *botrytis*): A preliminary study on greenhouse cultivation. *Horticulturae, 8*(9), 830.

Smit, L. (2023). *The use of a starch-based superabsorbent polymer to support and optimise potato production in the sandy soils of the Sandveld production region in South Africa*. Master's thesis, Stellenbosch University.

Sohi, S. P., Krull, E., Lopez-Capel, E., & Bol, R. (2010). A review of biochar and its use and function in soil. In *Advances in agronomy* (Vol. 105, pp. 47–82). Academic Press.

Spokas, K. A., Novak, J. M., & Venterea, R. T. (2012). Biochar's role as an alternative N-fertilizer: Ammonia capture. *Plant and Soil, 350*(1–2), 35–42.

Steiner, C., Teixeira, W. G., Lehmann, J., Nehls, T., de Macêdo, J. L. V., Blum, W. E. H., & Zech, W. (2007). Long-term effects of manure, charcoal, and mineral fertilization on crop production and fertility on a highly weathered Central Amazonian upland soil. *Plant and Soil, 291*, 275–290.

Sun, Z., Zhang, Z., Zhu, K., Wang, Z., Zhao, X., Lin, Q., & Li, G. (2020). Biochar altered native soil organic carbon by changing soil aggregate size distribution and native SOC in aggregates based on an 8-year field experiment. *Science of the Total Environment, 708*, 134829.

Tian, J., Kuang, X., Tang, M., Chen, X., Huang, F., Cai, Y., & Cai, K. (2021). Biochar application under low phosphorus input promotes soil organic phosphorus mineralization by shifting bacterial phoD gene community composition. *Science of the Total Environment*, 779, 146556

Ullah, A., Nafees, M., & Ahmed, N. (2023). Biochar and plant growth-promoting rhizobacteria improve drought resistance in *Brassica napus* L. *Environmental Science and Pollution Research, 30*(12), 4567–4579.

Vahedi, R., Rasouli-Sadaghiani, M. H., Barin, M., & Vetukuri, R. R. (2022). Effect of biochar and microbial inoculation on P, Fe, and Zn bioavailability in a calcareous soil. *Processes, 10*(2), 343.

Venugopalan, V. K., Nath, R., & Sarath Chandran, M. A. (2022). Smart fertilizers – A way ahead for sustainable agriculture. *Journal of Plant Nutrition, 45*(14), 2068–2076.

Voltr, V. (2012). Concept of soil fertility and soil productivity: Evaluation of agricultural sites in The Czech Republic. *Archives of Agronomy and Soil Science, 58*(sup1), S243–S251.

Woolf, D., Amonette, J., Street-Perrott, F. et al. Sustainable biochar to mitigate global climate change . Nat Commun 1, 56 (2010). https://doi.org/10.1038/ncomms1053

Wu, S., He, H., Inthapanya, X., Yang, C. (2017). Role of biochar on composting of organic wastes and remediation of contaminated soils—A review. *Environmental Science and Pollution Research, 24*, 16560.

Xi, J., Li, H., Xi, J., Tan, S., Zheng, J., & Tan, Z. (2020). Effect of returning biochar from different pyrolysis temperatures and atmospheres on the growth of leaf-used lettuce. *Environmental Science and Pollution Research, 27*, 35802–35813.

Xiang, Y., Deng, Q., Duan, H., & Guo, Y. (2017). Effects of biochar application on root traits: A meta-analysis. *GCB Bioenergy, 9*, 1563–1572.

Yu, X., Wang, X., Sun, M., Liu, H., Liu, D., & Dai, J. (2024). Cadmium immobilization in soil using phosphate-modified biochar derived from wheat straw. *Science of the Total Environment, 858*, 171614. https://doi.org/10.1016/j.scitotenv.2024.171614

Yu, Y. (2022). *Fate and transport of micro- and nanoparticles in the subsurface*. Doctoral dissertation, Washington State University.

Zhang, Y., Wang, J., & Feng, Y. (2021). The effects of biochar addition on soil physicochemical properties: A review. *CATENA, 202*, 105284.

Zheng, H., Wang, X., Chen, L., Wang, Z., Xia, Y., Zhang, Y., Wang, H., Luo, X., & Xing, B. (2018a). Enhanced growth of halophyte plants in biochar-amended coastal soil: Roles of nutrient availability and rhizosphere microbial modulation. *Plant, Cell & Environment, 41*(3), 517–532.

Zheng, H., Wang, X., Chen, L., Wang, Z., Xia, Y., Zhang, Y., et al. (2018b). Enhanced growth of halophyte plants in biochar-amended coastal soil: Roles of nutrient availability and rhizosphere microbial modulation. *Plant, Cell & Environment, 41*(3), 517–532.

Zulfiqar, F., Younis, A., & Chen, J. (2019). Biochar or biochar-compost amendment to a peat-based substrate improves growth of Syngonium podophyllum. Agronomy, 9(8), 460

Soil and Water Management for Agricultural Sustainability

Soil Conservation Techniques and Adaptive Management Practices for Climate-Resilient Agriculture

Ria Bhattacharya and Shyamal Kumar Mondal

Abstract In the contemporary context of swiftly growing human populations and dwindling arable land resources, sustainable agriculture is deemed an indispensable component of agricultural methodologies. Since soil is a natural substrate for plant growth, better soil is thought to be a crucial predictor of producing high-quality food. The resource of soil has been neglected and placed in danger due to increased food production, urbanization. Numerous issues, including soil fertility depletion and degradation, loss of soil organic carbon (SOC) and biodiversity, salinization, acidification, pollution, and soil erosion and degradation, have put "soil health" in jeopardy. Critical soil functions, ecological services, and food production are all significantly hampered by these processes of land degradation. Decreased soil organic carbon (SOC) content, deteriorating soil health, loss of soil resilience, and low crop productivity can be caused by management practices like over-tilling delicate soils and removing excessive residue, or by the loss of rich topsoil due to wind, water, or tillage erosion. The quantity of heavy metals, soluble salts, and hazardous compounds has a significant impact on the quality of soil. Additional factors that limit the amount of land that may be used for crop production include soil erosion, waterlogging, toxicity, deficiency in certain minerals, and incorrect tillage methods. In order to boost crop yield and satisfy the world's food demands, sustainable soil management (SSM) and conservation are crucial. Sustainable soil management (SSM) is therefore necessary to protect this valuable resource. Sustainable management aims to improve soil health and crop yields. However, the benefits of these techniques primarily depend on the soil and climate of the area. FAO states that "soil management is sustainable if the cultural, regulating, provisioning, and supporting services offered by soil are maintained or enhanced without significantly impairing either biodiversity or the soil functions that enable those services". In order to improve soil functions, this chapter discusses the definition, significance, difficulties, and different tactics and techniques of soil health and soil organic carbon. It also aims to lessen soil erosion and degradation.

R. Bhattacharya · S. K. Mondal (✉)
ICAR-Agricultural Technology Application Research Institute (ATARI), Kolkata, West Bengal, India

A. K. Singh et al. (eds.), *Sustainable Agriculture Management in Semi-Arid Climates*, https://doi.org/10.1007/978-3-031-94062-0_6

Keywords Sustainable agriculture · Soil health · Conservation · Crop production · SOC · SSM

1 Introduction

Higher food production and economic growth and development are the goals that lead to the usage, misuse, exploitation, and overexploitation of soils (UN, 2015; FAO, 2017). Food and nutritional security are ensured by soil resources (Amundson et al., 2015). Soil is related to farming and provides 98% of the world's food, either directly or indirectly, in addition to carrying out a variety of ecosystem activities and services (like carbon sequestration, maintain biodiversity, climate change mitigation, etc.) (FAO, 2020). The major problem is land degradation (~33% area), endangering soil resources and production of food (Lal, 2009, 2015a). The goal of sustainable soil management, or SSM, is to maintain soil health and food production while avoiding unfavourable environmental effects. Furthermore, it will take forward-thinking innovations—like those from the 1960s, which Nobel laureate Borlaug (2002)) revolutionized—to feed the world's 10 billion inhabitants. Subsequently, the world saw the "Green revolution", which caused the production of food grains to quadruple. However, the undernourished population and malnutrition remain a reason for concern. Therefore, the use of sustainable farming practices is crucial for achieving sustainable development. The health of the soil is essential for sustaining agricultural production, biodiversity, and the supply of food for a growing population (Jayaraman et al., 2023) (Fig. 1).

2 Sustainable Soil Management

A combination of techniques known as sustainable soil management preserves or enhances the services that soil offers without endangering the biodiversity or functionality of the soil. It is defined as soil management that provides for current requirements without affecting the ability of future generations to use that soil for their own needs Smith and Powlson (2007). According to the Intergovernmental Technical Panel on Soils, "the ability of the soil to sustain the productivity, diversity, and environmental services of terrestrial ecosystems" is the definition of soil health Intergovernmental Technical Panel on Soils (ITPS). "In managed systems, soil health can be maintained, promoted, or recovered through the implementation of sustainable soil management practices", according to the ITPS Soil Letters (UN, 2015; FAO/ITPS, 2020). "If the supporting, provisioning, regulating, and cultural services provided by soil are maintained or enhanced without significantly impairing either the soil functions that enable those services or biodiversity", states FAO (2020), then soil management is sustainable. In a similar vein, Lal (2009) put forth the "ten laws" of sustainable management techniques, which aim to minimize water

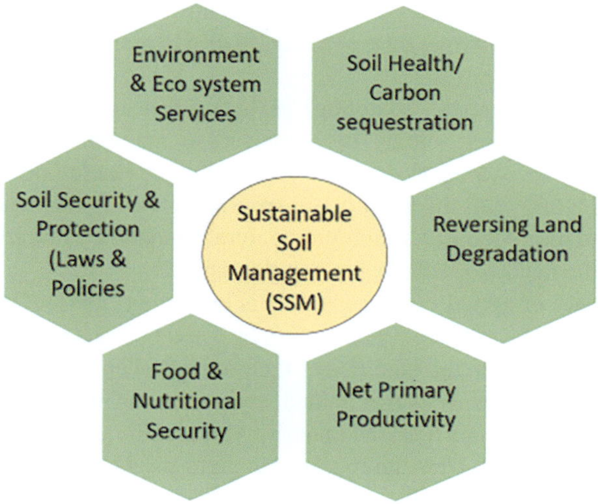

Fig. 1 Sustainable soil management, closely linked to soil functions and ecosystem services (Jayaraman et al., 2023)

and carbon footprints while also stopping soil degradation and restoration processes. "Management practices that threaten the soil biological community may also threaten soil sustainability by reducing the capacity of the soils to adapt in the future", according to Yachi and Loreau (1999). Many management practices put soil sustainability at risk. These include burning crop residue, over-cultivation, under-fertilization, and chemical use, as well as low additions of manures and crop residue etc. as a result the amount of organic matter in the soil is decreased (Smith & Powlson, 2007; Powlson et al., 2011a, b). Sustainable soil management is necessary to achieve both ecological services and food, nutritional security and maintain bio-diversity.

3 Sustainable Agriculture Principles

There are some general principles of sustainable agriculture can be listed as follows: (Soil Management in Sustainable Agriculture. DOI: https://doi.org/10.5772/intechopen.88319)

- Protecting and developing the soil is essential for producing high-quality, healthful goods. To enhance the quality of the soil, naturally occurring nutrients such as compost, organic, and green manure should be applied. Natural fertilizers are superior to chemical fertilizers for the land, plants, water, air, and people.
- Water and resources of water need to be preserved because, like life, agriculture depends entirely on water. The greatest strategy to preserve water in arid areas is to cultivate plants that are native to the area or that require water only in the rainy season. Retaining soil moisture can be aided by mulch and green manure. The

contour barriers safeguard water by halting its flow. Two further methods to save water are to schedule irrigation times and use drip irrigation instead of traditional irrigation method.

- Instead of using chemicals to control pests and diseases, natural or integrated protection management should be employed to maintain a balance between nature, products, diseases, weeds, and soil. Techniques including selecting hardy cultivars, planting with the appropriate spacing between plants, accurately timing agricultural procedures, employing natural predators, and crop rotation are crucial for the method's effectiveness in this aspect.
- Grow a variety of agricultural crops; this is known as "product rotation." Pests and illnesses can be avoided for 3–6 years by rotating crops or growing a variety of crops, depending on the qualities of the products. Thus, the soil retains nutrients, agricultural diversity is maintained, and nutritious food is produced.
- Make minor adjustments first: the majority of agricultural practices have taken a long time to develop. New approaches, though, might not always work. New concepts ought to be tested in limited settings first, then implemented when they prove to be effective.

4 Why Is it Necessary to Manage Soil Sustainably?

Since soil produces over 98% of the food that humans eat, either directly or indirectly, it is crucial to all terrestrial living things (Kopittke et al., 2019, 2021). This demonstrates the significance of soil for human survival and the survival of other living things. Additionally, soil has a right to appropriate management, restoration, and protection, just like any other living thing (http://www.fao-stories.org/en/c/1330050/). The cycling of nutrients, water retention and discharge, biodiversity, filtration and buffering, and physical stability and support are the five main roles of soil. One of the main risks to food production and sustainability in the world is land degradation. The Food and Agriculture Organization of the United Nations estimates one-third of the soil on Earth is categorized as "moderately to highly degraded". In addition, yields are decreased, crops are grown with less health, and soil minerals are reduced. As a result, the FAO (2017, 2019a, b) has recommended four more comprehensive groups of measures to reverse or control soil erosion: Group 1 is comprised of actions to minimize changes in land use, groups 2 and 3 are actions to lessen runoff velocity and soil erosion, and group 4 is an action to reduce the export of soil particles, related nutrients, and other contaminants from the soil. By restoring degraded lands that sustain activities and functions of ecosystem, with the goal of maintaining or improving the quantity and quality of land resources, neutrality of land degradation can be achieved through the right implementation of various management strategies (Fig. 2).

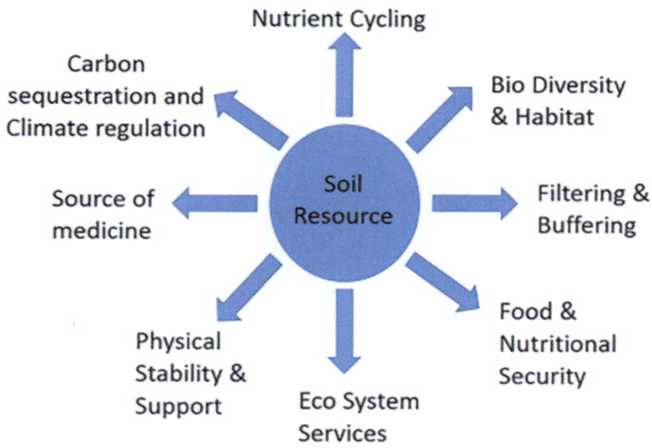

Fig. 2 Different functions of soil (Jayaraman et al., 2023)

5 Importance of Sustainable Soil Management

It is clear that agriculture contributes significantly to GDP in poor nations and only slightly to GDP in industrialized nations. Agriculture produces 4% of the world's GDP on average. The amount of land used for agriculture worldwide is 4.80 billion ha in 2021, a decrease of 0.13 billion ha from 4.93 billion ha in 2000. 3.2 billion hectares of this land are used for grazing and meadows, and 1.6 billion hectares are used for agriculture (FAO, 2021). Soil resources, in conjunction with optimal agronomic practices and sustainable soil management is the cornerstone of agriculture. Therefore, in order to attain the Sustainable Development Goals (SDGs), soil resources must be protected and conserved for future generations to ensure food and nutritional security.

The following are the main advantages of sustainable agriculture: (Soil Management in Sustainable Agriculture. DOI: https://doi.org/10.5772/intechopen.88319)

- Multiple products can be produced with great efficiency in small areas using sustainable agriculture methods.
- An environmentally conscious business will benefit the ecosystem. Not only will efficient soils support agricultural production, but they will also provide a home for animals.
- Soil fertilization ensures increased output.
- Promotes the development of new job sectors in addition to the advantages for agriculture.

6 Limitations, Prospects and Profits of Sustainable Soil Management

Sustainable Soil Management (SSM) stands for “One Stop Solution”, and it addresses concerns related to soil, plants, climate change, greenhouse gas (GHG) emissions, carbon sequestration, and greenhouse emissions. SSM covers several landscapes, including agriculture, horticulture, pasture, and integrated crop livestock systems (ICLS), with a focus on soil in order to increase the advantages in each land use system.

6.1 Relation between Soil and Crops

Soil is important for agricultural and food production. There have been pandemics, disasters, and extreme weather events including flooding, droughts, and water logging around the world. The soil is the most important component in providing food and nutrition when considering production of food grain and maintaining flora and fauna. Since the organic nutrient source is inadequate for the best possible food production and nutritional security, fertilizer inputs are also needed. FAO, 2021, there are around 190 million tonnes of fertilizer were used globally for production of food, with 57% of that amount being nitrogen-based fertilizer. In every region, the amount of cultivated land per person decreased (between 2000 and 2019) as a result of rising fertilizer costs; Africa saw the worst decline, followed by America and Asia. A number of other issues also pose a threat to soils, including soil erosion caused by wind and water, loss of soil organic carbon, an imbalance in soil nutrients, salinization, pollution, acidification, reduction in soil biodiversity, compaction, sealing, and waterlogging (FAO, 2017).

Agronomic and other management techniques should be used in conjunction with soil management to create sustainable soil management, as stated by White et al. (2012), Lal (2014), and Stagnari et al. (2019) selecting crop cultivars and cropping systems that will help farmers and producers maintain a stable income while also taking care of the soil system. Leguminous crops, for example, are soil-nourishing crops that can fix atmospheric nitrogen (by developing symbiotic relationships with bacterial groups, such as Rhizobium and Bradyrhizobium) and recycle nutrient residues back into the soil through leaf fall and roots. These crops serve as soil cover in addition to storing carbon. Similar to this, choosing crops with contrasting root systems will allow various crops to absorb water and nutrients from different strata (e.g., legume-cereal rotations, soybeans interplanted with maize, or pigeon peas interplanted with sorghum or maize) (Davies et al., 2011). Rainfed and water-scarce regions face a range of biophysical and socioeconomic challenges that impact crop and livestock productivity. Furthermore, choosing a cropping strategy that is climate resilient or conservation agriculture (CA) in rainfed regions gives farmers a source of income and helps maintain living mulch on the soil's surface and provides sustenance for fauna.

6.2 Role of Fertilizer/Manures/Amendments

Historically, farming was mostly done as "subsistence farming", which involved using more organic manure—such as crop residue, farm wastes and lime—and less chemical fertilizers. Ultimately, this led to lower crop yields. "Intensive farming", or high-input farming (greater fertilizer and nutrient application), was widely utilized in the early 1960s during the Green Revolution era to feed the expanding population (Sanchez & Swaminathan, 2005; Davies et al., 2011). Since then, the main practices for maintaining crop output, lowering hunger and malnutrition, and boosting economic prosperity have been applying fertilizers, manures, and amendments. FAO (2019a) created an "International Code of Conduct" for the Fertilizer Industry's Sustainable Use and Management. The "fertilizer code" is a voluntary guideline that focuses on preventing or minimizing nutrient imbalances and soil pollution in order to achieve SSM. Finally, in June 2019, the 41st session of the FAO Conference approved this "Fertilizer Code". Therefore, we must use extreme caution when applying fertilizers, insecticides, and manures at various sizes given the significance of soil biodiversity, food, and environmental safety. When applying manures, insecticides, and fertilizers on various scales, we must exercise extreme caution. Adopting integrated soil fertility management (ISFM) and the 4Rs (right rate, right source, right placement, and right timing) of nutrient stewardship entails using biofertilizers in conjunction with balanced/integrated nutrient management (INM) to increase crop yields and fertilizer use efficiency while also keeping nutrient levels within safe ranges.

Fertilizer application is based on soil tests (using sensors, instruments, and equipment) throughout the main cropping systems (site-specific nutrient management), recycling of nutrients (crop residues, farm waste), and careful handling of sewage water and sludge (circular nutrient management).

6.3 Weeds and Soil-Borne Disease Management

Both monocot and dicot weeds as well as soil-borne illnesses brought on by pathogens like Pythium, Rhizoctonia, and Fusarium species can be extremely problematic for crop management. Yield penalties can occasionally result from improper management. Thus, the application of appropriate pesticides to prevent soil-borne illnesses and weeds, crop rotation and intercropping, conservation tillage techniques, mulching (both organic and inorganic) and artificial materials, soil solarization, and summer ploughing are some of the proper agronomic and techniques for managing the soil that prevent weed growth and soil-borne fungi (Jayaraman et al., 2021a, 2021b, 2021c, 2021d).

7 Techniques and Approaches for Sustainable Management of Soil

7.1 Conservation Agriculture

Over 200 million hectares are used for conservation agriculture (CA) and no-till farming globally (Kassam et al., 2019). Conservation agriculture is a "set of management practices involving minimum soil disturbance (conservation tillage/No Tillage), maximum soil cover through residue cover or organic mulch and diversified cropping system/rotations" (FAO, 2021). An essential part of CA is the generation of residue from the growth of cover crops. This procedure not only reduces soil erosion but also inhibits weed growth and balances soil temperature (Davies et al., 2011), but it also increases soil health and soil organic carbon levels through the recycling and addition of residue (Dalal & Bridge, 2020; Lal, 2015a; Dalal et al., 2011, 2018). There are several advantages to incorporating cover crops into the rotation in semi-arid regions and habitats. It has been demonstrated that ground cover with living or dead mulch is an essential component of the CA system in perennial systems (Davies et al., 2011; Lal, 2015a, 2015b; Jayaraman et al., 2020, 2021a, 2021b, 2021c, 2021d; Dalal & Jayaraman, 2021). According to Jat et al. (2020), the use of CA is a sustainable intensification strategy that can lower global warming by 12–33% while increasing crop output (5.6%), net economic return (25.9%) and water usage efficiency (12.6%) (Fig. 3).

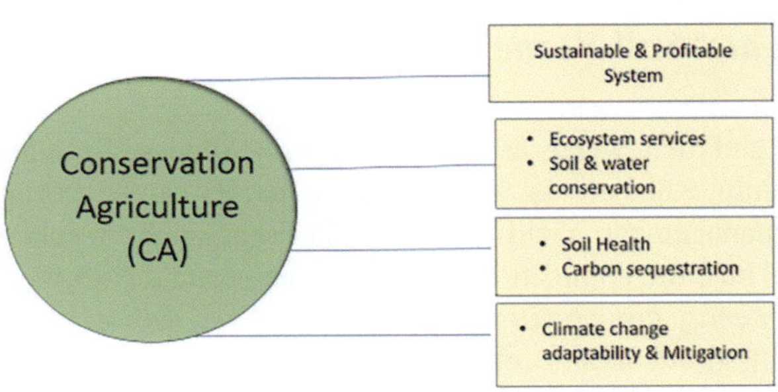

Fig. 3 Benefits of conservation agriculture (Jayaraman et al., 2023)

7.2 *Regenerative Agriculture*

The fundamental factor affecting soil health, local community livelihood, and production is acknowledged to be the degradation of land (soil) caused by soil erosion and loss of rich topsoil. Therefore, we must use sustainable soil management practices resource conservation technologies, sustainable intensification, and CA/precision farming to safeguard our natural resources.

Furthermore, it is imperative that current farming and management methods be changed in order to minimize soil disturbance and input utilization efficiency through the application of suitable SSM practices and procedures. For increased productivity, returns, sustainable production, and soil health, regenerative agriculture (RA) has thus been promoted (Lal, 2021; TAAS, 2021). Remedial agriculture refers to methods for improving soil health through the replenishment of soil organic carbon (SOC). It is an all-encompassing, principle-based farming method that contributes to the development and maintenance of ecosystems and community services. Regenerative agriculture provides multiple benefits:

- Soil carbon storage and maximize crop diversity.
- Maintain a cover crop or agricultural residue layer on the soil.
- Minimize soil disturbance.
- Understand the context of farm operations.
- Soil-crop-livestock interaction.
- Maintain living roots round the year.

This technique helps to reverse land degradation, increase agricultural yield per acre, improve soil health, soil biodiversity, and C storage, ultimately improving soil functions and ecosystem services. As a system-based strategy, regenerative agriculture encompasses a broad range of agricultural techniques designed to sequester SOC in order to monitor soil health sustainably and restore it. Instead of degrading or destroying the resources it uses, this agricultural technique enhances them. The RA concept advocates a number of practices and strategies, such as cover crops, integrated farming systems (IFS), integrated nutrient management (INM)/integrated plant nutrition systems (IPNS), pest management, agri-horticulture/agroforestry systems, and conservation agriculture/no-till farming with residue retention (Newton et al., 2020).

7.3 *Organic Farming*

According to FAO (2017), the sustainable development goals (SDGs) sought to identify and repair degraded soils as well as enhance soil health through SSM practices and technologies. One of the potential approaches and technologies that can increase agricultural productivity and sustainability sustainably without harming the environment or degrading soil quality is organic farming (OF) (Tully & McAskill, 2020).

According to Didawat et al. (2023), it is well recognized that organic farming (OF) is a whole approach to production control that supports and improves the total health of the agroecosystem, including soil biological activity, biodiversity, and biological cycles.

It is one of the ways that farmers may apply SSM technology to decrease their reliance on chemical-based inputs and increase their use of organic amendments, such as composts.

The 'Father of Modern Composting'—Albert Howard (1940), is credited with perfecting the "Indore method", a conventional Indian composting technique. According to Howard, "the connection between healthy soil and the villages' healthy populations, livestock, and crop" was one of the significant and intriguing features. In a book titled *An Agricultural Testament*, he wrote about his experiences and emphasized the value of humus preservation, soil moisture retention, the function of mycorrhizae, and the development of composting and organic farming methods (Howard, 2006). Compared to conventional farming, organic farming offers a multitude of potential advantages, such as enhanced soil aggregation, soil microbial biodiversity, porous structure, environmental protection, improved soil quality, improved food safety and quality, and higher crop output rates (Stinner, 2007). Therefore, where enough organic inputs are applied, organic farming may guarantee the long-term sustainability of agricultural systems. Increased SOC, soil health, and agricultural sustainability are all enhanced when a significant amount of biomass or organic manure is added through organic farming (Sharma et al., 2011; Schweizer et al., 2022). But it works better in specialized fields like vegetable farming, fruit trees, agri-horticulture, etc. Supplementing N and other nutrients to annual crops like rice and wheat with large amounts of biomass or manure is a difficult undertaking. Recently, there has been a push for natural farming, also known as zero budget natural farming (ZBNF), to address the growing issues with poor biodiversity in agriculture and cut back on fertilizer and pesticide inputs.

7.4 Four per Thousand (4PT) Programme

The "4 per thousand" program was created with the ambitious goal of increasing soil organic carbon (SOC) by 0.4% every year globally during the COP21 conference in Paris, 2015 (Lal, 2015b; https://4p1000.org/). The goal of the global "4 per 1000 (4PT)" program is to demonstrate how important agricultural soils may be for storing carbon in relation to ensuring food security and reducing the effects of climate change (Van Groenigen, 2018; Soussana et al., 2019; Rumpel et al., 2020). SOC sequestration remains off the international political agenda, and it was not discussed during the Bonn COP23 in Germany (2017), despite the importance of soil-carbon and these societal imperatives (Rumpel et al., 2018; Chenu et al., 2019). Through the adoption of best (recommended) carbon farming management practices, such as conservation agriculture (CA), crop residue mulching/retention, no-till farming, agroforestry, improved grazing, cover cropping, biochar application and reversing degraded soils through management, Lal (2016) has clearly indicated different strategies to promote SOC sequestration (Dalal et al., 2021a, b). Similar to this, Minasny et al. (2017) conducted a meta-analysis of the 4PT initiative and talked about different approaches and strategies of improving SOC (for example, increasing soil carbon in the upper 1 m of soil by 2–3 Gt C yr^{-1} can reduce greenhouse gas emissions by 20–35% (Soussana et al., 2019; Henry et al., 2023). Up to 5 Gt CO_2 yr^{-1} of soil carbon can be sequestered by adopting appropriate SOC practices; however, the effectiveness of this process varies amongst agro-ecological regions and depends on the extent to which different land holders adopt SOC practices (Fuss et al., 2018). In other words, it is specific to the region, site, and socio-economic context. Actually, raising C by atmospheric CO_2 capture and altering management techniques (by adding more nitrogen fertilizer) may cause other greenhouse gases, including N_2O, to rise as well. Consequently, it is suggested and advised to implement a suitable and appropriate SSM (involving nitrogen management) to reduce N_2O emissions into the atmosphere while reducing CO_2 emissions across major cropping systems that represent various agro-ecological regions of a country (Dalal et al., 2021a, b; Henry et al., 2023).

7.5 Soil, Water Conservation and Nutrient Management

Soil erosion and degradation pose a danger to the sustainability of agriculture (Lal et al., 2007; Lal, 2015b; Borrelli et al., 2017). For instance, farmers in rainfed parts of South Asia (such as India and Pakistan) and Africa (such as Vohland & Barry, 2009) receive all of their annual rainfall in fewer than 100 days. Site-specific nutrient management is an additional component of SSM that must be managed wisely to maintain crop output in addition to conserving water and soil. Furthermore, farmers with limited resources frequently lack the options to apply all necessary fertilizer nutrients in addition to nitrogen-based (N) fertilizer. Thus, in order to improve yield and maintain soil health, it is required to combine N-based fertilizer with organic input (farmyard manure, farm wastes, and crop residues) and biofertilizer. Furthermore, the implementation of microbial consortiums for in-situ recycling of crop residue will aid in decreasing residue burning and boosting SOC stock. Therefore, in order to attain nutritional and food security, soil-centric centred stewardship is necessary to reverse soil degradation, improve crop yield, and maintain soil health (Soussana et al., 2019; Dalal et al., 2021a, b).

7.6 *Reducing Erosion of the Soil and Reversing Land Degradation*

Increasing land degradation and erosion of soil worldwide as a result of improper or non-scientific management. Concerns are raised in many nations about the loss of top-fertile soils due to erosion, as well as about accelerated soil erosion and terrain distortion in agricultural areas. Numerous countries, including the USA (32% area) (Lal, 2003), China (30.7% area) (Guo et al., 2015), Africa (16%), Europe (17%) (Bai et al., 2008; Borrelli et al., 2017; Tamene et al., 2017), Australia (2/third area) and India (45%) (NAAS, 2010; FAO, 2019b) have already experienced significant land degradation due to one or more land degradation processes. It is imperative to take suitable conservation measures to reduce erosion of the soil and reverse land degradation. In order to offset the degradation, the neutrality program for land degradation has been implemented.

Out of 196 nations, 128 have already ratified Land Degradation Network (LDN), which is a paradigm change in the methods and regulation of land management (https://www2.unccd.int/actions/achieving-land-degradation-neutrality). The LDN strives to achieve SDGs 15.3, in particular, in order to reverse degraded land, stop desertification, and provide food security by 2030.

Three simultaneous actions comprise the Land Degradation Network (LDN): maintaining healthy land, preventing new land degradation, reducing existing land degradation by implementing sustainable land management techniques that can enhance biodiversity, improve soil health, increase food production, and slow down degradation.

7.7 *Climate-Resilient Soil Management Practices*

The SOC-centred strategy is among the possible choices for improving soil health, increasing food security and farm incomes, and mitigating climate change (Lal, 2004; Powlson et al., 2011a, b; Paustian et al., 2016; FAO, 2019a). Sustainable soil management techniques are comprised of several approaches. The SSM improves the physical, chemical, and biological properties of soil, which has an effect on carbon sequestration, greenhouse gas emission reduction, and soil health (Lal, 2015b). Organic C and N management greatly affects both the health of the soil and food grain yield. There are situations when raising SOC through increased N fertilizer application causes other greenhouse gas (GHG) emissions, like N_2O and CH_4.

8 Different Type of Soil Management Practices

8.1 Crop Rotation

In contrast to monoculture, crop rotation is cultivating a variety of crops in order to advance sustainable agriculture and enhance farming techniques. Crop rotation reduces compaction, supplies nitrogen, helps manage pests, lessens soil depletion, reduces the risks associated with farming, avoids using chemicals, and boosts soil activity.

8.2 Irrigation

The production of crops depends on irrigation is energy- and water-intensive. Using the least amount of water and energy feasible, sustainability maximizes plant hydration. Drip irrigation methods are more useful and effective than of furrow (flood) irrigation.

8.3 Cover Crop

To avoid erosion on their farms, farmers employ cover crops in the off-season. By utilizing green manure from cover crops to increase organic matter lowers fertilizer costs. Cover crops preserve soil moisture and suppress weed growth. The natural flowering of cover crops increases the population of honey bees.

8.4 Minimum or Zero Tillage

No-till farming, as opposed to traditional ploughing, stops soil erosion. Planting straight into crop residue reduces disturbance to the soil and biota, allowing for no-tilling. By using no-tillage techniques, time is saved, compaction is avoided, and ecological stability is increased.

8.5 Integrated Pest Management

IPM employs a variety of techniques that work better together than when used alone for chemical pest control. IPM reduces the harm that sustainable agriculture causes to all species. In integrated pest management, sustainable agriculture uses both biological and cultural control. Biological tactics include deploying predators, such as ladybugs or poultry, to get rid of pests and their eggs.

8.6 Integrated Weed Management

Sustainable weed control seeks to manage weeds while protecting natural resources by avoiding the use of chemicals and implementing eco-friendly methods. Crop rotation, allelopathic plants, insects, birds, cover crops, resilient crops, mechanical and manual weeding, and other organic techniques can all be used in integrated weed management.

8.7 Permaculture

Permaculture is the term used to describe the self-sufficient and sustainable development of agricultural ecosystems. Permaculture emulates harmony, stability, and diversity of natural ecosystems. By reducing waste, making use of renewable resources, eliminating pollution, and improving soil fertility with environmentally friendly approaches, this methodology promotes sustainability.

8.8 Composting

The process of dissolving organic materials into more fundamental inorganic and organic components yields compost. The process of composting is a natural way to create rich, nutrient-dense compost from organic materials like food scraps, yard clippings, and agricultural byproducts. Composting helps to maintain soil health by increasing the amount of organic matter in the soil.

8.9 Bio-Fertilizer

Utilizing bio-fertilizers allows plants to receive nutrients as they grow. The tiny living organisms that give plants essential nutrients and promote their growth are the origin of this phenomena.

9 Soil Management Practices for Sustainable Agriculture

The majority of living creatures get their mineral needs from soil. By managing the soil, one can prevent deficiencies in mineral elements or make it toxic to plants, as well as assure that the right minerals enter the food chain. Soil management affects human health, environmental sustainability, and crop productivity both directly and indirectly. Achieving future food security will need sustainable soil management, which includes appropriate nutrient management and soil conservation techniques. (FAO, 2017).

9.1 *Reduce Soil Erosion*

Due to deforestation and dry soil, topsoil naturally depletes faster due to human activity. Numerous human endeavours, such as farming, grazing, mining, building, and leisure activities, have been linked to soil erosion. It takes work to maintain soil cover, which can be in the form of gravel, plastic, or even plant or mulch, in order to control soil erosion. By preserving constant crop cover, a crop rotation system can help control erosion. Intercropping systems and perennial crops offer a practical cover that is consistent with profitable land use (FAO, 1983, 1989). One useful strategy for controlling soil erosion is to plant trees in windbreak positions. In areas where the soil texture is light and the moisture content is low, wind speed is important. The goal of contour bunds and channels is to redirect excess water in areas that are vulnerable to erosion caused by water. Crop rotation and ridge farming with intercropping are examples of alternative agricultural methods that can be used.

9.2 *Organic Matter Improvement*

To keep soil healthy and stop degeneration, SOM is necessary. The largest store of carbon on Earth is found in soils, which also balance greenhouse gas emissions to control and lessen climate change. Erosion, carbon emissions, and soil degradation can all result from loss of soil organic carbon. In addition to improving soil quality and increasing SOC, proper land use and soil management may also slow the increase of CO_2 in the atmosphere. Furthermore, soil organic matter is necessary to preserve the characteristics and functions of the soil. Thus, slight variations in SOC have a noteworthy impact on the characteristics of soil.

Therefore, to raise SOC, increase biomass production through techniques like drip irrigation or microsprayers, cover crops, balanced fertilizer use, organic amendments, improved vegetative stands, agroforestry, reforestation, and afforestation, while also minimizing soil erosion and nutrient leaching. By controlling agricultural residues, engaging in organic farming, utilizing compost and mulches, combining soil fertility and pest management, and protecting organic carbon-rich soils, one can raise the amount of organic matter in the soil (Lal, 2005).

To maintain soil functioning, burning crop leftovers after harvesting needs to be monitored. To make sure the soil has enough organic cover, use only organic inputs and implement techniques like cover crops, no-till farming, and live fences. Reduce the quantity of living things that decomposes in the soil by rotating your crops, adding legumes, and enhancing your crop mix without applying extra herbicides. Restoring organic matter to arable highland soils is essential for sustainable soil management—that is, if it is both economically and toxin-free.

9.3 Optimize Soil Nutrients and Cycles

One of the main barriers to food security and environmental sustainability is controlling the flow of nutrients to crops (Spiertz, 2009). Preserve or improve the soil organic matter to improve the fertility of the soil and the cycles of nutrients. Increased soil fertility can be attained by agroforestry, reduced/no-tillage with decreased pesticide usage, crop rotation, legume utilization, green and animal manures, cover crops, and other conservation techniques. Apply balanced, context-appropriate soil organic and inorganic amendments to maximize nutrient usage efficiency and manage the nutrient cycles in integrated systems. Utilize suitable fertilizer techniques to minimize losses and guarantee that crops absorb nutrients in a balanced manner. Analyses of the soil and plants over the long term are required rather than immediate action.

9.4 Reduce Soil Salinization and Alkalinization

Retention of soluble salts in the soil leads to salinization of the soil. The sources of salinization are human activity, seawater intrusion, and high rates of evapotranspiration. It lowers agricultural yields overall and deteriorates the health of the soil. Optimizing surface cover to minimize evaporation losses, applying water directly to the soil using low-pressure techniques, controlling irrigation to ensure proper plant growth and drainage, testing and monitoring irrigation water quality, and desalinating water when feasible are some strategies to lessen soil salinization. Use methods including leaching, planting salt-tolerant cultivars, utilizing local halophytes, and implementing amelioration strategies to restore damaged saline soils (Imadi et al., 2016).

9.5 *Minimize Soil Contamination*

Many things, including spills and agricultural waste, can pollute soil, and too much addition of certain substances can lead to buildup and contamination of the soil. Plant toxicities, water contamination, off-site pollution, and health concerns due to food-chain buildup are examples of negative outcomes. To protect public health, governments should control the build-up of contaminants and clean up contaminated soil that exceeds predetermined thresholds. Establish baseline levels and test, monitor, and evaluate levels for potentially contaminated locations in order to control local soil pollution. Evaluate the costs and hazards, then take corrective action to safeguard people and the environment. Furthermore, one of the essential actions to prevent soil pollution is to identify soils that are susceptible to the diffusion of contaminants (Hartmann & Six, 2023).

9.6 *Minimize Soil Acidification*

Depletion of base cations, decreased soil buffering, and increased input of nitrogen and sulphur are all associated with soil acidity brought on by humans. Low buffering capacity or a high aluminium content are typical characteristics of acidic soils, as is a low content of weatherable minerals. Keep an eye on the acidity of the soil and take steps to reduce it by adding lime, gypsum, clean ash, balanced fertilizer, organic amendments, and using acidifying fertilizers sparingly.

9.7 *Mitigate Soil Compaction*

Soil compaction by machinery and trampling damages the soil's structure. Higher runoff results from this reduction in aeration, water drainage, and infiltration. Compaction reduces soil diversity, causes surface crusting, and impedes the formation of seeds and roots, all of which lead to a decline in the health of the soil. Reducing vehicle traffic on bare soils by minimizing operations, putting in place controlled traffic systems, and carrying out agriculture and forestry operations with appropriate soil moisture are some strategies to prevent soil structure degradation caused by tillage. Utilize tire pressure management devices to lower surface pressure and adapt machinery and vehicles to the strength of the soil.

To loosen up compacted soil, use cropping systems that include pastures, crops, and agroforestry species with robust tap roots. In order to prevent trampling, keep the amount of organic matter in the soil sufficient to support microbial activity, stabilize the structure, and support grazing plants.

9.8 Improve Soil Water Management

Soils that are sustainably managed exhibit rapid water infiltration, excellent water retention, and effective drainage. In the absence of these prerequisites, waterlogging and shortage occur. Waterlogging increases the mobility of soil contaminants while having an impact on plant yields and roots. Drainage is essential for the healthy operation of roots in moist regions. High water retention soils need to be taken into consideration. To prevent waterlogging and regulate rising groundwater, install and maintain drainage systems. To improve plant irrigation water efficiency, make improvements to the conveyance, distribution, and field application processes. Water loss can also be reduced by using water management strategies, plant selection, and irrigation techniques. Manage soil cover to maximize water-use efficiency in dryland agriculture. Crop rotation, water harvesting, and lowering runoff and evaporation are some of the techniques that can maximize soil water availability. Encourage the best possible soil water extraction through the use of suitable cultivars, timely agronomic practices, and routine irrigation water quality monitoring.

10 Conclusion

In order to fulfil the food security, soil plays as an important resource that serves a variety of ecosystem services and functions including water purification, nutrient cycling, mitigation and regulation of climate change, retention and degradation of pollutants and pesticides, environmental cleaning, flood prevention, preservation of cultural heritage, and aesthetic provisioning. To maintain soil quality, every department like forestry, fisheries, agriculture and others should come up with sustainable development goals. The use of sustainable agricultural techniques is essential for finding answers to the problems of growing populations and unhealthy soil. It is critical to preserve soil, water, and atmosphere by implementing sustainable ecosystem practices. Examine sustainable management strategies to satisfy food demand without endangering the environment. Poor soil management, excessive fertilizer use, and tillage all negatively impact the health of the soil and crops. It is necessary to pay attention to long-term fertility for sustainable soil management. Greenhouse gas emissions can be decreased, crop yield can be increased, and soil organic carbon can be increased through innovative management techniques.

References

Amundson, R., Berhe, A. A., Hopmans, J. W., Olson, C., Sztein, A. E., & Sparks, D. L. (2015). Soil and human security in the 21st century. *Science, 348*(6235), 1261071.

Bai, Z. G., Dent, D. L., Olsson, L., & Schaepman, M. E. (2008). Proxy global assessment of land degradation. *Soil Use and Management, 24*(3), 223–234.

Borlaug, N. E. (2002). Feeding a world of 10 billion people: The miracle ahead. In Proceedings of the National Academy of Sciences, 100(15), 8032–8037. https://doi.org/10.1073/pnas.172403799
Borrelli, P., Robinson, D. A., Fleischer, L. R., Lugato, E., Ballabio, C., Alewell, C., Meusburger, K., Modugno, S., Schutt, B., Ferro, V., & Bagarello, V. (2017). An assessment of the global impact of 21st century land use change on soil erosion. *Nature Communications, 8*(1), 1–13.
Chenu, C., Angers, D. A., Barré, P., Derrien, D., Arrouays, D., & Balesdent, J. (2019). Increasing organic stocks in agricultural soils: Knowledge gaps and potential innovations. *Soil and Tillage Research, 188*, 41–52.
Dalal, R. C., Allen, D. E., Wang, W. J., Reeves, S., & Gibson, I. (2011). Organic carbon and total nitrogen stocks in a Vertisol following 40 years of no-tillage, crop residue retention and nitrogen fertilisation. *Soil and Tillage Research, 112*(2), 133–139.
Dalal, R. C., & Bridge, B. J. (2020). Aggregation and organic matter storage in sub-humid and semi-arid soils. In *Structure and organic matter storage in agricultural soils* (pp. 263–307). CRC Press.
Dalal, R. C., & Jayaraman, S. (2021). Soil organic carbon sequestration and turnover in semi-arid tropics and subtropics in relation to agronomic yield and sustainability. In *Soil organic carbon and feeding the future* (pp. 287–311). CRC Press.
Dalal, R. C., Strong, W. M., Weston, E. J., Cooper, J. E., Lehane, K. J., & King, A. J. (2018). Farming systems' productivity and soil organic carbon stocks following fertilisers, no-tillage or legumes on a fertility-depleted soil in a semi-arid subtropical region. *Soil Research, 56*(4), 429–440.
Dalal, R. C., Thornton, C. M., Allen, D. E., & Kopittke, P. M. (2021a). A study over 33 years shows that carbon and nitrogen stocks in a subtropical soil are increasing under native vegetation in a changing climate. *Science of the Total Environment, 772*, 145019.
Dalal, R. C., Thornton, C. M., Allen, D. E., Owens, J. S., & Kopittke, P. M. (2021b). Long-term land use change in Australia from native forest decreases all fractions of soil organic carbon, including resistant organic carbon, for cropping but not sown pasture. *Agriculture, Ecosystems & Environment, 311*, 107326.
Davies, W. J., Zhang, J., Yang, J., & Dodd, I. C. (2011). Novel crop science to improve yield and resource use efficiency in water-limited agriculture. *The Journal of Agricultural Science, 149*(S1), 123–131.
Didawat, R. K., Sharma, V. K., Nath, D. J., Patra, A., Kumar, S., Biswas, D. R., et al. (2023). Soil biochemical properties and nutritional quality of rice cultivated in acidic inceptisols using long-term organic farming practices. *Archives of Agronomy and Soil Science, 69*(8), 1282–1297.
FAO. (1983). *Keeping the land alive: Soil erosion-its causes and cures*. Soils Bulletin no. 50.
FAO. (1989). *Soil conservation for small farmers in the humid tropics*. Soils Bulletin no. 60.
FAO. (2017). *Voluntary guidelines for sustainable soil management*. Food and Agriculture Organization of the United Nations, Rome.
FAO. (2019a). *Soil erosion: The greatest challenge to sustainable soil management* (100 pp). Rome. Licence: CC BY-NC-SA 3.0 IGO.
FAO. (2019b). *The international code of conduct for the sustainable use and management of fertilizers* (pp. 1–57). Rome.
Food and Agriculture Organization of the United Nations. (2020). The State of Food Security and Nutrition in the World 2020: *Transforming food systems for affordable healthy diets*. FAO. https://doi.org/10.4060/ca9692en
FAO. (2021). *World food and agriculture – Statistical yearbook 2021*. Rome. https://doi.org/10.4060/cb4477en.
Food and Agriculture Organization of the United Nations & Intergovernmental Technical Panel on Soils. (2020). State of knowledge of soil biodiversity: Status, challenges and potentialities. FAO. https://www.fao.org/documents/card/en/c/CB1929EN
Fuss, S., Lamb, W. F., Callaghan, M. W., Hilaire, J., Creutzig, F., Amann, T., Beringer, T., de Oliveria Garcia, W., Hartmann, J., Khanna, T., Luderer, G., Nemet, G. F., Rogelj, J., Smith, P., Vicente, J. L. V., Wilcox, J., & del Mar Zamora Dominguez, M. & Minx, J. C. (2018).

Negative emissions—Part 2: Costs, potentials and side effects. *Environmental Research Letters, 13*(6), 063002.

Guo, Q., Hao, Y., & Liu, B. (2015). Rates of soil erosion in China: A study based on runoff plot data. *Catena, 124*, 68–76.

Hartmann, M., & Six, J. (2023). Soil structure and microbiome functions in agroecosystems. *Nature Reviews Earth & Environment, 4*(1), 4–18.

Henry, B., Dalal, R., Harrison, M. T., & Keating, B. (2023). *Creating frameworks to foster soil carbon sequestration. BurleighDodds series in agricultural sciences* (pp. 1–43). BurleighDodds Scientific Publication.

Howard, A. (2006). *The soil and health: A study of organic agriculture*. University Press of Kentucky.

Imadi, S. R., Shah, S. W., Kazi, A. G., Azooz, M. M., & Ahmad, P. (2016). Phytoremediation of saline soils for sustainable agricultural productivity. In *Plant metal interaction* (pp. 455–468). Elsevier.

Jat, M. L., Chakraborty, D., Ladha, J. K., Rana, D. S., Gathala, M. K., McDonald, A., & Gerard, B. (2020). Conservation agriculture for sustainable intensification in South Asia. *Nature Sustainability, 3*(4), 336–343.

Jayaraman, S., Dalal, R., Patra, A. K., & Lal, R. (2023). Sustainable soil management: Challenges, prospects and benefits. In *Sustainable soil management: Beyond food production*. Cambridge Scholars Publishing.

Jayaraman, S., Dalal, R. C., Patra, A. K., & Chaudhuri, S. K. (2021d). Conservation agriculture: A sustainable approach for soil health and food security. In *Conservation agriculture for sustainable agriculture*. Springer Nature. https://doi.org/10.1007/978-981-16-0827-8

Jayaraman, S., Dang, Y. P., Naorem, A., Page, K. L., & Dalal, R. C. (2021b). Conservation agriculture as a system to enhance ecosystem services. *Agriculture, 11*(8), 718. https://doi.org/10.3390/agriculture11080718

Jayaraman, S., Naorem, A. K., Lal, R., Dalal, R. C., Sinha, N. K., Patra, A. K., & Chaudhari, S. K. (2021c). Disease-suppressive soils—Beyond food production: A critical review. *Journal of Soil Science and Plant Nutrition, 21*, 1437–1465.

Jayaraman, S., Sinha, N. K., Mohanty, M., Hati, K. M., Chaudhary, R. S., Shukla, A. K., & Dalal, R. C. (2021a). Conservation tillage, residue management, and crop rotation effects on soil major and micro-nutrients in semi-arid Vertisols of India. *Journal of Soil Science and Plant Nutrition, 21*, 523–535.

Kassam, A., Friedrich, T., & Derpsch, R. (2019). Global spread of conservation agriculture. *International Journal of Environmental Studies, 76*(1), 29–51.

Kopittke, P. M., Menzies, N. W., Dalal, R. C., McKenna, B. A., Husted, S., Wang, P., & Lombi, E. (2021). The role of soil in defining planetary boundaries and the safe operating space for humanity. *Environment International, 146*, 106245.

Kopittke, P. M., Menzies, N. W., Wang, P., McKenna, B. A., & Lombi, E. (2019). Soil and the intensification of agriculture for global food security. *Environment International, 132*, 105078.

Lal, R. (2003). Soil erosion and the global carbon budget. *Environment International, 29*(4), 437–450.

Lal, R. (2004). Soil carbon sequestration to mitigate climate change. *Geoderma, 123*(1–2), 1–22.

Lal, R. (2005). Influence of soil erosion on carbon dynamics in the world. In *Soil erosion and carbon dynamics* (pp. 23–35). CRC Press.

Lal, R. (2009). Soil degradation as a reason for inadequate human nutrition. *Food Security, 1*, 45–57.

Lal, R. (2014). Climate strategic soil management. *Challenges, 5*(1), 43–74.

Lal, R. (2015a). Restoring soil quality to mitigate soil degradation. *Sustainability, 7*(5), 5875–5895.

Lal, R. (2015b). Cover cropping and the "4 per thousand" proposal. *Journal of Soil and Water Conservation, 70*(6), 141A.

Lal, R. (2016). Beyond COP 21: Potential and challenges of the '4 per thousand' initiative. *Journal of Soil and Water Conservation, 71*(1), 20A–25A.

Lal, R. (2021). Feeding the world and returning half of the agricultural land back to nature. *Journal of Soil and Water Conservation, 76*(4), 75A–78A.

Lal, R., Reicosky, D. C., & Hanson, J. D. (2007). Evolution of the plow over 10,000 years and the rationale for no-till farming. *Soil and Tillage Research, 93*(1), 1–12.

Minasny, B., Malone, B. P., McBratney, A. B., Angers, D. A., Arrouays, D., Chambers, A., Chaplot, V., & Winowiecki, L. (2017). Soil carbon 4 per mille. *Geoderma, 292*, 59–86.

NAAS. (2010). *Degraded and wastelands of India: Status and spatial distribution* (pp. 1–158). Directorate of Information and Publications of Agriculture, Indian Council of Agricultural Research, New Delhi.

Newton, P., Civita, N., Frankel-Goldwater, L., Bartel, K., & Johns, C. (2020). What is regenerative agriculture? A review of scholar and practitioner definitions based on processes and outcomes. *Frontiers in Sustainable Food Systems, 4*, 577723.

Paustian, K., Lehmann, J., Ogle, S., Reay, D., Robertson, G. P., & Smith, P. (2016). Climate-smart soils. Nature, 532(7597), 49–57. https://doi.org/10.1038/nature17174

Powlson, D. S., Gregory, P. J., Whalley, W. R., Quinton, J. N., Hopkins, D. W., Whitmore, A. P., Hirsch, P. R., & Goulding, K. W. (2011b). Soil management in relation to sustainable agriculture and ecosystem services. *Food Policy, 36*, S72–S87.

Powlson, D. S., Whitmore, A. P., & Goulding, K. W. (2011a). Soil carbon sequestration to mitigate climate change: A critical re-examination to identify the true and the false. *European Journal of Soil Science, 62*(1), 42–55.

Rumpel, C., Amiraslani, F., Chenu, C., Garcia Cardenas, M., Kaonga, M., Koutika, L. S., & Wollenberg, E. (2020). The 4p1000 initiative: Opportunities, limitations and challenges for implementing soil organic carbon sequestration as a sustainable development strategy. *Ambio, 49*, 350–360.

Rumpel, C., Lehmann, J., & Chabbi, A. (2018). '4 per 1,000' initiative will boost soil carbon for climate and food security. *Nature, 553*(7686), 27.

Sanchez, P. A., & Swaminathan, M. S. (2005). Cutting world hunger in half. *Science, 307*(5708), 357–359.

Schweizer, S. A., Graf-Rosenfellner, M., Bhat, N. A., Kayser, G., Sisodia, B. S., Kirchhof, G., Zikeli, S., Cadisch, G., & Bhullar, G. S. (2022). Responses of soil organic carbon, aggregate diameters, and hydraulic properties to long-term organic and conventional farming on a Vertisol in India. *Land Degradation & Development, 33*(5), 785–797.

Sharma, K. L., Grace, J. K., Mishra, P. K., Venkateswarlu, B., Nagdeve, M. B., Gabhane, V. V., Sankar, G. M., & Ramachandran, K. (2011). Effect of soil and nutrient-management treatments on soil quality indices under cotton-based production system in rainfed semi-arid tropical vertisol. *Communications in Soil Science and Plant Analysis, 42*(11), 1298–1315.

Smith, P., & Powlson, D. S. (2007). Sustainability of soil management practices-a global perspective. In *Soil biological fertility* (pp. 241–254). Springer.

Soussana, J. F., Lutfalla, S., Ehrhardt, F., Rosenstock, T., Lamanna, C., Havlík, P., & Lal, R. (2019). Matching policy and science: Rationale for the '4 per 1000-soils for food security and climate' initiative. *Soil and Tillage Research, 188*, 3–15.

Spiertz, J. H. J. (2009). Nitrogen, sustainable agriculture and food security: A review. In *Sustainable agriculture* (pp. 635–651). Springer.

Stagnari, F., Galieni, A., D'Egidio, S., Pagnani, G., & Pisante, M. (2019). Sustainable soil management. In *Innovations in sustainable agriculture* (pp. 105–131). Springer.

Stinner, D. H. (2007). The science of organic farming. In *Organic farming: An international history* (pp. 40–72). CABI.

TAAS. (2021). Regenerative agriculture for soil health, food and environmental security. In: *Proceedings and recommendations. Trust for advancement of agricultural sciences (TAAS), Avenue II, Pusa Campus, New Delhi* (vii+30 p).

Tamene, L., Adimassu, Z., Ellison, J., Yaekob, T., Woldearegay, K., Mekonnen, K., Thorne, P., & Le, Q. B. (2017). Mapping soil erosion hotspots and assessing the potential impacts of land management practices in the highlands of Ethiopia. *Geomorphology, 292*, 153–163.

Tully, K. L., & McAskill, C. (2020). Promoting soil health in organically managed systems: A review. *Organic Agriculture, 10*(3), 339–358.

United Nations. (2015). *Sustainable development goals: The 17 goals*. UN Department of Economics and Social Affairs, Sustainable Development Goals. Retrieved from https://www.sdgs.un.org/goals

Van Groenigen, J. W. (2018). The '4 per 1000' initiative: A credibility issue for the soil science community? *Geoderma, 309*(Supplement C), 118–123.

Vohland, K., & Barry, B. (2009). A review of in situ rainwater harvesting (RWH) practices modifying landscape functions in African drylands. *Agriculture, Ecosystems & Environment, 131*(3–4), 119–127.

White, P. J., Crawford, J. W., Díaz Álvarez, M. C., & García Moreno, R. (2012). Soil management for sustainable agriculture. *Applied and Environmental Soil Science, 2012*, 850739.

Yachi, S., & Loreau, M. (1999). Biodiversity and ecosystem productivity in a fluctuating environment: The insurance hypothesis. *Proceedings of the National Academy of Sciences, 96*(4), 1463–1468.

Soil Nutrient Analysis for Sustainable Agriculture in Palakkad Gap Region, South India

P. P. Jincy, Richard Scaria, Harikrishnan T. Menon, P. Pankajakshan, B. S. Jeena, M. C. Muhsin, C. Satheesh, and Dhanya Vijayan

Abstract A research on soil fertility in Palakkad Gap region enables analysis of the nutrient status with respect to its geology as well. The Palakkad Gap, being located in the Palakkad Cauvery Shear Zone, constitutes probably one of the oldest regions in the Southern Granulite Terrain and has undergone repeated events of tectogenesis besides it is one of the major breaks of the Western Ghats. Politically, the study area falls in the portions of Palakkad district of Kerala and the Coimbatore district of Tamil Nadu with 116,677 ha. Soil nutrient analysis help in soil management by checking soil degradation for a sustainable future of agricultural practices. In this respect, a comprehensive approach associating physio-chemical analysis of soil, geostatistical and ArcGIS platforms has been utilised. The spatial distributions of soil nutrient parameters have been represented by an interpolation method in the ArcGIS software. Soil sampling results along with supportive secondary sources have been incorporated as the database for this research. The conditions of soil pH and electrical conductivity (EC), the distribution of available macro and micro-nutrients, have been analysed with the test results from Kerala Forest and Research Institute (KFRI), Govt. of Kerala. The results show that almost 45% of the total area

P. P. Jincy (✉)
Department of Geography, Govt. Arts and Science College, Tholanur, Palakkad, Kerala, India

R. Scaria · P. Pankajakshan · B. S. Jeena · M. C. Muhsin
Department of Geography and Research Center, University of Calicut, Govt. College, Chittur, Palakkad, Kerala, India
e-mail: pankajchirayil@chitturcollege.ac.in

H. T. Menon
SLUSI, Ministry of Agriculture and Farmer's Welfare Government, New Delhi, India

KRISDAM, Thrissur, Kerala, India

C. Satheesh
Department of General and Applied Geography, Doctor Harisingh Gour Vishwavidyalaya (A Central University), Sagar, Madhya Pradesh, India

D. Vijayan
Leibniz Centre for Agricultural Landscape Research (ZALF), Muncheberg, Germany
e-mail: Dhanya.Vijayan@zalf.de

A. K. Singh et al. (eds.), *Sustainable Agriculture Management in Semi-Arid Climates*, https://doi.org/10.1007/978-3-031-94062-0_7

is normal for pH and electrical conductivity falls in non-saline category for 98% of the region. The nitrogen, organic content and boron levels are deficient while all other macro and micro-nutrients are sufficient in the region. The Soil Nutrient Index has been employed to evaluate the fertility status of the region as a whole and reached the inference that Palakkad Gap region is unique from the rest of the region of Kerala and Tamil Nadu. Besides, proper attention for deficient soil nutrients can ensure outstanding agricultural productivity in the region.

Keywords Nutrient status · Macro and micro-nutrients · Soil fertility · Deficient · Sufficient

1 Introduction

Soil is the imperative constituent of any cultivation. Studies on the spatial distribution of soil nutrients are always a hot topic, especially for promoting agricultural sustainability in recent years (Li et al., 2021). Thus, an analysis of soil nutrients is crucial in improving the soil productivity through proper management practices. The success of a soil management practice in order to sustain the soil fertility is subject to the soil response to the agricultural practices (Thakare, 2020). The soil nutrient analysis evaluates the nutrient status of the given soil and provides all essential information required to set the target of nutrient applications. It also facilitates in detection and monitoring of all possible changes in the soil parameters (Sonikajha & Suneetha, 2015). Soil macro and micro-nutrients play predominant role in the plant growth by supplying nitrogen, phosphorous and potassium (NPK), organic content (OC) and other micro-nutrients to the topsoil (Krishnaveni, 2015) and also they are the key attributes that regulates the plant growth, soil microbial activities and the potential environmental pollution. As the constant usage of the land has increased, farmers are increasingly using chemical fertilisers and pesticides which rise the chance of diminishing return (USDA-ERS, 1997). Equally, due to crop cultivation and harvesting, substantial quantities of mineral nutrients are lost from soils in alarming rate. Therefore, soil quality checking and conservations can definitely give most relevant and optimistic impact on the economy which reflects in productivity, water and air quality, and optimal use of nutrients and pesticides (Ravikumar, 2013). The available soil nutrients are associated with the absorption capacity of the plants and the transformations, migrations and microbial decompositions of fertilisers in the soil. Hence, a balance of soil nutrients is highly connected to the amount of fertilisers applied to the soil (Pan et al., 2021). When the nutrient levels in soil are very high, it can be a sign of the nutrient imbalances and consequently there are chances of environmental pollution from the field. In contrast, when there is low nutrient level, the soil should be applied with required nutrients and in case of very low conditions, it should be checked and changed from the existing fertilisation practices (Yost & Uchida, 2000). Proportionate to primary and secondary nutrients, crops require smaller quantities of micronutrients (Shukla, 2015).

Soil variability determination and soil health management is very much important for ecological modelling, environmental predictions, precise agricultural management of natural resources as well (Hangsheng et al., 2005; Wang et al., 2009).

2 Context of the Study

Due to an unbalanced and insufficient fertiliser usage along with low efficiencies of other attributes, the effectiveness level of the chemical fertilisers declines enormously under intensive agricultural scenario. Difference in available nutrients can also be natural; therefore, some can be deficient and the others can be in sufficient level. Therefore, judicious usage of both macro and micro-nutrient supplements is inevitable to overcome both deficiencies and imbalances (Singh & Mishra, 2012). An updated and area-specific information on the availability of soil nutrients in the study area is meagre.

Soil nutrient analysis is essential for farmers, gardeners, and agronomists to make informed decisions about fertilisation and crop management. The analysis provides data on essential nutrients such as soil pH, electrical conductivity (EC), boron, copper, iron, manganese, organic carbon (OC), available nitrogen, zinc, sulphur, phosphorus, potassium, and other elements necessary for plant growth. Soil nutrient analysis is a valuable tool for sustainable agriculture, helping to optimise nutrient management, reduce environmental impact, and improve overall crop productivity.

Understanding the complex interactions of land use, land capability and nutrient variability are vital for sustainable agriculture (Yagaso et al., 2024). The variation of nutrient resources across landscapes has become the key point of many of the researches nowadays (Benning & Seastedt, 1995). Several studies identified that the net primary productivity is primarily controlled by the availability of soil nutrients (Pastor & Post, 1986; Seastedt et al., 1991). Consequently, distribution and spatial variability of soil nutrients along with climate, land use and other variables are critical for predicting the effects of future land use change on nutrients (Kosmas et al., 2000).

Soil characteristics are highly influenced by soil-forming factors, such as climate, topography as well as the nature of parent materials from which it is weathered (Jenny, 1941). The correlation among soil properties and soil-forming factors is an important subject matter that has been researched globally (Li et al., 2018). Parent material (Kosmas et al., 1993), climate and geological history have a marked influence on soil properties in regional as well as continental scale. The physical properties of soil, for instance distribution of clay content with depth, sand content and pH show high correlation with landscape (Ovalles & Collins, 1986). Tectonically active regional expanses possess complex landscapes of soils with a heterogeneous physico-chemical nature. Correspondingly, spatial variability of nutrient sources enriches regional biodiversity and forms diverse habitats possibly attractive for animals and humans (Kübler et al., 2021). The productivity of crops is proportional to the soil health which in turn results in better livelihood of the people inhabiting the region (Muhsina et al., 2021).

Moreover, the soils of the Palakkad region (Table 1) are characterised by inceptisols (minimal soil profile development) and entisols (no profiles but just formed from parent materials) alfisols (rich in fertility) and ultisols (ultimately weathered). Inceptisols are young soils, in which the soil parent material has a major influence on the nutrient element concentrations, which are more pronounced and less understood in the *ultisols*, which undergo continuous weathering of minerals. Hence, this

Table 1 Soil classifications of Palakkad gap region

Soil series	Soil reaction	Soil classification	Order
Alagapuri	Calcareous	VerticHaplustepts	Inceptisols
Anakkal	Non-calcareous	TypicHapludults	Ultisols
Anamalai	Non-calcareous	FluventicHaplustepts	Inceptisols
Annur	Non-calcareous	Lithic Rhodustalfs	Alfisols
Attipalayam	Calcareous	VerticHaplustepts	Inceptisols
Chuduvalattur	Non-calcareous	TypicHapludults	Ultisols
Chullimada	Non-calcareous	TypicUdorthents	Entisols
Dasarapatti	Calcareous	GypsicHaplustepts	Inceptisols
Ilayaimuthur	Non-calcareous	Lithic Haplustepts	Inceptisols
Irugur	Non-calcareous	TypicHaplustepts	Inceptisols
Kalkalam	Non-calcareous	TypicUdipsamments	Entisols
Kallivalasu	Calcareous	FluventicHaplustepts	Inceptisols
Kanjampatti	Non-calcareous	TypicHaplustepts	Inceptisols
Kattampatti	Calcareous	VerticUstorthents	Entisols
Kumbamala	Non-calcareous	TypicUdorthents	Entisols
Kundur	Non-calcareous	TypicDystrudepts	Inceptisols
Kunnathur	Non-calcareous	Lithic Rhodustalfs	Alfisols
Kunpulli	Non-calcareous	TypicDystrudepts	Inceptisols
Malampuzha	Non-calcareous	TypicUdipsamments	Entisols
Mandur	Non-calcareous	RupticAlficDystrudepts	Inceptisols
Mannur	Non-calcareous	TypicDystrudepts	Inceptisols
Mettupalaiyam	Non-calcareous	RhodicPaleustalfs	Alfisols
Muppamkulam	Non-calcareous	OxicDystrudepts	Inceptisols
Palathurai	Calcareous	TypicHaplustepts	Inceptisols
Peringottukurussi	Non-calcareous	AquicDystrudepts	Inceptisols
Pichanur	Non-calcareous	TypicUdipsamments	Entisols
Pilamedu	Non-calcareous	UlticHaplustepts	Inceptisols
Pudur	Non-calcareous	TypicHaplustepts	Inceptisols
Salaiyur	Calcareous	FluventicHaplustepts	Inceptisols
Somayyanur	Non-calcareous	TypicHaplustepts	Inceptisols
Valipparamlat	Non-calcareous	TypicDystrudepts	Inceptisols
Varapatti	Calcareous	VerticUstorthents	Entisols
Velladikunnu	Non-calcareous	TypicDystrudepts	Inceptisols

is primary step to understand the uniqueness of Palakkad Gap region with respect and relation to its soil formation and soil fertility.

3 Objectives of the Study

- To comprehend the *uniqueness* of soil types in Palakkad Gap for agricultural perspective in relation to its present soil nutrient status in comparison to other parts of the world.
- To analyse the relationship between the present soil nutrients with the parent material, which can be the primary step laid to decipher the *geological history* of Palakkad Gap region.

4 General Description of the Study Area: Palakkad Gap Region

The Palakkad Gap, also known as the Palghat Gap, is a low mountain pass in the Western Ghats of Kerala, India. It is one of the most significant mountain passes in the Western Ghats, allowing for a natural corridor between the states of Kerala and Tamil Nadu. The gap serves as a vital communication link between the two states. Geographically, this area lies between 76°38′40″ to 77°2′21″ east longitude and 10°29′29″ north latitude in parts of Palakkad and Coimbatore districts of Kerala and Palakkad. It is situated between the Nilgiri Hills to the north and the Anaimalai Hills to the south. The gap is part of the Western Ghats, a mountain range running parallel to the western coast of India.

4.1 Delineation of Palakkad Gap Region

For demarking the study area of Palakkad Gap region, based on the hierarchical approach (Level-III of 1:50,000 scale), four determinants such as slope classes, dissection index, relative relief and drainage density were considered. Hence, the present study area is a region of slope <5%, relative relief <5 m, low dissected and drainage density <1. The capacity to add data at different levels and the possibility to increase subunits at any levels is the peculiarity and merit of this classification. Due to internal homogeneity, this demarcation has sharp/clear cut unit boundaries. Here, limited or no generalisations can be done for areal classes in which relief, lithology, and genesis are the unit determinants. Specific origin and landform associations are important demarcating criteria of the study area. This delineation is highly helpful in developmental planning at the district or block level, for soil erosion potential mapping, prioritisation, etc. Thus, it is applied to demarcate and delineate the study area for carrying out both physical and cultural studies in Palakkad Gap region which spread both in portions of Palakkad District of Kerala and Coimbatore District of Tamil Nadu (Figs. 1 and 2).

The origin and evolution of Palakkad Gap is still a controversy with contradicting hypothesis postulated by various geologists. The Palakkad Gap, being located in the Palakkad Cauvery Shear Zone, constitutes probably one of the oldest regions in the Southern Granulite Terrain and has undergone repeated events of tectogenesis. Palghat Gap is the only major break within the Western Ghats. Topographic analysis of the Palghat area has brought to light remarkable linear, consequent rivers of high orders in the Gap proper, as compared to closely spaced, lower order streams of trellis, dendritic, obsequent and subsequent patterns in the hill ranges. The Gap region represents an area of low undulating relief, with convex gently graded inter-stream tracts, sloping down to broad valley floors consisting of local erosional remnants. These erosional landforms are often seen inter-fingered with alluvial plains and lateritic hummocks, and the terrain as a whole represents a dissected pediment.

5 Database and Methodology

5.1 Data Collection and Soil Sampling

The soil samples were collected from various locations across the study area. Then, the soil samples were tested from Kerala Forest and Research Institute (KFRI), Peechi for essential nutrients such as soil pH, electrical conductivity (EC), boron, copper, iron, manganese, organic carbon (OC), available nitrogen, zinc, sulphur, phosphorus and potassium.

5.2 Spatial Interpolation Using IDW

The ArcGIS 10.2 software has been used for IDW interpolation to generate the soil fertility map. With the IDW interpolation, the unknown values of unsampled locations are estimated based on the inverse of the distance to the nearest sampled points, weighted by their distances. Specified parameters such as the power parameter (which controls the influence of nearby points on the interpolated values and the search radius (the maximum distance within which neighbouring points) are considered for interpolation.

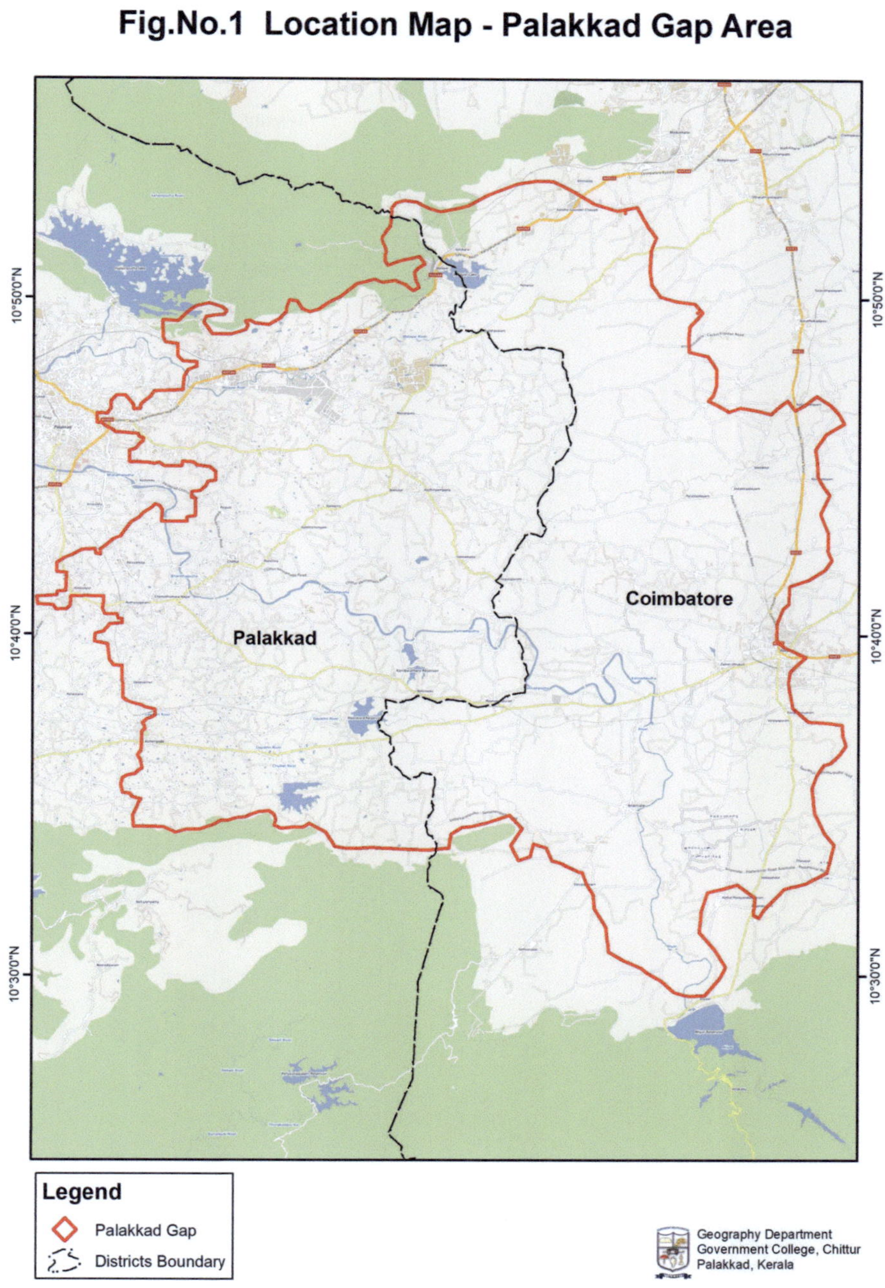

Fig. 1 Location map—Palakkad Gap region

Fig. 2 Index Map—Palakkad Gap region

5.3 *Validation and Adjustment*

The interpolated soil fertility map is then validated by comparing it with the original soil fertility data collected from the sampling points. Assess the accuracy of the interpolated map using statistical measures such as root mean square error (RMSE) or cross-validation techniques.

5.4 *Secondary Data Sources*

The soil characteristics of Palakkad Gap region are analysed with the help of soil data collected from Tamil Nadu Agricultural University (TNAU) for Tamil Nadu portion and Soil and Land Use Survey of India (SLUSI).

6 Results and Discussions

6.1 *Soil Reaction (pH)*

Soil pH is a key factor that determines nutrient availability, microbial activities and crop development in the soil. However, researches on the soil pH variations in regional level remains inadequate (Zhang et al., 2019). It is a key index of soil properties, which is considered as one of the main influencing variables other than any soil properties (Bolan & Kandaswamy, 2005). Government of India Ministry of Agriculture and Farmers Welfare initiated soil health card (SHC) scheme all across all India on village level by collection soil samples from field level since 2014–2015. Soil health cards (SHCs) scheme, a massive programme of soil sampling, testing and generation of SHCs, was launched to assist the farmers to know the fertility status of their land holdings. State Governments have intended to issue soil health cards to all farmers in the country. As per its guidelines of the ministry of agriculture and farmers welfare, the soil pH ranges are given in Table 2 and in this report pH analysis made using the same data and techniques.

Table 2 Limits of pH concentration in soil

Class	pH
Strongly acidic	<4.5
Moderately acidic	4.5–5.5
Slightly acidic	>5.5–6.5
Normal	>6.5–7.5
Slightly alkaline	>7.5–8.5
Moderately alkaline (Sodic)	>8.5–9.5
Strongly alkaline (Highly Sodic)	>9.5

Table 3 Distribution of soil reaction (pH) in Palakkad gap area

Legend	Area (ha.)	AREA_PER
2. Moderately acidic (4.5–5.5) 10.2 ha.	10.2	0.01
3. Slightly acidic (>5.5–6.5) 35051.89 ha.	35051.89	30.04
4. Normal (>6.5–7.5) 51723.16 ha.	51723.16	44.33
5. Slightly alkaline (>7.5–8.5) 27708.16 ha.	27708.16	23.75
6. Moderately alkaline (Sodic) (>8.5–9.5) 372.05 ha.	372.05	0.32
7. Strongly alkaline (Highly Sodic) (>9.5) 162.86 ha.	162.86	0.14
8. Misc.	1648.68	1.41
Total	**1,16,677**	**100**

Soil pH in Palakkad Gap area ranges from moderately acidic (4.5–5.5) which constitutes an area of 10.2 ha (0.01%), 10.2, slightly acidic soils (>5.5–6.5) covers an area of 35,051.89 ha (30.04%) followed by normal soils (>6.5–7.5) with an area of 51,723.16 ha (44.33%) which predominately dominates majority of the area. Slightly alkaline soils (>7.5–8.5) are mainly seen in Coimbatore district of the study area with an area of 27,708.16 ha (23.75%). In the same study area of Coimbatore part, moderately alkaline (sodic) (>8.5–9.5) soils were seen which need immediate attention to reclaim. These soils occupy an area of 372.05 ha (0.32%), further strongly alkaline (highly sodic) (> 9.5) occupies an area of 162.86 ha (0.14%) needs immediate reclamation by adding more gypsum and organic matter. A table showing the area and percentage of its distribution in the Palakkad gap area is given below for ready reference (Table 3, Fig. 3).

6.2 *Electrical Conductivity (EC)*

The Ministry of Agriculture and Farmers Welfare, Integrated Nutrient Management (INM Division) Government of India, in its soil health cards (SHCs) programme defined electrical conductivity as in Table 4.

In a total area of 116,677 ha. of Palakkad Gap region, 114,860. 87 ha. (98.44%) of soils are non-saline (0–1.68 dS/m) which is a favourable condition for soil fertility while an expanse of 167.45 ha. (0.14%) of the gap soils is in low-saline category (1.68–3.36 dS/m) which can also be tolerated for plant growth. Irrespective of Kerala and Tamil Nadu portions of gap region, except for the pocket region of Sokkannur Village in Coimbatore District, shows non-saline soils. The Table 5 shows the distribution of EC in Palakkad Gap region in area and percentage (Fig. 4).

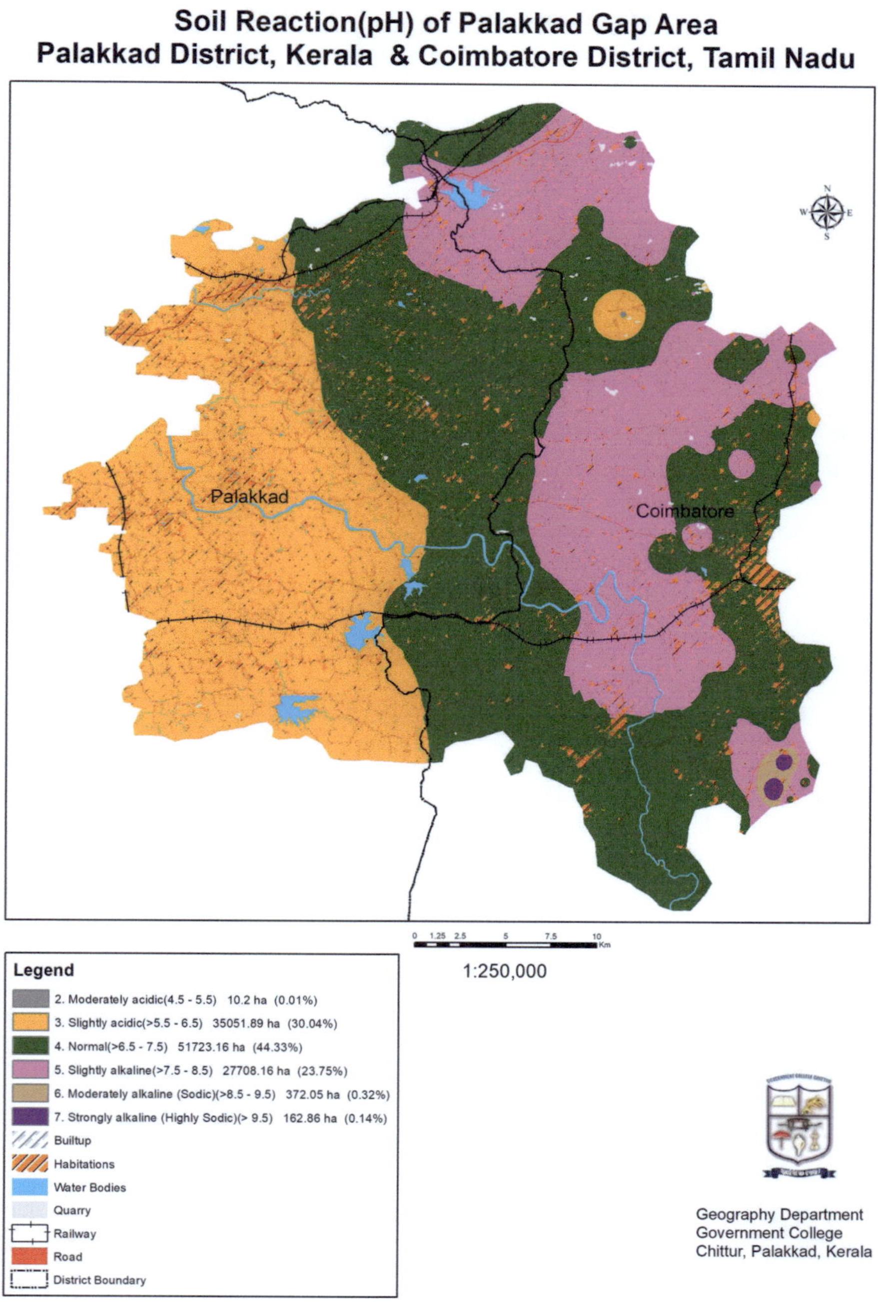

Fig. 3 Distribution of potential hydrogen (pH) in Palakkad Gap area

Table 4 Limits of electrical conductivity (EC) in soil

Class	EC (dS/m) in 1:2.5 extract at 25 °C	Electro-conductivity of soil saturated extract ECe at 25 °C (dS/m)
Non-saline	0–1.68	0–2
Low salinity	1.68–3.36	2–4
Mild salinity	3.36–6.72	4–8
High salinity	6.72–13.44	8–16
Sever salinity	>13.44	>16

Table 5 Area of electrical conductivity (EC) distribution in Palakkad gap region

Legend	Area	Area in percentage
1. Non-saline (0–1.68 dS/m)	1,14,860.87	98.45
2. Low saline (1.68–3.36 dS/m)	167.45	0.14
3. Misc.	1, 648.68	1.41
Total	**1,16,677**	**100**

6.3 *Boron (B)*

As per the soil health cards (SHCs) scheme of Ministry of Agriculture and Farmers Welfare, under Integrated Nutrient Management (INM Division), Government of India, the available boron (B) concentration has been classified as in Table 6.

It is evident that the available boron in Palakkad Gap region comes under deficient category (<0.5 ppm), falls under an area of 115,028 ha. (98.59%) out of 116,677 ha. of the total area of interest. Both, Kerala and Tamil Nadu portions of the study area, are deficient in the boron distribution. The distribution of boron in Palakkad Gap region is shown in the Table 7 and pie chart (Fig. 5).

6.4 *Copper (Cu)*

In the soil health card scheme formulated by Ministry of Agriculture and Farmers Welfare, Integrated Nutrient Management (INM) Division, Government of India classified the limits of Copper as in Table 8.

The study area, Palakkad Gap region, shows that the availability of copper is sufficient (>0.2 ppm) in 93,507.7 ha (80.14%). Hence, the region comes under sufficient category. But 21,520.62 ha (18.44%) of this region falls in deficient category with copper availability <0.2 ppm. The concentration of deficient copper

Fig. 4 Distribution of electrical conductivity (EC) in Palakkad Gap area

Table 6 Limits of boron concentration in soil

Category	Boron (B) ppm
Deficient	<0.5
Sufficient	>0.5

Table 7 Area of boron concentrations in Palakkad gap region

Legend	Area in hectares	Area in percentage
1. Deficient (<0.5 ppm)	1,15,028.32	98.59
2. Misc.	1648.68	1.41
Total	**1,16,677**	**100**

can be identified in parts of western gap region i.e. in the Palakkad portions of the study area and rest of the major portions of the study area have sufficient copper concentration especially in Tamil Nadu portions of the study area. The Table 9 shows the distribution of Copper in Palakkad Gap region in area and percentage (Fig. 6).

6.5 Iron (Fe)

The soil health card scheme initiated by Ministry of Agriculture and Farmers Welfare, Integrated Nutrient Management (INM) Division, Government of India, classified the limits of iron as in Table 10.

An area of 971,23.97 ha has sufficient available iron which constitutes 83.24% of the study area with >4.5 ppm which is concentrated mainly in Kerala portion of Palakkad Gap region. An area of 179,04.35 ha (15.35%) shows deficient iron concentration with <4.5 ppm which is mainly dominated in the Tamil Nadu region of the study area. The Table 11 shows the distribution of Iron in the study area and its percentage (Fig. 7).

6.6 Manganese (Mn)

Integrated Nutrient Management (INM) Division introduced soil health card programme initiated by Ministry of Agriculture and Farmers Welfare, Government of India has classified the limits of manganese (Mn) as in Table 12.

Out of the total land area of 116,677 ha of Palakkad Gap region, it is found that manganese (Mn) deficiency (<2.0 ppm) in 1676.79 ha (1.44%) but then 97.15% of the study region has the sufficient distribution of manganese (Mn) in an area of 113,351.63 ha. Hence, the distribution of Mn in the Palakkad Gap region is said to be sufficient especially in Kerala portions while in a few pocket spaces have slight

Fig. 5 Distribution of boron (B) in Palakkad Gap Area

Table 8 Limits of copper (Cu) in soil

Category	Copper (Cu) ppm
1. Deficient	<0.2
2. Sufficient	>0.2

Table 9 Area of copper (Cu) concentration in Palakkad gap region

Legend	Area	Area in percentage
1. Deficient (<0.2 ppm)	21,520.6	18.44
2. Sufficient (>0.2 ppm)	93,507.7	80.15
3. Misc.	1648.6	1.41
Total	**1,16,677**	**100**

deficiency which mainly distributed in the Coimbatore parcels of the study area. The Table 13 shows the distribution of Mn in the Palakkad Gap region.

Manganese, when exists in low level is an essential soil nutrient for all crops, while when it is in excess it acts as a toxicant. Therefore, managing Mn for its deficiency and toxicity is significant for optimum crop development in the agricultural soils. The Fig. 8 represents the distribution of Mn in the soils of Palakkad Gap region.

6.7 *Organic Carbon (OC)*

Soil organic carbon (SOC) refers to the carbon stored in soil in the form of organic matter. This organic matter comes from the decomposition of plant and animal residues, as well as the synthesis of organic compounds by soil organisms. SOC plays a crucial role in soil health, fertility, and overall ecosystem functioning. Here are some key points about soil organic carbon.

Organic matter composition: soil organic carbon is a component of soil organic matter (SOM), which also includes other organic substances like plant and animal residues in various stages of decomposition. OC content in soil is a key indicator of soil fertility. It improves soil structure, water retention, and nutrient availability. Soils with higher organic carbon content generally support better plant growth. Carbon sequestration: soil organic carbon contributes to carbon sequestration, which is the process of capturing and storing atmospheric carbon dioxide in the soil. This is important in mitigating climate change by reducing the amount of carbon dioxide in the atmosphere.

Microbial activity: SOC serves as a food source for soil microorganisms. Microbes break down organic matter, releasing nutrients in a form that plants can use. This microbial activity is essential for nutrient cycling in the soil. Soil with higher organic carbon content tends to have better structure and stability, reducing the risk of erosion. The organic matter acts like a glue, binding soil particles together.

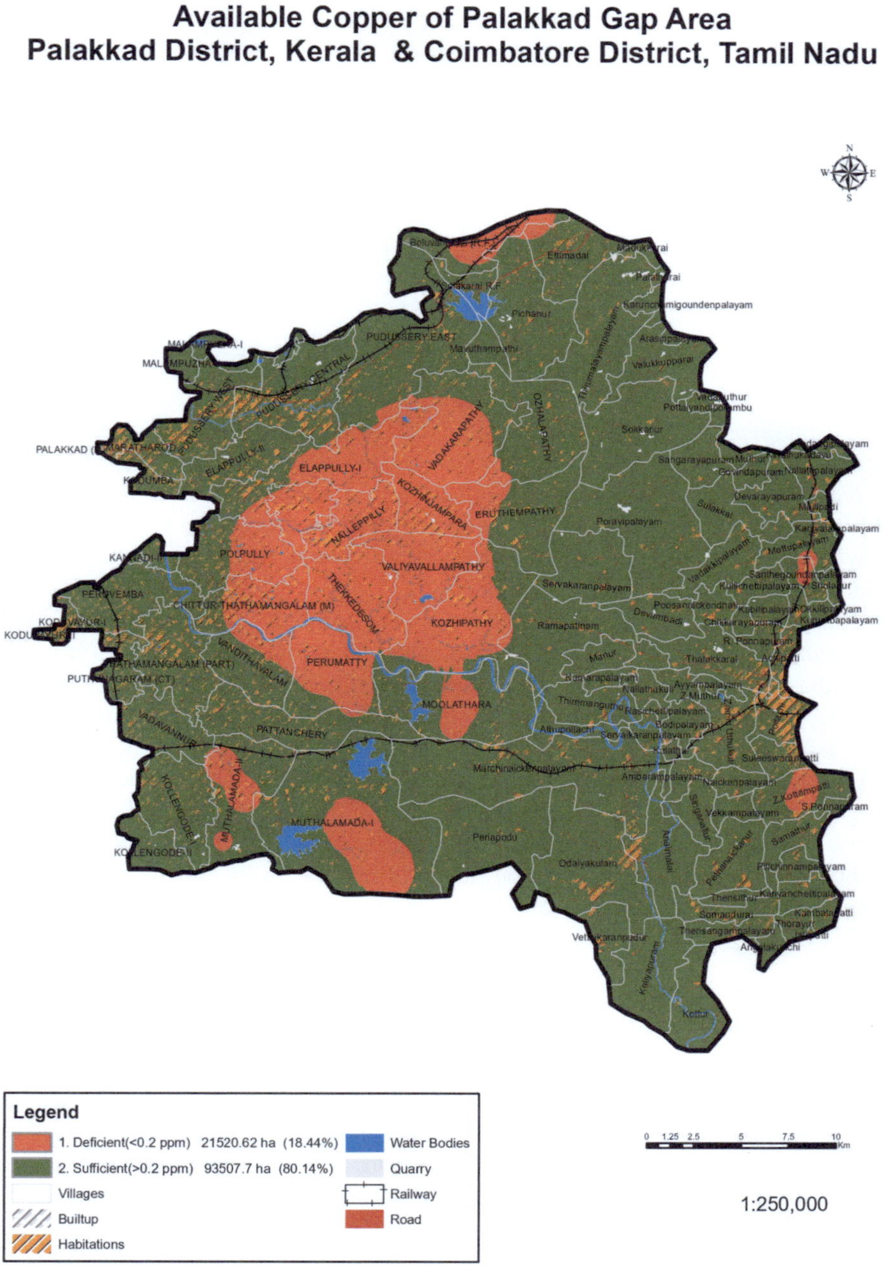

Fig. 6 Distribution of copper (Cu) in Palakkad Gap area

Table 10 Limits of iron (Fe) in soil

Category	Iron (ppm)
1. Deficient	<4.5
2. Sufficient	>4.5

Table 11 Area of iron (Fe) concentration in Palakkad gap region

Legend	Area	Area in percentage
1. Deficient (<0.2 ppm)	1,79,04.35	15.35
2. Sufficient (>0.2 ppm)	9,71,23.97	83.24
3. Misc.	1648.6	1.41
Total	**1,16,677**	**100**

Agricultural practices, land use changes, and deforestation can influence soil organic carbon levels. Sustainable land management practices, such as cover cropping, reduced tillage, and agroforestry, can enhance SOC and promote overall soil health. Measuring soil organic carbon is a common practice in soil quality assessments and environmental monitoring. It helps researchers, farmers, and land managers understand the health and productivity of soils.

Integrated Nutrient Management (INM) Division introduced soil health card programme initiated by Ministry of Agriculture and Farmers Welfare, Government of India has classified the limits of organic carbon is given in Table 14.

Out of the total land area of 116,677 ha of Palakkad Gap region, the majority of the availability of organic carbon falls under the low category with an area of 112,421.46 ha which occupies (96.35%) of the total area. The medium range of organic carbon constitutes 2350.41 ha (2.01%) of the total area. In the high category, available organic carbon is very less with an area of 256.46 ha only. The distribution of organic carbon is given in the Table 15 with its graphic presentation also. This table is without miscellaneous area (Fig. 9).

6.8 Nitrogen (N)

The available soil nitrogen usually represents the fertility and highly correlated with the crop yield (Zhang, 2018). Soil health card programme initiated by Ministry of Agriculture and Farmers Welfare, Integrated Nutrient Management (INM) Division, Government of India has defined the available limits of nitrogen as in Table 16.

Out of the total area of 116,677 ha of Palakkad Gap region, 115,028.32 ha (98.59%) falls in <280 kg/ha of available nitrogen which comes under low category irrespective of Kerala and Tamil Nadu portions. The Table 17 depicts the nitrogen concentration in Palakkad Gap region (Fig. 10).

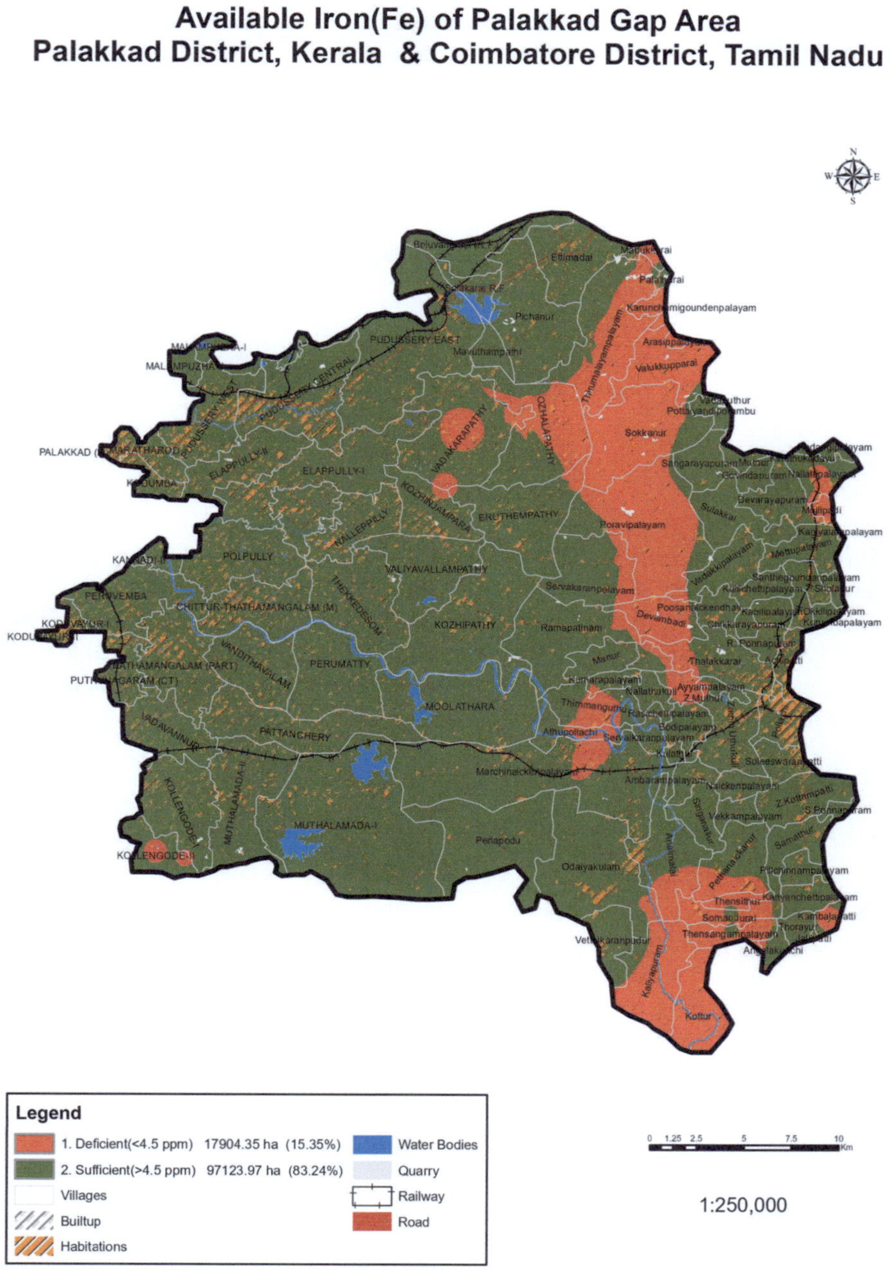

Fig. 7 Distribution of iron (Fe) in Palakkad Gap region

Table 12 Limits of manganese (Mn) in soil

Category	Manganese (ppm)
1. Deficient	<2.0
2. Sufficient	>2.0

Table 13 Area of manganese (Mn) distribution in Palakkad gap region

Legend	Area	Area in percentage
1. Deficient (<0.2 ppm)	1676.69	1.44
2. Sufficient (>0.2 ppm)	1,13,351.63	97.15
3. Misc.	1648.6	1.41
Total	**1,16,677**	**100**

6.9 Zinc (Zn)

Based on the definitions given by soil health card scheme initiated by Ministry of Agriculture and Farmers Welfare, Integrated Nutrient Management (INM) Division, Government of India, the limits of zinc in soils are as in Table 18.

Out of 116,677 ha. of total land area of Palakkad Gap region, the available zinc has 112,716.65 ha. (96.6%) area in sufficient and 2311.66 ha. (1.98%) area in deficient category. The Table 19 depicts the area and percentage of zinc availability in Palakkad Gap region (Fig. 11).

6.10 Sulphur (S)

As stated by the soil health card scheme introduced by Ministry of Agriculture and Farmers Welfare, Integrated Nutrient Management (INM) Division, Government of India, the limits of sulphur in soils are defined as in Table 20.

Palakkad gap has a total area of 116,677 ha. in which sulphur distribution is sufficient in 102,211.57 ha (87.61%) whereas deficient in 12,816.75 ha (10.98%) of the study area. The Table 21 represents the area and percentage of sulphur distribution in Palakkad Gap region (Fig. 12).

6.11 Potassium (K)

The Ministry of Agriculture and Farmers Welfare, in their soil health card scheme, Integrated Nutrient Management (INM) Division, Government of India, defined the potassium limits in soils as in Table 22.

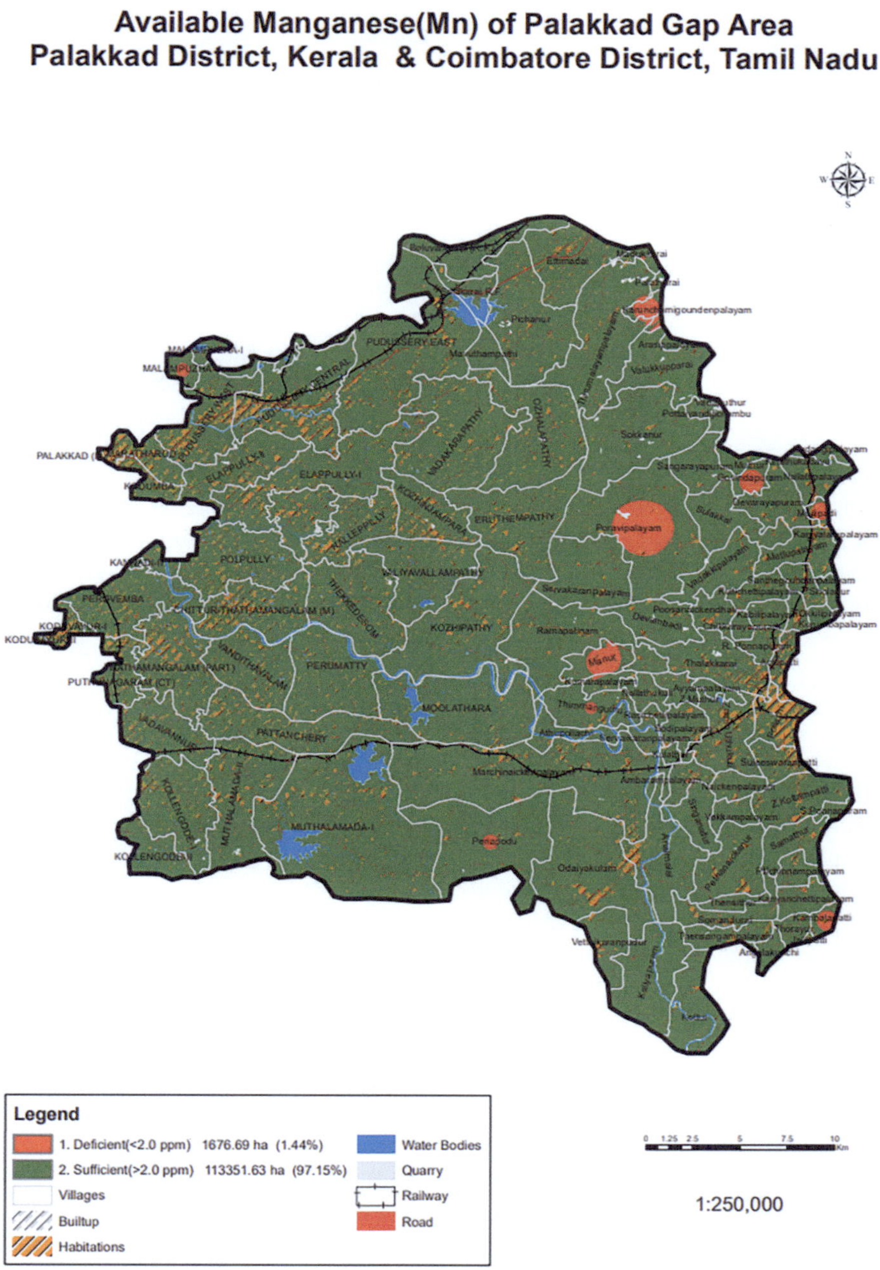

Fig. 8 Distribution of manganese (Mn) in Palakkad Gap region

Table 14 Limits of organic carbon (OC) in soil

Organic carbon (%)	Category
<0.50	Low
0.50–0.75	Medium
>0.75	High

Table 15 Area of organic carbon (OC) distribution in Palakkad gap region

Legend	Area	Area in percentage
Low (<0.50%)	112421.5	96.36
Medium (0.50–0.75%)	2350.41	2.01
High (>0.75%)	256.46	0.22
Miscellaneous	1648.6	1.41
Total	**1,16,677**	**100.00**

It is evident that the potassium distribution is high in 64,042.69 ha (54.89%), medium in 38,685.54 ha (33.16%) and low in 12,300.09 ha (10.54%) of the study area. Tamil Nadu portion of the gap region is dominated by high concentration of potassium while in the Kerala portions the distribution is medium and low (Table 23, Fig. 13).

6.12 *Phosphorous (P)*

According to soil health card scheme initiated by Ministry of Agriculture and Farmers Welfare, Integrated Nutrient Management (INM) Division, Government of India, the limits of phosphorous in soils are as in Table 24.

Out of 116,677 ha. of total land area of Palakkad Gap region, the available phosphorous is high in 73,985.27 ha (63.41%), medium in 40,978.92 ha (35.12%), and low in (0.05%). The Palakkad portions of the study area are dominated by high phosphorous availability while the Coimbatore portion shows the medium level of phosphorous availability as a dominant class. The Table 25 depicts the area and percentage of zinc availability in Palakkad Gap region (Fig. 14).

7 Soil Nutrient Index

Soil fertility assessment of any region is important especially in the context of sustainable agricultural production (Kalambe, 2021). In order to understand a real difference in soil fertility, it is necessary to obtain a single value for each nutrient. Calculating the nutrient index of each nutrient is critical to analyse the soil fertility status in each location and interpreting fertility status based on that single result. It

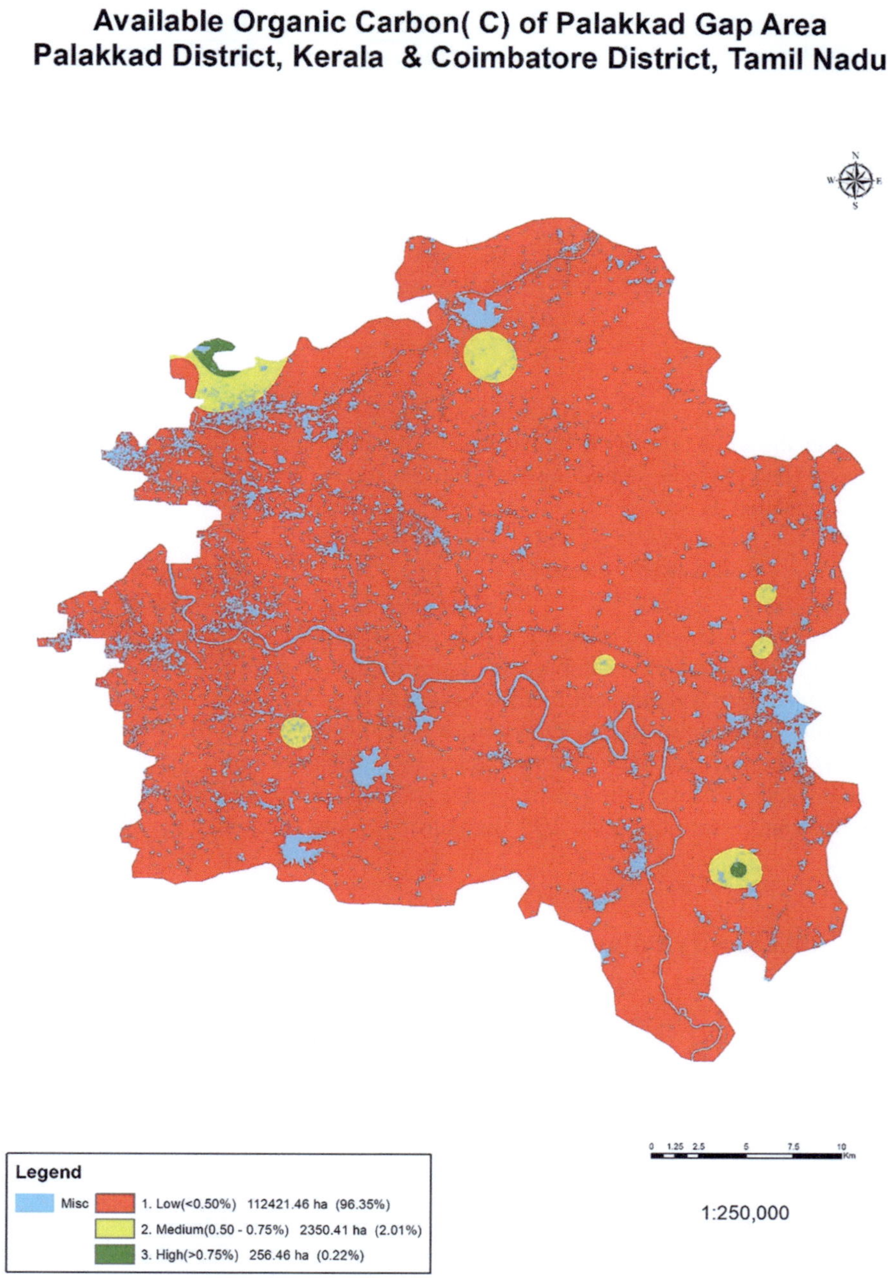

Fig. 9 Distribution of organic carbon (OC) in Palakkad Gap region

Table 16 Limits of nitrogen (N) in soil

Category	Nitrogen (kg/ha)
1. Low	<280
2. Medium	280–560
3. High	>560

Table 17 Area of nitrogen (N) distribution in Palakkad gap region

Legend	Area	Area in percentage
1. Low (<280 kg/ha)	1,15,028.32	98.59
2. Miscellaneous	1648.68	1.41
Total	**1,16,677**	**100**

assesses the soil's capacity to deliver plant nutrients (Singh et al., 2016). Parker and others First (1951) presented the Soil Nutrient Index (SNI) method was presented. It is calculated as

$$\text{Soil Nutrient Index} = \left\{\left(1^{*}\ \text{NL}\right) + \left(2^{*}\ \text{NM}\right) + \left(3^{*}\ \text{NH}\right)\right\} / \text{TNS}$$

where NL = number of samples in low category; NM = number of samples in medium category; NH = number of samples in high category, TNS = total number of samples. The Soil Nutrient Index was used to determine the fertility status of the soil. Based on the number of samples that were classified into the three categories of low, medium, and high, and were rated by a particular rating chart (Patil et al., 2017). Table 26 shows rating chart for soil nutrient index values. Soil Nutrient Index (SNI) Value Description Low 2.33 High fertility status (Parker et al. 1952).

8 Soil Quality Potential and Crop Peculiarities of Palakkad Gap Region

Soil quality can be defined as the capacity of the soil to perform within natural as well as managed ecosystem boundaries and to sustain plant productivity and reduce the soil degradation. In the present situation, plants are supposed to meet the demand for food supply for the increasing population. In order to get high yield, it is essential for the plantations to be cultivated with soils comprising appropriate amount of nutrients with respect to nitrogen (N), phosphorus (P) and potassium (K) and other micro-nutrients (Potdar et al., 2021). Nevertheless, constantly cultivating high-yield varieties without appropriate alternatives for inorganic fertilisers, no addition of micronutrients and no sufficient supply of organic manures definitely leads to excessive subtraction of necessary soil nutrients from the soil that eventually results in deficiencies of micronutrients in soils (Amar & Shanmugasundaram, 2020).

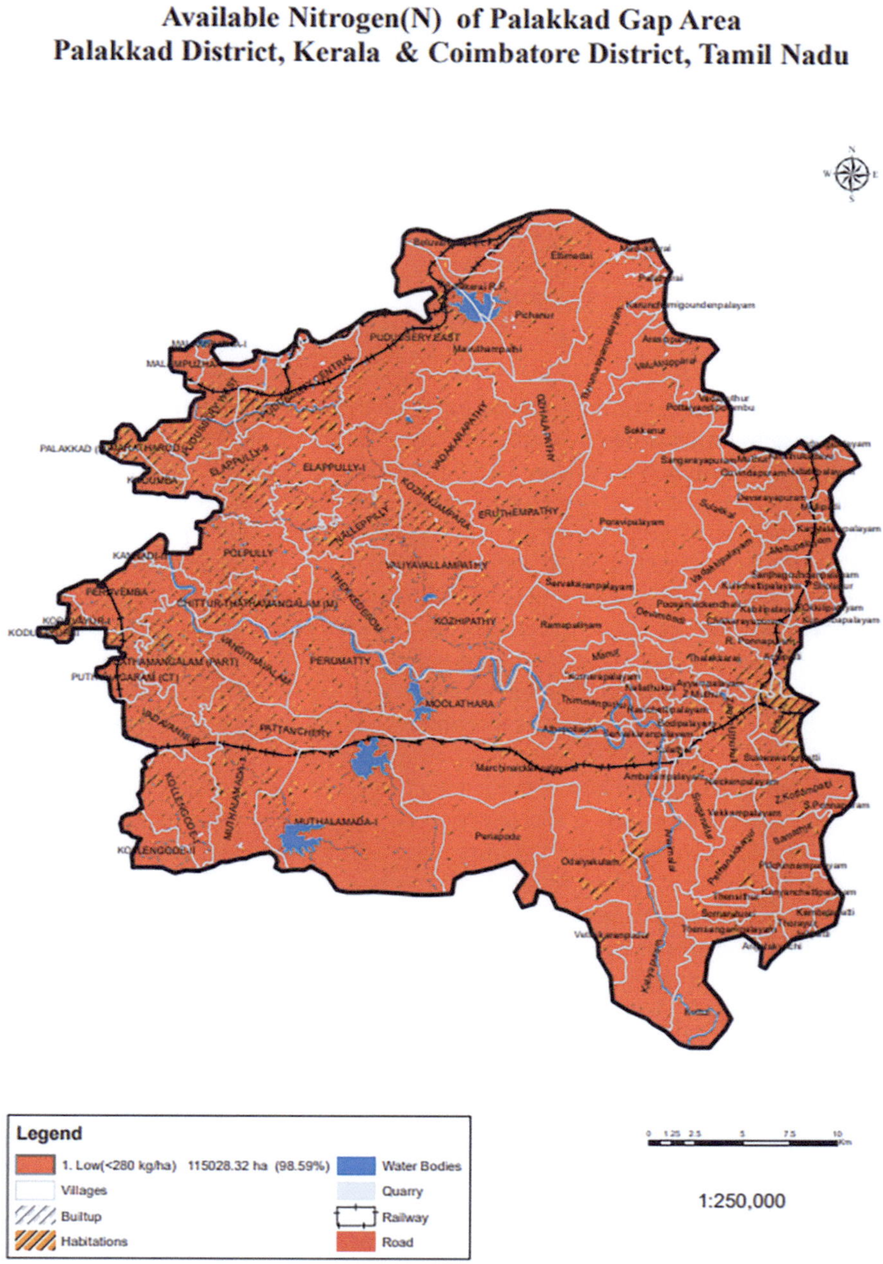

Fig. 10 Distribution of nitrogen (N) in Palakkad Gap region

Table 18 Limits of zinc (Zn) in soil

Category	Zinc (ppm)
1. Deficient	<0.6
2. Sufficient	>0.6

Table 19 Area of zinc (Zn) distribution in Palakkad gap region

Legend	Area	Area in percentage
1. Deficient (<0.6 ppm)	2311.67	1.98
2. Sufficient (>0.6 ppm)	1,12,716.65	96.61
3. Miscellaneous	1648.68	1.41
Total	**1,16,677**	**100**

Tables 3 and 4 clearly show that majority of region is coming under normal pH and non-saline soils which evidently indicates the region supports wide variety of plants and vegetation. It is evident that the available boron in Palakkad Gap region comes under deficient category (<0.5 ppm), falls under an area of 115,028 ha. (98.59%) out of 116,677 ha. of the total area of interest. Both, Kerala and Tamil Nadu portions of the study area, are deficient in the boron distribution. The availability of potassium, phosphorous, copper, iron, manganese, zinc, and sulphur is sufficient even though a few tracts show deficiencies which is the regional uniqueness that strongly supports agricultural activities at present as well shows much higher potentials in future with more varieties of crops. More varieties of vegetables and fruits can be grown in this soil condition which may support the economic status of the society.

The land use structure have a tremendous role in soil availability, nutrient cycle and may also have impact on secondary successions and biomass productions (Lu et al., 2002). Natural forests of the tropics are self-sustaining, efficient, and closed nutrient cycling ecosystems with fairly little loss or gain of the actively cycling nutrients and with high amounts of nutrient turnover. On the other hand, most of the agricultural systems are open or leaky systems with comparatively high losses of its soil nutrients. Nutrient cycling in agroforestry systems falls between these extremes as well (Nair, 1993; Nair et al., 1995).

The agricultural land use of Palakkad Gap region is occupied by paddy, coconut, banana, mango, mixed cultivation, seasonal crops, areca nut, ginger, rubber, gooseberries, and palm farms. Almost 63.3% of the study area i.e. 740.31 sq.km of Palakkad Gap region is utilised for agricultural and plantation purposes. The net sown area has a trend of paddy cultivation in the West, coconut in the East and Mango production in the Southern part of the Gap.

Fig. 11 Distribution of zinc (Zn) in Palakkad Gap region

Table 20 Limits of sulphur (S) in soil

Category	Sulfur (ppm)
1. Deficient	<10.0
2. Sufficient	>10.0

Table 21 Area of sulphur (S) distribution in Palakkad gap region

Legend	Area	Area in percentage
1. Deficient (<10.0 ppm)	12,816.75	10.98
2. Sufficient (>10.0 ppm)	1,02,211.57	87.61
3. Miscellaneous	1648.68	1.41
Total	**1,16,677**	**100**

9 Major Findings and Recommendations

- Palakkad Gap region constitutes an area of low undulating relief, with convex gently graded inter-stream tracts, sloping down to broad valley floors consisting of local erosional remnants.
- The Gap area, which is the only major breaks in Western Ghats, constitutes probably one of the oldest regions in the Southern Granulite Terrain and has undergone repeated events of tectogenesis which increases its research value of focusing on its geological history.
- The soil characteristics of the Gap region are comprised of four soil types according to USDA classifications, inceptisols in larger share, followed by entisols, alfisols and ultisols.
- The larger share of inceptisols and entisols indicates the age of soils here as young with less profile development.
- The moderately acidic soils of the study area should be applied with lime while moderately and strongly alkaline soils should be reclaimed by adding gypsum and organic matter into those soils (Xin et al., 2023).
- Systematic monitoring of soil salinity with remedial measures may definitely lead to healthier plant growth, enhanced crop yields and sustainable practices of agriculture in the gap region. Even though only a few fractions of the gap area show low saline soils, it tends to be very low in future, hence monitoring and managing for an ideal balance of salts and nutrients in the region is vital. Appropriate irrigation practices, for example, using low salt content water and avoiding overwatering can be maintained. Also, in order to ensure healthy root development and prevent salt accumulation, adequate drainage is very essential. In addition, making the soil rich with organic matter can improve the soil structure and decrease the risk of soil bulk density (compaction) which helps in enhancing plant nutrient availability and overall soil fertility (Tsigoida & Argyrokastritis, 2020).

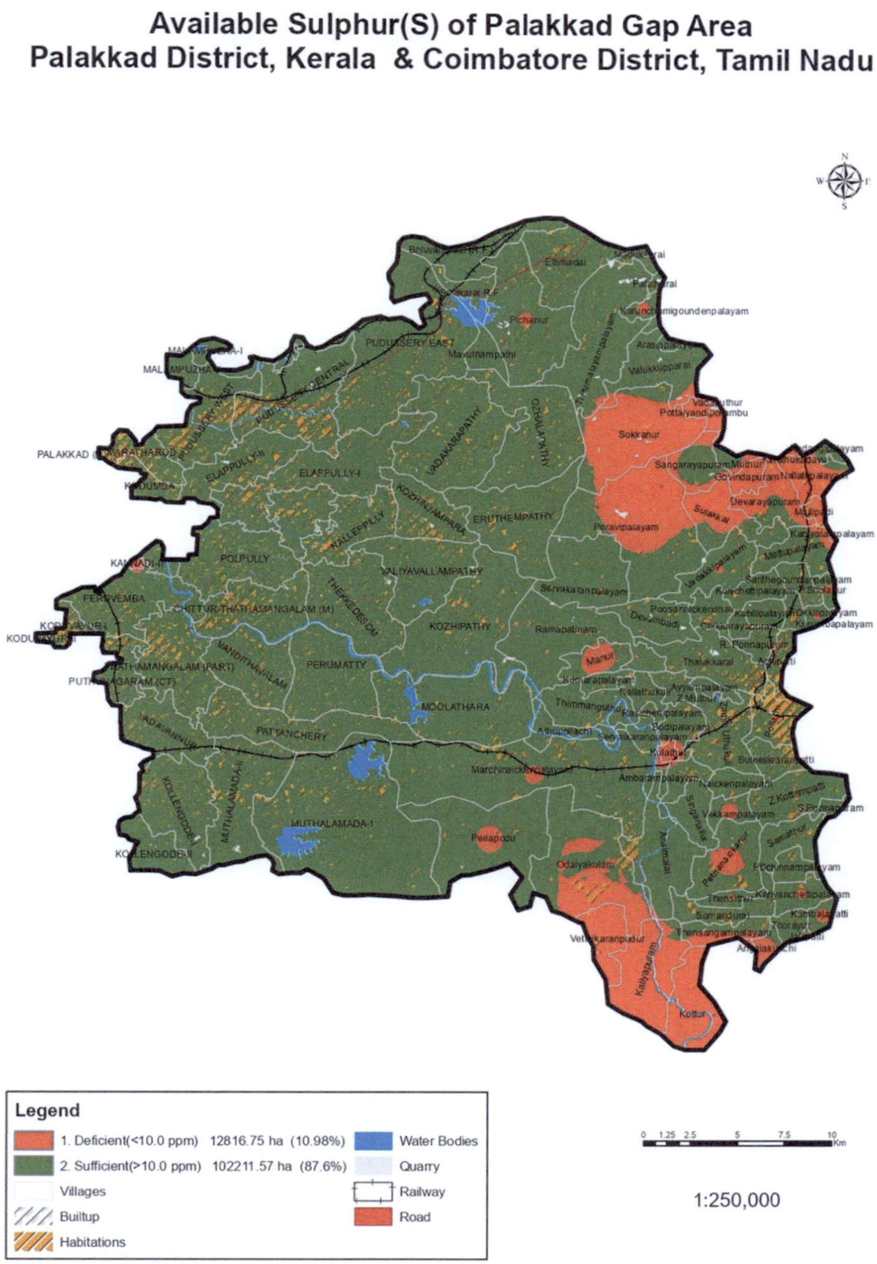

Fig. 12 Distribution of sulphur (S) in Palakkad Gap region>

Table 22 Limits of potassium (K) in soil

Category	Potassium (kg/ha)
1. Low	<120
2. Medium	120–280
3. High	>280

Table 23 Area of potassium (K) distribution in Palakkad gap region

Legend	Area	Area in percentage
1. Low (<120 kg/ha)	12,300.09	10.54
2. Medium (120–280 kg/ha)	38,685.54	33.16
3. High (>280 kg/ha)	64,042.69	54.89
3. Miscellaneous	1648.68	1.41
Total	**1,16,677**	**100**

- Boron fertilisers such as solubor, boric acid and borax can be supplemented to boron-deficient soils according to crop needs based on soil test results. Therefore, deep watering will relieve heavy boron soil concentrations by leaching the nutrient away from the roots (Singh et al., 2020).
- Increasing soil organic carbon (SOC) is a valuable goal for improving soil fertility, structure, and overall health. Here are several practices that can help increase SOC levels. Adding compost to the soil is an effective way to introduce organic matter. Compost is rich in nutrients and helps improve soil structure. Organic manures from animals, such as cow or chicken manure, can contribute organic matter and nutrients to the soil. Growing cover crops during periods when the main crop is not present can protect the soil from erosion and add organic matter when the cover crop is incorporated into the soil. Leaving crop residues on the field after harvest can increase organic matter content. No-till or reduced tillage practices help preserve crop residues and reduce the rate of decomposition. Planting-specific crops that are grown and then incorporated into the soil while still green is known as green manuring. This practice adds organic matter and nutrients to the soil.
- Since the available nitrogen is low in the study area, in the first phase the organic manures may be added in to soil and usage of nitrogen fertilisers can be applied for the crops to improve their quality and yield. In organic manuring, application of biosolids, animal wastes such as fish emulsions, blood meals; cultivation of legume plants which can fix nitrogen in their roots will help to increase the available nitrogen such as peas, beans, clover, vetch and alfalfa, etc. are the effective ways to increase the nitrogen concentration in agricultural soils. On the other hand, if the nitrogen is excessive or in toxic situation, the crop productivity may decline. Hence, nitrogen fertilisation should only be applied in required quantity after checking the soils.
- Zinc availability can be maintained in soil in several ways. Zinc sulphate can be recommended when dry fertilisers are used while the zinc–ammonia complex

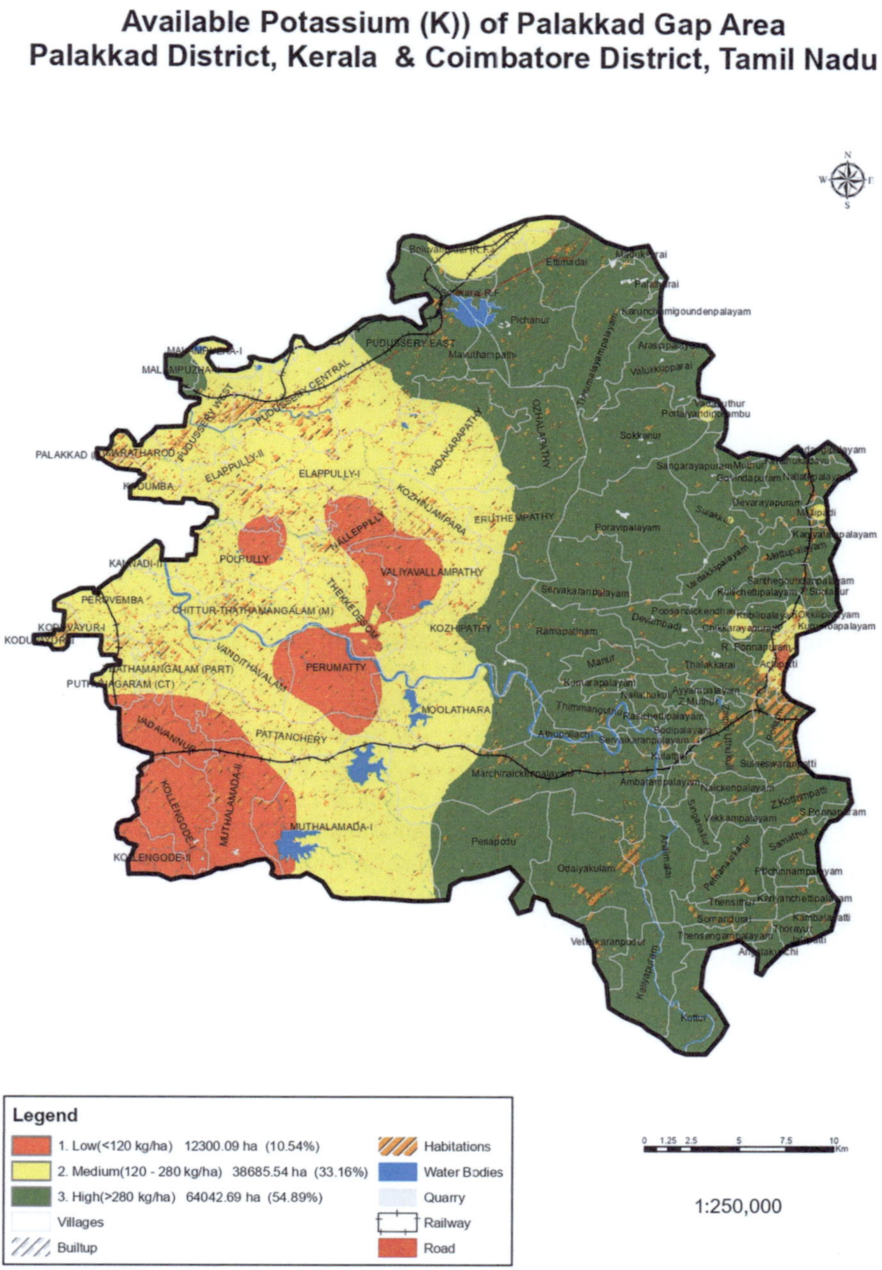

Fig. 13 Distribution of potassium (K) in Palakkad Gap region>

Table 24 Limits of phosphorous (P) in soil

Category	Phosphorous (kg/ha)
1. Low	<10
2. Medium	10–25
3. High	>25

Table 25 Area of phosphorous (P) distribution in Palakkad Gap region

Legend	Area	Area in percentage
1. Low (<10 kg/ha)	64.13	0.05
2. Medium (10–25 kg/ha)	40,978.92	35.12
3. High (>25 kg/ha)	73,985.27	63.42
4. Miscellaneous	1648.68	1.41
Total	**1,16,677**	**100**

can be suitable when fluid fertilisers are used. Application of poultry manure can improve the availability of zinc in the soil. Application of zinc as a starter fertiliser is the most effective and economical method to zinc fertilisation which provides the nutrient to the whole year.

- Sulphur can be supplied to the soils through manures, mineral fertilisers and compost. But excessive applications of sulphur in soils may result in decreasing soil pH and consequently rises the problems with the lower soil pH decrease. In effect, uptake of sulphur is reduced with the decrease of soil pH.
- Potassium deficiency can be reduced and improve the soil fertility by adding organic wastes such as green or manure wastes, application of wood ash and granite dusts or else, the addition of chemical fertilisers such as Muriate of Potash (MOP) will increase the potassium availability quickly in soils.
- Available phosphorous uptake can be managed by practices such as liming of acidic soils, application of organic amendments for both acidic and alkaline soils, tillage practices and regulation of time and methods for fertiliser applications.

10 Conclusions

The soil nutrient status of Palakkad Gap region is found to be outstanding since the soil here possesses a sufficient amount of macro- and micro-nutrients except for low organic contents in soil, corresponding decrease in available nitrogen, and the boron deficiency. When compared to Kerala and Tamil Nadu as whole, the data from soil survey department of Kerala and Tamil Nadu Agricultural University shows that there is an excess deficiency of micro-nutrients in these regions. In contrast, the research outcomes Palakkad Gap region clearly show that it is well sufficient with primary, secondary and tertiary nutrients. This makes the region agriculturally

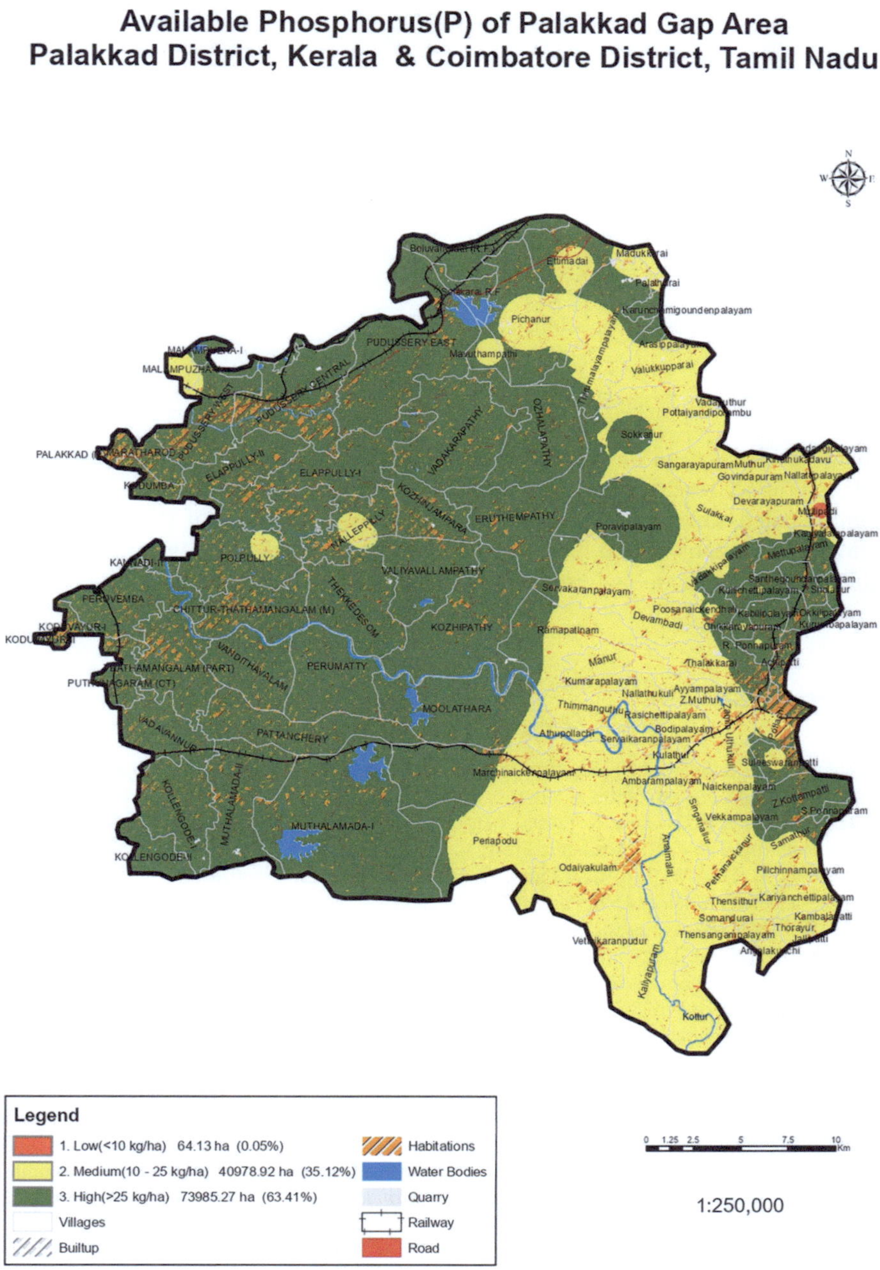

Fig. 14 Distribution of phosphorous (P) in Palakkad Gap region

Table 26 Soil nutrient index of Palakkad gap region

Study area	No. of soil samples falling in the category of			Nutrient index value	Nutrient index rating	Soil fertility status
	Low	Medium	High			
Palakkad gap region	**Available Nitrogen (N)**					
	22	0	0	1.00	Low	Low fertility status
	Available Phosphorous (P)					
	0	3	19	2.86	Medium	Medium fertility status
	Available Potassium (K)					
	8	9	5	1.86	Low	Low fertility status

sound and leads to further research on this uniqueness. Hence, it is concluded that this is primary research on Palakkad Gap which may be continued in future with semi-detailed soil survey for updating soil series, collection of profile-wise soil samples from entisols and inceptisols for stepping into the geological history and alfisols with respect to its further scope in agricultural interventions.

References

Benning, T. L., & Seastedt, T. R. (1995). Landscpae-level interactions between topoedaphic features and nitrogen limitation in tallgrass prairie. *Landscape Ecology, 10*, 337–348.

Bolan, N. S., & Kandaswamy, K. (2005). pH. In D. Hillel (Ed.), *Encyclopedia of soils in the environment* (pp. 196–202). Academic Press.

Hangsheng, L., Dan, W., Jay, B., & Larry, W. (2005). Assessment of soil spatial variability at multiple scales. *Ecological Modelling, 182*, 271–290.

Jenny, H. (1941). *Factors of soil formation*. McGraw-Hill.

Kalambe, N. (2021). Determination of nitrogen in soil samples of Tiwasa Region in Amravati District. *International Journal of Scientific Research in Science and Technology, 9*, 109–110. https://doi.org/10.32628/IJSRST221119

Kosmas, C., Danalatos, N., Moustakas, N., Tsatiris, B., Kallianou, C., & Yassoglou, N. (1993). The impacts of parent material and landscape position on drought and biomass production ofwheat under semi-arid conditions. *Soil Technology, 6*, 337–349.

Kosmas, C., Gerontidis, S., & Marathianou, M. (2000). The effect of land use change on soils and vegetation over various lithological formations on Lesvos (Greece). *Catena*, 40(1), 51–68.

Krishnaveni, M. (2015). Nutrient analysis of soil collected from Panuchakuli Village, Kanyakumari District, Kanyakumari. *Research Journal of Pharmacy and Technology., 8*, 857–859. https://doi.org/10.5958/0974-360X.2015.00140.7

Kübler, S., Rucina, S., Aßbichler, D., Eckmeier, E., & King, G. (2021). Lithological and topographic impact on soil nutrient distributions in tectonic landscapes: Implications for pleistocene human-landscape interactions in the southern Kenya rift. *Frontiers in Earth Science, 9*. https://doi.org/10.3389/feart.2021.611687

Li, C., Wang, X., & Qin, M. (2021). Spatial variability of soil nutrients in seasonal rivers: A case study from the Guo River Basin, China. *PLoS ONE, 16*(3), e0248655. https://doi.org/10.1371/journal.pone.0248655

Li, X., McCarty, G. W., Karlen, D. L., & Cambardella, C. A. (2018). Topographic metric predictions of soil redistribution and organic carbon in Iowa cropland fields. *Catena, 160*, 222–232.
Lu, D., Moran, E., & Mausel, P. (2002). Linking Amazonian secondary succession forest growth to soil properties. *Land Degradation and Development, 13*, 331–343.
Muhsina, A., Joseph, B., & Kumar, V. (2021). Principal component analysis on soil fertility parameters in vegetable growing locations of Kottayam District of Kerala. *International Research Journal of Pure and Applied Chemistry, 22*, 64–72. https://doi.org/10.9734/irjpac/2021/v22i430404
Nair, P. K. R. (1993). *An introduction to agroforestry*. Kluwer.
Nair, P. K. R., Kang, B. T., & Kass, D. B. L. (1995). Nutrient cycling and soil erosion control in agroforestry system. In *Agriculture and environment: Bridging food production in developing countries* (ASA Special Publication No. 60, ch. 7). American Society of Agronomy.
Ovalles, F. A., & Collins, M. E. (1986). Soil-landscape relationships and soil variability in north-central Florida. *Soil Science Society of America Journal, 50*, 401–408.
Pan, X., Lv, J., Dyck, M., & He, H. (2021). Bibliometric analysis of soil nutrient research between 1992 and 2020. *Agriculture, 11*(3), 223. https://doi.org/10.3390/agriculture11030223
Parker, D. I., Sowden, F. J., & Atkinson, H. J. (1952). The nitrogen distribution and amino acid content of certain soil organic matter fractions. *Scientific Agriculture, 32*(3), 163–169.
Pastor, J., & Post, W. M. (1986). Influence of climate, soil moisture and succession on forest carbon nitrogen cycles. *Biogeochemistry, 2*, 3–27.
Patil, P. M., Shashikant, A., Roy, S., & Momoniat, E. (2017). Unsteady mixed convection over an exponentially decreasing external flow velocity. *International Journal of Heat and Mass Transfer, 111*, 643–650.
Potdar, R., Shirolkar, M., Verma, A., More, P., & Kulkarni, A. (2021). Determination of soil nutrients (NPK) using optical methods: A mini review. *Journal of Plant Nutrition, 44*, 1–14. https://doi.org/10.1080/01904167.2021.1884702
Pratik, B., Aru, S., Lokade, V., & Deshmukh, N. (2023). The remarkable neem tree: a comprehensive review of its botanical pharmacological and therapeutic properties. *IJCRT, 11*, a738–a744.
Ravikumar, P. (2013). Evaluation of nutrient index using organic carbon, available P and available K concentrations as a measure of soil fertility in Varahi River basin, India. *Proceedings of the International Academy of Ecology and Environmental Sciences, 3*(4), 330.
Seastedt, T. R., Briggs, J. M., & Gibson, D. J. (1991). Controls of nitrogen limitation in tall grassprairie. *Oecologia, 87*, 72–79.
Shukla, G. (2015). Estimation of micronutrients in the soil. *The Bioscan, 10*, 4.
Singh, A., Singh, A., & Singh, J. (2020). Boron in crop production from soil to plant system: A review. *Archives of Agriculture and Environmental Science, 5*, 218–222. https://doi.org/10.26832/24566632.2020.0502020
Singh, J. S., Koushal, S., Kumar, A., Vimal, S. R., & Gupta, V. K. (2016). Book review: microbial inoculants in sustainable agricultural productivity-Vol. II: functional application.
Singh, R. K., & Mishra, S. K. (2012). Available macro nutrients (N, P, K and S) in the soils of Chiraigaon block of district Varanasi (U.P.) in relation to soil characteristics. *Indian Journal of Scientific Research, 3*, 97–100.
Sonikajha, & Suneetha, V. (2015). Nutrient analysis of soil samples from various places. *Journal of Chemical and Pharmaceutical Research, 7*, 291–293.
Thakare, P. B. (2020). Physiochemical analysis and nutrient assessment of farmland soil samples in the village of Wardha District, Maharashtra, India. *Asian Journal of Research in Chemistry, 13*, 415–418.
Tsigoida, A., & Argyrokastritis, I. (2020). Electrical conductivity, pH and other soil chemical parameters after sub-irrigation with untreated and treated municipal wastewater in two different soils. *Global Nest Journal, 22*, 55–66. https://doi.org/10.30955/gnj.003217
USDA-ERS. (1997). *Agricultural resources and environmental indicators 1996–1997*. Agricultural Handbook, 712. U.S. Gov. Print. Office, Washington DC, USA.

Wang, Y., Zhang, X., & Huang, C. (2009). Spatial variability of soil total nitrogen and soil total phosphorus under different land uses in a small watershed on the Loess Plateau, China. *Geoderma, 150*, 141–149.

Xin, X., Guo, L., Wang, S., Wang, X., Ren, M., Zhao, P., Huang, Z., Jia, H., Wang, J., & Lin, A. (2023). Effective strategies for reclamation of saline-alkali soil and response mechanisms of the soil-plant system. *Science of The Total Environment, 905*, 167179. https://doi.org/10.1016/j.scitotenv.2023.167179

Yagaso, Z. S., Bayu, T. Y., & Bedane, M. D. (2024). The dynamics between soil erosion, nutrient variability, land use, topography, and implications for food security in Southern Ethiopia. *Cogent Social Sciences, 10*(1). https://doi.org/10.1080/23311886.2024.2307178

Yost, R. S., & Uchida, R. (2000). Interpreting soil nutrient analysis data. Plant Nutrient Management in Hawaii Soils. JA Silva, and RS Uchida (eds.). *College of Tropical Agriculture and Human Resources* (pp. 87–94). University of Hawaii, Honolulu, HI.

Zhang, J. (2018). Research on the soil nutrient characteristics of tea plantation. *IOP Conference Series: Earth and Environmental Science, 208*, 012079.

Zhang, Y.-Y., Wu, W., & Liu, H. (2019a). Factors affecting variations of soil pH in different horizons in hilly regions. *PLoS One, 14*(6), e0218563. https://doi.org/10.1371/journal.pone.0218563

Agroforestry Microbiome Diversity and Function for Climate Resilience and Biotic Stress Tolerance in Agriculture

Gyanaranjan Sahoo, Asim Chandra Dash, Meenakhi Prusty, Madhumita Jena, Afaq Majid Wani, Antony Joseph Raj, and Kalyan Sundar Das

Abstract Agroforestry is a sustainable land-use strategy that aims to boost overall productivity and revenue by maintaining crops, trees, and livestock on the same plot of land. The soil's microclimate beneath tree canopy can be changed via agroforestry. Enhancing farm production, mitigating climate change, sequestering carbon, protecting biodiversity, phytoremediation, conserving water, and improving soil quality through the addition of plant and animal waste are all significant benefits of it. Increased microbiological activity in the soil, habitat for beneficial insects, microclimate adjustment, nitrogen fixation, and other benefits are all facilitated by plant species diversity. By destroying the monoculture structure, this agroecological method promotes intricate interactions between different species of microflora, fauna, crops, and tree species. This strategy improves biomass production, nutrient cycling, and the natural control of hazardous species, leading to ecological sustain-

G. Sahoo · M. Jena
Krishi Vigyan Kendra, Odisha University of Agriculture and Technology, Nayagarh, Odisha, India
e-mail: gsahoo.forestry@ouat.ac.in; madhumitajena@ouat.ac.in

A. C. Dash
College of Agriculture, Odisha University of Agriculture and Technology, Bhubaneswar, Odisha, India
e-mail: acdash@ouat.ac.in

M. Prusty
AICRP on LTFE, College of Agriculture, Odisha University of Agriculture and Technology, Bhubaneswar, Odisha, India
e-mail: meenakhiprusty@ouat.ac.in

A. M. Wani
Department of Forest Biology and Tree Improvement, College of Forestry, SHUATS, Prayagraj, Uttar Pradesh, India

A. J. Raj
College of Forestry, SHUATS, Prayagraj, Uttar Pradesh, India

K. S. Das (✉)
ICAR-Agricultural Technology Application Research Institute, Kolkata, West Bengal, India

A. K. Singh et al. (eds.), *Sustainable Agriculture Management in Semi-Arid Climates*, https://doi.org/10.1007/978-3-031-94062-0_8

ability. Throughout the world, improper management techniques lead to the degradation of soils. Therefore, more conservationist agriculture systems must be developed. An alternative approach called agroforestry permits the continued sustainable agriculture growing practices on lands and livestock while also aiding in the prevention of land degradation. By adding cover crops and permanent vegetation, this strategy improves the buildup of organic carbon in soils, which is anticipated to raise the biomass of soil microorganisms. Legumes can only be used successfully if proper attention is paid to the development of productive symbioses with root nodule bacteria. Agroforestry may prove to be a financially viable venture and climate-smart farming method that helps smallholders adapt to the extremes of the changing environment by improving and preserving soil quality, reducing wind erosion, and diversifying food supplies.

Keywords Agroforestry · Biodiversity climate change · Soil microorganisms · Soil nutrients · Soil productivity

1 Introduction

Climate change is one of the foremost threats to the productivity of agricultural systems. Changes in the mean climate alone are unlikely to have as much of an impact on yield output as extreme events and increased climate variability (Burgess et al., 2022). In fact, repeated heat waves and protracted droughts have caused crop production to stagnate in recent years (Beillouin et al., 2020; Chavez et al., 2015; Ray et al., 2012). A global-scale meta-analysis of studies published from 1980 to 2015 revealed that drought-related reduced wheat produce (*Triticum aestivum* L.) and maize (*Zea mays* L.) were 21% and 40%, respectively (Daryanto et al., 2016).

For every degree Celsius that the mean temperature rises, the production of wheat worldwide is expected to fall by 6% (Zhao et al., 2017). There is more evidence that climate unpredictability has a detrimental impact on grasslands, with cascading animal welfare and productivity will be strongly impacted by changes in the thermal environment, particularly by increased heat stress (Cheng et al., 2022). Additional difficulties that interact with climate change and have the potential to worsen its consequences are being faced by agricultural systems (Aguilera et al., 2020). Threats to soils include dehydration, wind erosion, decomposition of organic matter (OM), nitrogen loss, compaction, salinization, and pollution. For instance, soil deterioration has caused farmland soils to lose 20%–60% of their C before cultivation (Olsson et al., 2019).

Udawatta et al. (2017) state that soil material loss from erosion is one of the primary variables driving productivity. Agriculture productivity is also impacted by biodiversity loss, especially from the soil (Aguilera et al., 2020). These ecological functions have been steadily lost as a result of agricultural intensification, making agricultural systems more susceptible to disturbance and environmental change and eventually losing their ability to self-regulate (Barrios et al., 2015). In light of this, modifications to agricultural systems are necessary to lessen the influence of climate change. According to Vermeulen et al. (2012), this second "green revolution" will necessitate the adoption of new crop types as well as adjustments to farming systems and cropping techniques. The majority of adaptation efforts have been directed at changing aboveground components (such as cultivating cultivars that are stress-tolerant; Arenas-Corraliza et al., 2018).

However, there is a rising interest, particularly in the agroecology community, in metrics that concentrate on subsurface components (Altieri et al., 2015; Hatfield et al., 2017). Thus, in order to sustain crop production in the face of current climate change, farmers are encouraged to pursue four interconnected objectives: (1) adopting practices and systems with a high available water capacity (AWC) and water-use efficiency (WUE); (2) maintaining soil fertility and productivity; (3) safeguarding against new pests and pathogens; and (4) promoting carbon sequestration (Rao et al., 2007). The negative consequences of dry spells may be lessened by investments that support green water storage, or precipitation that is buried in the ground and may be accessible to vegetation (Falkenmark & Rockström, 2008).

In addition to regulating C-climate feedbacks, soil organic matter (SOM) and the physiological function it supports are essential for soil conservation (Crowther et al., 2019). These elements work together to influence how ecosystem functions and populations react to environmental change, especially extreme weather events (de Vries et al., 2018; Delgado-Baquerizo et al., 2017). Thus, in the context of global change, methods or structures that support the accumulation of SOM and biological activity ought to be supported. One tactic for mitigating the effects of climate change is system diversification (Altieri et al., 2015). Agroforestry (Cardinael et al., 2021), livestock integration with crops in mixed farms (Bonaudo et al., 2014), and functionally varied crops (Lakhran et al., 2017) are a few examples of diversification. In order to increase agricultural and forest productivity, agroforestry systems (AFS) intentionally mix trees (or shrubs) with crops or pastures (livestock). Agroforestry includes a variety of techniques, such as silvopastoralism and alley cropping. According to Schoeneberger et al. (2012) and Mustafa et al. (2022) vegetation layers (such as trees and crops) interact with soil resources in a complementary or facilitative rather than competitive manner, increasing the application of renewable resources and acting as a buffer against environmental stress.

Research conducted worldwide over a minimum of 40 years has demonstrated that AFS can enhance the three-dimensional (chemical, biological) quality of soil while increasing effective soil depth (deeper and more dense rooting systems) and water-use efficiency (the system uses more rainfall water; Dollinger & Jose, 2018). Thus, compared to other agricultural options, agroforestry has been promoted as a technique that avoids negative climate impacts (Thomas et al., 2019).

2 Enhancement of Soils and Improvement of Microclimate by Agroforestry

Depending on the crop type, temperature, and geography, agroforestry can modify the condition of soil by alterations to environmental services and functions brought about by both the immediate and indirect impacts of trees. Trees are an important part of the nutrient cycle because their deep roots allow them to bring back elements that have been leached, acting as a 'safety net' opposed to the cycle's nutritional deficits. In tree-based systems, trees aid in dry deposition and atmospheric nutrient absorption (Sahoo & Wani, 2019). Because of the increased use of fertilizer, deep ploughing, and enhanced agriculture, agroforestry presents a potential means of accumulating and storing carbon in the soil (UNDP, 2012).

Because the elements obtained in a combination farming cover have a greater variety of form and function than those obtained in monocropping, agroforestry encourages the more economical use of resources (Dagar, 2012). The addition of plants to fields improves field capacity (FC), organic matter (OM), available potassium, accessible phosphorus, soil carbon stocks, and decreased bulk density (BD) by raising the water holding capacity (WHC) and delivering water to plants gradually like a sponge (Islam & Quli, 2016). In order to promote soil consolidation and lower bulk soil density, OM is applied to the soil. In dry and semi-arid regions, this decreased soil BD improves groundwater replenishment, ventilation, soil nutrient quality, and water dispersion in the rhizosphere.

In agroforestry systems, the primary supply for minerals and organic carbon (OC) is the buildup of litter from fallen leaves and twigs. Both directly and indirectly, the amount of soil organic carbon (SOC) affects how well fertilizers are used in agriculture. Because of their higher availability and absorption as well as their active deep root system, nutrients will be used more effectively in high (OM) soils (Islam et al., 2015). Moreover, mycorrhizae most likely benefit from the greater microbial variety that follows the addition of OM and the release of P for crop use. Through the breakdown of leaf litter, the nitrogen fixed by the trees encourages soil and N cycling and raises the long-term soil N content through OM infusions.

Moreover, a higher rate of N cycling is caused by the nutrient pumping of N from underground soils to the underground zone of maize through profound plant roots (Garrity, 2004). The inclusion of non-leguminous and leguminous species, such as rambutan and durian, enhances the nutritional value of the soil in cocoa-based agroforestry (Ofori et al., 2014). In the southwest of Cameroon, the views of farmers on agroforestry were investigated by Sinha et al. (2018). They found that about 53% of farmers believed that the agro-silvopastoral system was the most effective at increasing soil fertility, while 57% said that it did so in the greatest quantity.

Furthermore, compared to monocropping, agroforestry systems have increased accessible P and K, according to Mohanty et al. (2016). A Bangladeshi study by Riyadh et al. (2018) found that in different crops connected with jackfruit-based agroforestry, soil temperature decreased by 3.37%–9.25%, soil moisture increased by 10%–20%, and total nitrogen increased by 9%–19% in contrast to open seasonal crop fields. Soil fertility may be increased by using jackfruit-based agroforestry systems, according to the theory, by enhancing the soil's chemical and physical characteristics. When compared to monocropping, guava and poplar-based agroforestry systems displayed higher SOC (Sahoo et al., 2023a). The addition of field trees may enhance the Earth's physical properties (Richard & Munishi, 2013).

3 Agroforestry's Effects on Soil Communities

The microorganisms that live there and the physical and chemical characteristics of the soil are affected when trees are planted close to crops. The variety of soil microorganisms promotes fertility and production by indirectly stimulating plant development. Plant growth is indirectly influenced by the soil microbial population since it contributes significantly to increased fertility and productivity (Sahoo et al., 2023b). Microorganisms in particular are significant for the health and productivity of plants as well as the biochemical process in the soil. When comparing agroforestry systems to solo cropping, it may be observed that the soil contains a greater number of diverse and functioning soil bacteria. In these situations, it is anticipated that this will lead to increased biological soil fertility. Agroforestry systems have higher rates of microbial productivity and abundance because of the deposition of organic matter, the influence of trees, quantities and root transudes, and a variety of litter quality. Earthworms, fungi, acari, diplopoda, nematodes, and other insects affect the cycling of nutrients and the conversion of carbon to oxygen (Thomas et al., 2019). Ants, termites, and earthworms are examples of soil engineers that are crucial to aggregate development and soil structure maintenance. Important biological control agents include centipedes, ground or rove beetles, predatory mites, collembola, and carnivorous nematodes (Cardinael et al., 2021). Microbial communities receive the essential energy source from falling litter and root exudation in the form of amino acids, carbohydrates, organic acids, and other compounds (Schoeneberger et al., 2012). Agroforestry can enhance soil biodiversity in comparison to monocropping by offering specific habitat, sanctuary for epigenic microbes, diversity of environments, buffering capacity, soil moisture content, and humidity. This is predicted to have an impact on the related ecosystem services (Mustafa et al., 2022).

The soil fauna determines the health of the soil through a variety of soil-related processes. *P. reticulatum* increases the diversity and activity of nematode microorganisms, which improves nutrient mineralization, OM breakdown, and nutrient enrichment in soil (Olsson et al., 2019). As integrating trees creates ideal circumstances for soil microflora to grow, agroforestry and agro-silvopastoral systems have higher rates of microbial diversity than monoculture, according to research from Rajasthan (Lakhran et al., 2017). But with Germany's alley cropping, growth in the byproducts of microbes was limited to the rows of trees (Raj et al., 2020). Higher bacterial biomass rates were seen in the vicinity of tree rows in an Alnus rubra–maize system in a different study conducted in a temperate agroforestry setting (Sahoo et al., 2022). Dollinger and Jose (2018) discovered that the different nutrient availability from root exudation and litter fall, along with the low fertilization and tillage activities in alley cropping, caused the microbial populations in tree rows of poplar-wheat alley farming systems in cooler regions to be different from those in cultivated land. Integrated tree row farms provide soil microbes associated with the rows, increasing the total variety of alley cropping and possibly increasing functional diversity. The nitrifying Nitrosospira and Nitrospira abundance rates were lower under the tree monoculture than in agroforestry systems, despite the fact that the trees promoted the N-fixing bradyrhizobium and mesorhizobium.

4 Prospects for Reducing Soil Pollution with Agroforestry

Inadequate handling of agricultural chemicals and the widespread use of agrochemicals, like herbicides and agricultural products, in farming are major worldwide contributors to non-point source groundwater and surface water contamination, and can lead to the degradation of soil (Cardinael et al., 2021; Hernández-Morcillo et al., 2018). For crops to be able to use subsurface water to combat dry spells and to absorb nutrients leached from the deeper soil layer, they need a more substantial soil structure with a significant root length frequency. Agroforestry improves soils and shallow groundwater, speeds up mineralization, and uses less agrochemicals (Sahoo et al., 2021). In agroforestry systems, tree roots can decrease soil N and P deposits by 20%–100%, with pesticide transmission reaching 100% of the total, according to a review of various research (Mishra et al., 2016). It is commonly believed that agroforestry trees, like Populus, Willow, and Platanus species, can absorb pollutants, including pesticides, and help break them down so they become immobile in the woody parts of the tree or on their leaves (Mustafa et al., 2022). This capability is also dependent on the agrochemicals' physicochemical properties. Remarkably, tree roots in AFS only take up to 60%–90% of the agricultural chemicals injected into the soil's upper layers due to their depth; this reduces the adverse environmental effects of the fertilizers and keeps them from competing with the crops for basic nutrients (Dollinger & Jose, 2018) (Fig. 1).

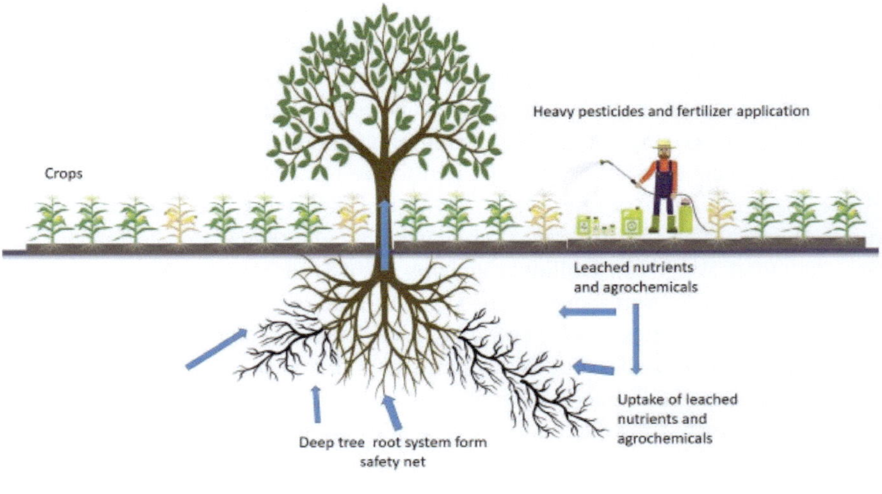

Fig. 1 Absorption and immobilization of leached chemicals in agroforestry systems

5 Soil Microbial Biomass, Composition and Functioning

According to Sinha et al. (2018), soil microorganisms are essential to soil fertility and ecosystem productivity. The relationship between the makeup of rhizosphere microbes and nutrient uptake, drought tolerance, disease resistance, and ultimately plant fitness has been extensively recognized in the literature (Mbow et al., 2014a). AFS is said to have a favourable impact on microbial C and N as well as soil microorganisms (Bijalwan et al., 2011; FAO, 2002). AFS tends to have more soil microbial biomass (SMB) than treeless systems, which is a general measure of soil health (Sinha et al., 2018). Matocha et al. (2012) found that changes in FOM brought about by trees in AFS and the nutritional specificity of the resulting microorganisms can be important factors affecting SMB. Regarding whether environmental driver or combination of causes influences SMB and community composition patterns, however, little is still known (Sinha et al., 2018). The size of the microbial biomass pool, substrate availability, and temperature all affect microbial enzyme activity, which is employed as a biological soil quality indicator (Mbow et al., 2014b). Trees should be associated with higher microbial activity because of the increased FOM inputs, SMB, and more regulated microclimates in AFS. In fact, agroforestry alley cropping has been shown by various authors to have considerably higher microbial diversity and enzyme activity than mono-crop agriculture (Sharma & Sahoo, 2021; Ravindranath et al., 2012).

The quantity, variety, and functioning of soil microbes in subtropical alley-cropping regimes were examined in Cheng et al. (2022) study of the literature. In comparison to farmland monocultures, they discovered that in the majority of instances, soil microbial diversity, abundance, and functions increased in AFS. Enzyme activity in an experimental Missouri watershed dropped from agroforestry shields to row crops, according to Udawatta et al. (2008). Richard and Munishi (2013) discovered that unmanaged shrub species had higher rates of residue decomposition and microbial reactions beneath them than beside them in the Senegalese Sahel. Additionally, they discovered that the presence of more macrofauna enhanced microbial biomass C. The potential mechanism of enzyme activities, or symbiogenic phenomena, may provide significant benefits to AFS, but additional study is required to properly comprehend this phenomenon and determine the best tree-crop pairings.

6 Farmers' Perspectives on Agroforestry Systems

Agroforestry was viewed differently in the communities. In general, respondents believe that agroforestry has the ability to meet their fuelwood demands, enhance soil fertility, and offer a positive climate for crops. Farmers have traditional information on tree use in farming systems as well as environmental issues like as watershed preservation and climate control (Sahoo & Wani, 2020). In general, a significant majority of farmers (62.5%) were familiar with agroforestry, and 68.3% of those polled were eager to use the system. The major cause for the system's adoption was the many profits that the crop-tree amalgamation provides to farmers. Other requirements for implementing the technique were the availability of land and the separation of trees from crops, or the crops not being destroyed or impacted by tree development (Sahoo et al., 2019). Trees were cited by a few responders as a source of fuelwood. Those who were unwilling to embrace agroforestry (8.3%) explained their position by pointing out that the benefits of agroforestry take time to notice, which is not their aim in farming. Contractors would also devastate their farms by cutting wood species and exploiting the little land space available (Barrios et al., 2015).

In addition, 65% of defendants said they would be capable of managing the network's many elements since it requires the same degree of attention as their current farming approach. Farmers said that the agroforestry framework was likewise sensible with legitimate arranging since pruning as well as weeding was adequate to keep the framework in astounding condition (Matocha et al., 2012). Three farmers indicated that they needed additional labour to run the system, while three others stated that they could only handle it on a modest scale. Following crop damage caused by timber merchants during extraction, farmers were hesitant to keep certain types of timber on their properties. They also think the plan was designed to keep them out of the forest. (Leakey, 2010).

Farmers' attitudes toward agroforestry imply that:

- To avoid crop devastation during felling, timber species should be kept apart from farmlands.
- During tree harvesting, the tree/crop combination should be such that there is little or no harm to the crops.
- Farmers' chosen tree species, particularly those with dual advantages, such as medicinal value, should be considered.
- Existing taboos should be considered when incorporating animal components into agroforestry.

7 Agroforestry System Constraints and Issues

Agroforestry as a substitute land use system in the area is hampered by a number of issues. These elements are listed below in order of importance.

7.1 Inadequate Understanding of Logging Regulations/ Procedures

The majority of respondents had insufficient knowledge of tree tenure, logging regulations, and mechanisms for compensating rights lost as a result of tree destruction. As a result, chainsaw operators and wood businesses take advantage of farmers to their harm. A number of farmland wood species are felled without the farmers' permission (Sahoo et al., 2020a). As a result, many farmers are unmotivated to either leave or grow such trees on their fields.

7.2 *Farm Size*

The amount of land that the farmers had access to acted as a limitation, even if the respondents were aware of the idea of agroforestry and its advantages. The majority of farmers contended that the size of their farms prevented tree integration. They further asserted that this endeavour would shade their crops, lowering crop production (Sahoo et al., 2020b). Because of the limited amount of the property, 47.5% of the respondents were unwilling to fully and deliberately implement this method.

7.3 *Lack of Seedlings*

Another barrier to agroforestry adoption in the research region is a scarcity of seedlings. The Forestry Services Division (FSD), which had previously given complimentary seedlings for ranchers to sow, has now ceased to do so. Farmers want to grow trees on their fields, but they cannot since they do not have access to seedlings. Farmers do not plant trees for a variety of reasons, according to Kumar (2006), including outside ownership concerns and a lack of acreage:

- Lack of information about which species to plant and how to plant them, as well as a tree-planting culture
- Lack of seeds

7.4 *Seasonal Disease Incidence in Plants and Animals*

Agroforestry is threatened by the seasonal appearance of plant and animal diseases in the research region. Animal illnesses were prevalent throughout various times of the year, according to focus group talks done in all of the communities. The bulk of the animals died, according to farmers, during this period, implying that the Ministry of Food and Agriculture's veterinary services division, which is in charge of preventing and controlling some of these illnesses, was not up to the task in the study location (Sahoo et al., 2020b). As a result, the respondents were uninterested in raising animals. Plantain sigatoka illness was also found to be prevalent. If a beneficial interface among trees, crops, and animals is to be established, then these issues must be addressed in order for AF systems to be used effectively (Sahoo et al., 2020a).

7.5 Inadequate Restitution for Lost Crops

Contractors harvesting timber species generate agricultural losses to farmers in the farming region, which are not appropriately reimbursed. When their crops are harmed, they are not usually compensated appropriately. As a result of this scenario, a number of farmers have destroyed seedlings of various wood species on their properties in order to prevent crop damage during future felling.

7.6 Inadequacy of Loan Amenities

Farmers in the farming area claimed they could not acquire finance to develop their farms or buy more farm supplies because credit was scarce. Credit is said to be a major issue for farmers as well as project implementation; thus, it is recommended that farmers' financial needs be handled in order for project implementation to be successful (Goli et al., 2015).

7.7 Tree Tenure

Most farmers were unaware of their rights to grow trees on their properties and so were unwilling to plant trees. According to Garrity (2004), agroforestry is dependent on people's rights to grow and utilize trees, which are dependent on the current land and tree tenure structure. Since the government owns the trees on farms, farmers do not feel required to maintain or care for them, and the compensation for crops destroyed due to tree removal is minimal. As a result, community education about tree tenure is required. Islam and Quli (2016) proposed that an ethical foundation of right and ownership be established to represent the labour and expertise involved in forest management outside of forest reserves. This is because farmers in Ghana's forest zone controlled the majority of the soil resources, and the final fate of trees and forest resources is determined by their management techniques (Islam et al., 2015).

7.8 Others

Because product transportation will be a limiting issue when AF is adopted, the transportation network connecting villages to market centres, as well as the absence of storage facilities, will diminish AF's impact. Due to the aforementioned issues, as well as the perishability of specific commodities, a lot of agricultural food deteriorates during a big harvest (Sahoo & Wani, 2019).

8 Conditions and Local Behaviours that Encourage the Use of Agroforestry

The ensuing is known local situations/opportunities that favour agroforestry implementation:

Farmers leave trees on farms: farmers are more most likely to take up a superior a variant of this approach due to the fact that agroforestry is a recognized drive and people are already cognizant of the value and advantages of trees. Farmers value trees' soil and crop-enhancing abilities, according to Islam and Quli (2016), since most farmers keep some trees to preserve soil structure, increase soil fertility and nutrient cycling, and interact favourably with crops. As a result, all that is required is to educate and encourage farmers to maximize the system's potential. There is no prohibition against tree planting in the research area. When agroforestry is brought into a community, the new tree species are likely to be welcomed as long as the uses and advantages of the trees are described.

Mixed cropping: this is an agroforestry practice that varies depending on the components. The integration of maize, cassava, and other food crops with cocoa trees, as seen in the fields, is one example.

Animal numbers: animals are being grown in all communities, and farmers are willing to expand the number of animals now being kept. As a result, it is critical that this system should be implemented with the goal of preventing the spread of animal illnesses.

Understanding the significance of trees: trees are employed in the communities for fuel, shade, windbreaks, feed, and charcoal, implying that local applications for the woody perennial already exist and are in great demand. However, pressure on these trees has resulted in the loss of several useful tree species for fuelwood, such as Esa (*Celtis mildbraedii*). Farmers have expressed their discontent about a lack of fuelwood due to the fact that women who are in charge of gathering firewood must go considerable distances to obtain it. As a result, it is predicted that when agroforestry is implemented, the community will be able to access a conveniently accessible supply of fuel. Some trees have also been recognized by farmers, which can be used as fodder, as Nyankyerene (*Ficus exasperata*), Sesea (*Trema orientalis*), and cassava (*Manihot esculenta*), avocado pear (*Persea americana*) leaves. Tree species that work well with food crops have also been found by farmers. They comprise simul (*Ceiba pentandra*), Kumnini (*Lanneawelwitschii*), and Odum (*Milicia excelsa*). Farmers in Anyimaye were found to be aware of trees as a means of watershed conservation. Because some communities are already aware of the necessity for a water management system, it is expected that the adoption of tree planting for enhanced watershed management and weather would be easier (Dhyani et al., 2003; Mishra et al., 2016). As a result, it is predicted that farmers would support future initiatives to improve agroforestry systems. According to the data, fruit trees are also grown in home gardens as well as agricultural fields. Farmers labelled Wama (*Ricinodendronheudelotii*) and other trees as medicinal plants, suggesting that these tree species were valuable to them.

Home fence and boundary demarcation: almost all communities had live fence, indicating that this form of agroforestry was already in existence. Trees are also used to demarcate land divides, and farmers recognize the prominence of using various tree kinds for this purpose.

Presence of nurseries: some villages have a basic understanding of how to cultivate tree seedlings. Some nursery development facilities are likely to be provided, and nurseries in general may be encouraged.

A favourable climate for tree and agricultural development: farmers have reported no extra meteorological or environmental limitations limiting crop or tree development, other from the decrease in rainfall.

Existence of NGOs: NGOs are assisting the community in virtually every hamlet by training farmers, which will ultimately augment the practice of agroforestry.

Fruit trees for subsistence and income: fruit trees are grown not only in the neighbourhood, but also in the field by farmers. When it comes to integrating food crops with non-traditional tree species in the future, this technique is viewed as highly beneficial (timber species). Mango (*Mangifera indica*), avocado pear (*Persea americana*), and sweet orange (*Citrus sinensis*) are the most frequent fruit trees grown in the region, albeit on a modest balance.

Farmers' interest in beekeeping, mushroom farming, and snail farming: farmers' interest in beekeeping, mushroom farming, and snail farming might be beneficial when improved agroforestry systems are offered.

Soil fertility: the fact that soil fertility is deteriorating is well-known. Farmers expressed their dissatisfaction with decreased food output as a result of diminished soil fertility. Increased production is one of the major benefits that agroforestry can provide to farmers, since it maintains soil fertility while also having beneficial impacts on plants. According to Sahoo and Wani (2019), the loss in soil fertility provides several agroforestry trees, provide a possibility for agroforestry to be introduced to communities, such as *Luecaena*, *Glyricidia*, *Cassia*, *Calliandra,* and *Alchornea*, are known to improve soil fertility. Farmers have long seen fallowing as a means of preserving and restoring soil fertility while also supporting the farming system's profitability. Farmers are likely to take recommended steps (e.g. enhanced fallow systems) to boost soil fertility when installing agroforestry systems.

Farmland tree regeneration: farmers on farmland let the stumps of chopped trees to regrow naturally. This phenomenon may be beneficial to agroforestry systems that incorporate native tree species.

Costs of farm inputs: farmers that cannot afford the high costs of agricultural inputs like fertilizer and pesticides might profit from low-input systems like agroforestry. According to Bijalwan et al. (2011), agricultural infrastructure, fertilizers, and pesticides are more expensive and out of reach for most farmers. As a result, it was claimed that the benefits of agroforestry include higher potential stability in reacting to drought and heat stress, enhanced nutrient cycling, and improved moisture balance, as well as a larger diversity of crops.

9 Women Participation in Agroforestry

As a lone farmer, some females were found to be interested in agroforestry (19.1%), implying that agroforestry has enhanced their social and financial situation. It was discovered that 33% of female agroforestry farmers were unmarried, causing them to become enmeshed in the industry in order to sustain themselves and their children (FAO, 2019). Approximately 67% of married women had old or infirm husbands (none had never married); therefore, they took up farm work. People in the family comprising men and women contribute significantly to agroforestry by cooperating in tree development and propagation. Tree and crop protection, irrigation, weeding, and annual crop harvesting are all shared responsibilities between men and women. Women are in charge of post-harvest food storage and processing, including handmade food prepared from fruits, vegetables, and cereal crops for consumption and sale by each social unit.

10 The Impact of Drought Stress on Agriculture

- Decreased growth of seedlings and proliferation of embryos
- Inadequate growth of vegetation
- Sexual development is negatively impacted
- Plant height and leaf surface decreased
- Leaf weight substantially decreased
- Respiration decreased; stomatal activity decreased and
- Overall moisture content drastically decreased

Drought-related crop destruction in Tamil Nadu: the drought not only had a negative impact on crop husbandry but also on rural livelihoods. Lack of fodder caused cattle grazers to take the uncommon practice of slicing palmyra leaves to nourish their livestock. Unexpectedly, even those strong, well-rooted, and drought-tolerant trees were beginning to exhibit symptoms of decline. The livelihoods of the palm sugar manufacturers in these areas, who are primarily female, as well as the palm leaf basket and mat weavers and tree climbers (sometimes called toddy tappers) have also suffered. Additionally, the proprietors of the trees in these groves have nothing (Aguilera et al., 2020).

Palmyra do not have crop insurance coverage, in contrast to coconut palms. In the drought-prone Salem district's villages of Attur taluk-Thammampatti, Veeraganur, and Ulipuram-areca nuts, another perennial crop are also in danger. This taluk is well known for growing cash crops including sugarcane, areca nuts, papaya, and betel wine because it has an abundance of groundwater and lush, fertile soil. This region, which is at the base of the Servarayan hills, has been severely damaged by drought. Local progressive farmers have indicated that the area's groundwater table has been declining and wells have been drying up as a result of the previous 2 years' inadequate rainfall.

The lack of water in the area has caused areca crops and papaya gardens to wither this summer (2014). This year, almost 50,000 of the high-yielding Red Lady papaya saplings that were planted on 230 acres in the Salem district and surrounding areas Pethanaickkanpalayam, Manivizhundhan, and Kallanatham clusters had already wilted. Farmers erected government-subsidized drip irrigation systems to irrigate their crops, but there wasn't enough water available to run them.

11 Effects of Flooding Stress on Plants

Anaerobic circumstances lead to substantial iron toxicity. An increase in polyphenol oxidase activity triggers the synthesis of oxidized flavonoids. Furthermore, it leads to a reduction in the power of rooted degradation and leaf bronzing. Symptoms of iron intoxication include-tiny brown spots on lower leaves that begin at the tip and extend to the root, whole leaves that are coloured orange, yellow, and brown. When patches form on leaf intervenes, the leaves turn dark and eventually die. Narrow leaves that frequently stay green, certain types of leaves have orange-yellow tips that eventually dry up. If iron poisoning is severe, leaves turn purple-brown, stunted growth and severely constrained tillering. A coarse, sparse, damaged root system with many dead roots and a coating on the outermost layer of its root system is dark brown to black. Recently, uprooted rice hills frequently contain numerous black roots and weak root systems. To reduce stress, apply a 500 ppm cycocel growth retardant spray to stop apical dominance and encourage the growth of laterals; apply 2% DAP + 1% KCl foliar spray (MOP); apply 0.5 parts per million brassinolide spray to boost photosynthetic activity; apply a 100 ppm salicylic acid foliar spray to increase the use of stem reserves in situations with extreme moisture stress; during the most crucial phases of the stress, a foliar spray of 0.3% boric acid + 0.5% $ZnSO_4$ + 0.5% $FeSO_4$ + 1.0% urea. Use fertilizers (NPK or NPK + lime) in moderation and use enough fertilizer containing K. Apply lime to acidic soils; avoid adding too much organic matter (straw, manure, etc.) to soils with a lot of organic matter.

12 Fatigue Related to Salinity

Soil salinization is one of the most frequent and important causes of environmental harm. Soils on almost every continent are salinized. According to Kumar (2006), salinity is the primary factor influencing dry and semi-arid parts of the planet, where the root zone loses soluble salts due to low rainfall and transpiration capacity exceeding rainfall. Seven million hectares of land are damaged by salt in India alone. One of India's top 371 rice-growing regions, Tamil Nadu, is vulnerable to salinity stress. Salinity significantly reduces the economic worth of using land for agriculture and forestry. According to an early Food and Agriculture Organization (FAO) research, 7% of the world's land area is harmed by soils damaged by salt.

Salinity is another issue that is growing quickly, having more than doubled in the previous 20 years. Three hectares of the minimum 10 ha of arable land that the world is losing every minute are thought to be lost due to salinization. Thus, each year, 1.6 million hectares of arable land are lost. Most people concur that secondary salinization is one of the primary causes of the degradation and desertification of the world's arid regions. Surprisingly, negative effects from irrigation can affect up to 20% of irrigated land, and these effects greatly contribute to land loss due to salt accumulation (Matocha et al. 2012). Plants' ability to tolerate salt stress is essential for both agriculture and the natural dispersion of species of vegetation (Garrity, 2004).

13 Government Programmes Involving Tree Planting

The National Climate Change Action Plan's Greening India Mission aims to bring 1.5 million hectares of agriculture and fallow lands that have been damaged due to agroforestry, including 0.8 million hectares under enhanced agroforestry techniques on existing lands and 0.7 million hectares under agroforestry (FAO, 2019). Because the vast majority of India's productive land is farmed, agroforestry on agricultural areas offers a significant prospect to stock carbon through afforestation (FAO, 2019). By 2025, the overall potential for agroforestry is projected to be 25.36 million hectares, with tree-borne oilseeds and silvipasture accounting for more. The Government of India's National Forest Policy was introduced in 1988 with the goal of expanding the amount of forest and tree cover everywhere in the country and, consequently, the ecosystem services provided by forests to nearby populations (NRCAF, 2007). Some examples of these services include carbon sequestration, timber, and non-technical fishpond products (NTFPs). Nowadays, a number of programs are used to carry out reforestation and afforestation, such as the Farm Forestry Program, which was started in the late 1970s; the Social Forestry Program, which started in the early 1980s; and the Joint Forest Management Program, which started in 1990. Additionally, afforestation is carried out as part of the National Afforestation and Eco-development Board programs, which started in 1992; and plantation forestry initiatives are personal, initiated by farmers and industry. The purpose of the Farm Forestry Program is to raise awareness of the benefits of planting trees. More trees were planted in rural and commercial areas as part of this project. (Torquebiau et al., 2012). In the semi-arid areas of the province, agricultural forestry has shown especially promising results. Tree culture blossomed because the government handed land unsuitable for producing crops to impoverished rural communities.

14 Livelihood Security by Agroforestry

14.1 Food Security

Although many people refer to Asia as the "continent of the twenty-first century", other analysts think that many Asian countries would not be able to feed their anticipated populations by the year 2021 (Sahoo et al., 2022). Comparatively speaking, Asia has less land per person than other parts of the world, but civilization is also displacing productive land. The FAO states that historically, food production in Asia has increased at a pace that has kept pace with the region's population growth. Nonetheless, population growth has outpaced trends in food production during the past 10 years, indicating the need for an increased food supply. Over 800 million people in the developing world are hungry, according to the FAO (FAO, 2019). Adding insult to injury, improvements in wheat yields are declining globally, especially in Asia, as a result of what is referred to as "technological weariness". Securing food security for people can be achieved through agroforestry and other woody perennial-based agricultural methods. Though small and marginal farmers have historically employed agroforestry to meet their needs for fuel, fodder, and food, tropical farmers generally rely on annual crops. Such methods would increase the financial returns to producers in addition to guaranteeing food production. (Sahoo et al., 2021). People with higher incomes have more "buying power" when it comes to food, particularly in cases when agriculture is not practiced or crops do not succeed. Additionally, diversified output is a risk-avoidance tactic, which is crucial in light of the current agricultural issues in a number of South and Southeast Asian nations. Additionally, agroforestry has been emphasized as a possible revenue streams and jobs for people living in rural areas (Sahoo et al., 2021). In addition to providing food variety, year-round systems like home gardens that produce a range of fruits, vegetables, spices, and other items also help provide food security during "lean" times.

14.2 Poverty Reduction

Agroforestry contributes a larger percentage of a farmer's total annual revenue. This total includes agricultural crops, forestry (timber), and animals. Agricultural crops such as cocoa, coffee, cloves, rice, and fruits provide the majority of their revenue from crop harvest, which includes plants that develop quickly and have economic value, allowing farmers to make a regular income to fulfil daily needs (Sahoo et al., 2021). Timber is grown primarily for long-term savings; but, if households require significant sums of money, the wood is chopped down. The cost of timber varies significantly depending on the kind, age, size, and quality of the wood. Commercialize cattle producers as a source of forthcoming savings. Many households have cattle that are sold or redeemed for cash and food on a yearly basis as part of their routine activities (Sahoo et al., 2021; Raj et al., 2020). Farmers most often raise cows and goats, in addition to using manure as a fertilizer for crops and as a biogas fuel. The cost of fertilizer, pesticides, labour, and seed are all included in the cost of production in agroforestry management. Land management is not done in a systematic way (Sahoo et al., 2023b).

15 Comparison of Traditional and Commercial Farming Strategies for Sustainable Opportunities

The expenses of tree care are minor in traditional agroforestry systems since trees grow organically, particularly in traditional agroforestry zones, and farmers just let them to thrive. The only indirect expenses connected with tree care are those related to shade and competition caused by moisture and fertilizer requirements. Because of this, only the advantages of trees that result from their harvest and sale were evaluated, even though the industrial agroforestry system is characterized by trees that are planted adjacent to plants on farm bunds protects, boundaries, or inside pitches (Sharma & Sahoo, 2021). Some researchers conducted a socioeconomic analysis of traditional and commercial agroforestry techniques used by farmers in western Uttar Pradesh, and they discovered that tree species like *Azadirachta indica, Acacia nilotica, Dalbergia sissoo and Eucalyptus* spp. were dominant species in the traditional system whereas *Populus deltoides* and *Eucalyptus* spp. were commercial agroforestry's most important species (Dhyani et al., 2003). As illustrated in figure, the primary factors influencing the implementation of agroforestry in the traditional agroforestry region were fuel wood (50.6%), increased revenue (24.4%), and shade (17.5%). In the commercial agroforestry province, surplus revenue (71.3%) was the most important factor. Although traditional agroforestry appears to be less viable than commercial agroforestry, it is nevertheless significant to the farmers (Fig. 2).

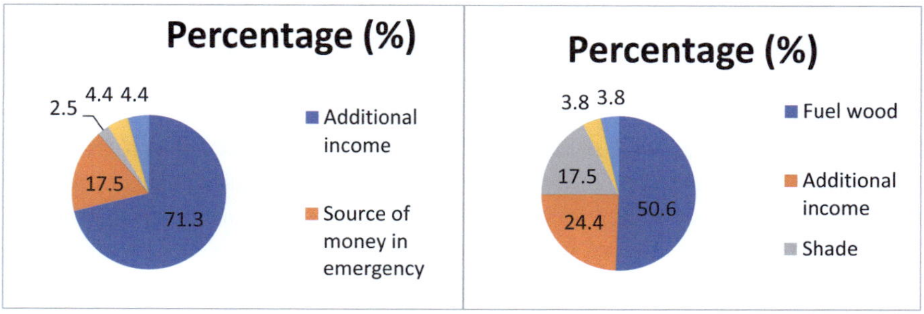

Traditional agroforestry system Commercial agroforestry system

Fig. 2 Determinants of traditional vs. commercial agroforestry system

16 Overcoming Challenges

Agroforestry systems' efficacy is exaggerated by a number of variables, including the choice and management of tree species, soil properties, topography, downpours, farming methods, objectives for food safety, opportunities for financial and community development, and more. Not all activities have good impacts; in fact, some practices may have more negative or neutral effects than positive ones. To begin, creating the right agroforestry system through the selection of the most appropriate plants and management practices is critical for attaining desired social and environmental benefits (de Vries et al., 2018; Delgado-Baquerizo et al., 2017). Species must be chosen with care, taking into consideration local priorities as well as biophysical circumstances. Research indicates that, even though shade trees compete with other plants for water, additional resources like water can be optimized by understanding the unique root characteristics of intercropped trees, such as differential geographic growing structure, root anatomy, and mycorrizal condition (Rao et al., 2007; Sahoo and Wani, 2022).

Farmers should be encouraged to participate in agroforestry by guaranteeing land security and tree tenure, as agroforestry takes longer than other conventional agricultural systems to realize benefits. While applying community-inclusive stewardship concepts, land should be regarded as a collective benefit. Farmers' capacity to develop their optimal management methods may be harmed if they continue to get low returns (World Bank, 2006). As a result, enhancing local microcredit, providing suitable and adequate funding at the grassroots level, and promoting agroforestry as a viable and lucrative venture are all critical (Crowther et al., 2019). To meet the ideal location social and financial boundaries, maintain social conditions like sex cooperative energies, improvement of more planned, organized and reasonable market frameworks with aggregate expecting the two sources of info and yields of agroforestry, better vehicle foundation, contribution of slighter delegates in the production network, improved interest in portrayal of tree food sources, arrangement of impetuses from instalment for eco framework administrations and certificate plans ought to be taken on (Dollinger & Jose, 2018).

Furthermore, given the growing urban population, demand-side concerns such as diet modification should be further investigated (Arenas-Corraliza, 2018). Scaling up effective tree domestication procedures is another imperative challenge. It is also essential to increase understanding of the positive features of existing indigenous plants utilized in agroforestry, as well as the creation of new cultivars from those indigenous species, as well as the availability and ease of access to high-quality planting materials and the ability to cross borders safely transfer of exceptional varieties developed over the globe. In addition, small-scale farmers play an indisputable role in the agricultural industry (Arenas-Corraliza, 2018; Tscharntke et al., 2011). Around 80% of agricultural areas in South Asia are smaller than 0.6 hectares in size (Kumar, 2006). Because the majority of ecological and economic research is done independently, it is hard to combine the data, making it difficult to analyse the financial benefits, extent, and trends of agroforestry, which are not included in official inventories, in real time (FAO, 2019). Even the FAO's Global Forest Assessment does not acknowledge agroforestry as a separate land use since it falls under the categories of forestry or agriculture. Agroforestry is, nevertheless, growing increasingly common in various places of the globe, based on the latest studies. Emergent recognition in affluent countries will increase empathetic and sustenance for its spread in poor countries, safeguarding essential savings (Dollinger & Jose, 2018).

17 Conclusion

The incorporation of trees alongside agricultural crops, in conjunction with appropriate species selection and management strategies, can contribute to improving the biological, chemical, and physical characteristics of degraded soil, given that soil can influence the microclimate. Farm trees promote infiltration through litter fall and canopy impacts, and they have a good impact on hydrologic functions. Agroforestry improves soil quality by increasing microbial activity in the soil due to the effects of trees, organic matter deposition, root exudates, and various types of litter. Agroforestry methods mitigate soil erosion and nutrient loss while mitigating the impact of precipitation. This evergreen could be used in farming techniques as a practical solution to reduce contamination from pollutants and agrochemicals. To encourage smallholders to adopt agroforestry practices, farmers need to be informed about the long-term non-cash environmental benefits of these systems. The restoration of soil-based ecosystem services in degraded soil, a feasible route for intensification to increase agricultural sustainability, and its impact on the soil biota, microclimate, litter fall, and nutrient pumping and cycling are some advantages of agroforestry. Smallholders' livelihoods and soil conditions are significantly impacted by climate-related variability, which increases their susceptibility to disease and climate change in the current climate scenario.

Agroforestry systems are being called for to be used globally as more resilient and sustainable farming methods. To better understand how crops and soils in the AFS will react to potential climate change, manipulative experiments and long-term monitoring should be combined with modelling research. The goals of the research should take into account the traditional knowledge held by farmers. In order to clarify the connections between, among other things, forestry, crop production, microclimate, ecohydrology, and soil biology, future research will need to use a multidisciplinary approach.

18 Recommendations

Policy suggestions will be required if significant changes in agroforestry practice occur in the country. Whether or not there is an agroforestry policy in place, there are still issues that need to be updated or handled, including research objectives and techniques, extension strategies and correspondences, socioeconomic challenges, agroforestry education and training, and organizational issues. Instead of fragmented efforts marked by rivalry, overlap, and lack of coordination, the objective of this strategy should be to integrate and improve a multidisciplinary and inter-institutional strategy is used to establish agroforestry research and development programmes.

References

Aguilera, E., Diaz-Gaona, C., Garcia-Laureano, R., Reyes-Palomo, C., Guzmán, G., Ortolani, L., Sanchez-Rodriguez, M., & Rodriguez-Estevez, V. (2020). Agroecology for adaptation to climate change and resource depletion in the Mediterranean region. A review. *Agricultural Systems, 181*, 102809.

Altieri, M. A., Nicholls, C. I., Henao, A., & Lana, M. A. (2015). Agroecology and the design of climate change-resilient farming systems. *Agronomy for Sustainable Development, 35*(3), 869–890.

Arenas-Corraliza, M. G., López-Díaz, M. L., & Moreno, G. (2018). Winter cereal production in a Mediterranean silvoarable walnut system in the face of climate change. *Agriculture, Ecosystems and Environment, 264*, 111–118.

Barrios, E., Shepherd, K., & Sinclair, F. (2015). Soil health and agricultural sustainability: The role of soil biota. In *Agroecology for food security and nutrition, proceedings of the international FAO symposium, 18–19 September 2014* (pp. 104–122). Food and Agriculture Organization of the United Nations (FAO).

Beillouin, D., Schauberger, B., Bastos, A., Ciais, P., & Makowski, D. (2020). Impact of extreme weather conditions on European crop production in 2018. *Philosophical Transactions of the Royal Society B, 375*, 20190510.

Bijalwan, A., Sharma, C. M., & Kediyal, V. K. (2011). Socioeconomic status and livelihood support through traditional agroforestry systems in hill and mountain agro-ecosystem of Garhwal Himalaya, India. *Indian Forester, 138*(12), 1423–1431.

Bonaudo, T., Bendahan, A. B., Sabatier, R., Ryschawy, J., Bellon, S., Leger, F., Magda, D., & Tichit, M. (2014). Agroecological principles for the redesign of integrated crop–livestock systems. *European Journal of Agronomy, 57*, 43–51.

Burgess, A. J., Cano, M. E. C., & Parkes, B. (2022). The deployment of intercropping and agroforestry as adaptation to climate change. *Crop and Environment, 1*, 145–160.

Cardinael, R., Cadisch, G., Gosme, M., Oelbermann, M., & van Noordwijk, M. (2021). Climate change mitigation and adaptation in agriculture: Why agroforestry should be part of the solution. *Agriculture, Ecosystems and Environment, 319*, 107555.

Chavez, E., Conway, G., Ghil, M., & Sadler, M. (2015). An end-to-end assessment of extreme weather impacts on food security. *Nature Climate Change, 5*, 997–1001.

Cheng, M., McCarl, B., & Fei, C. (2022). Climate change and livestock production: A literature review. *Atmosphere, 13*(1), 140.

Crowther, T. W., Riggs, C., Lind, E. M., Borer, E. T., Seabloom, E. W., Hobbie, S. E., Wubs, J., Adler, P. B., Firn, J., Gherardi, L., Hagenah, N., Hofmockel, K. S., Knops, J. M. H., McCulley, R., MacDougall, A., Peri, P. L., Prober, S. M., Stevens, C. J., & Routh, D. (2019). Sensitivity of global soil carbon stocks to combined nutrient enrichment. *Ecology Letters, 22*, 936–945.

Dagar, J. C. (2012). Utilization of degraded lands/habitats and poor-quality water for Livelihood security and mitigating climate change. *Indian Journal of Agroforestry, 14*(1), 1–16.

Daryanto, S., Wang, L., & Jacinthe, P.-A. (2016). Global synthesis of drought effects on maize and wheat production. *PLoS One, 11*, e0156362.

de Vries, F. T., Griffiths, R. I., Bailey, M., Craig, H., Girlanda, M., Gweon, H. S., Hallin, S., Kaisermann, A., Keith, A. M., Kretzschmar, M., Lemanceau, P., Lumini, E., Mason, K. E.,

Oliver, A., Ostle, N., Prosser, J. I., Thion, C., Thomson, B., & Bardgett, R. D. (2018). Soil bacterial networks are less stable under drought than fungal networks. *Nature Communications, 9*, 3033.
Delgado-Baquerizo, M., Eldridge, D. J., Ochoa, V., Gonzalo, B., Singh, B. K., & Maestre, F. T. (2017). Soil microbial communities drive the resistance of ecosystem multifunctionality to global change in drylands across the globe. *Ecology Letters, 20*, 1295–1305.
Dhyani, S. K., Sharda, V. N., & Sharma, A. R. (2003). Agroforestry for water resources conservation: Issues, challenges and strategies. In P. S. Pathak & R. Newaj (Eds.), *Agroforestry: Potentials and opportunities*. Agribios.
Dollinger, J., & Jose, S. (2018). Agroforestry for soil health. *Agroforestry Systems, 92*, 213–219.
Falkenmark, M., & Rockström, J. (2008). Building resilience to drought in desertification prone savannas in sub-Saharan Africa: The water perspective. *Natural Resources Forum, 32*, 93–102.
FAO. (2002). *Forest-based poverty reduction: A brief review of facts, figures, challenges and possible ways forward.* Paper prepared for the International Workshop on "Forests in poverty reduction strategies: Capturing the potential, 01–02 October, Tuusula, Finland.
FAO. (2019). *CIFOR. FAO framework methodology for climate change vulnerability assessments of forests and forest dependent people.*
Garrity, D. P. (2004). Agroforestry and the achievement of the millennium development goals. *Agroforestry Systems, 61*, 5–17.
Goli, S., Maurya, N. K., & Sharma, M. K. (2015). Continuing caste inequalities in rural Uttar Pradesh. *International Journal of Sociology and Social Policy, 35*, 252–272.
Hatfield, J. L., Sauer, T. J., & Cruse, R. M. (2017). Chapter one-soil: The forgotten piece of the water, food, energy nexus. In D. L. Sparks (Ed.), *Advances in agronomy* (pp. 1–46). Academic Press.
Hernández-Morcillo, M., Burgess, P., Mirck, J., Pantera, A., & Plieninger, T. (2018). Scanning agroforestry-based solutions for climate change mitigation and adaptation in Europe. *Environmental Science & Policy*, 80, 44–52.
Islam, M. A., & Quli, S. M. S. (2016). Motivation strategy for agroforestry intensification among small holders. *Advances in Life Sciences, 5*(10), 3878–3883.
Islam, M. A., Quli, S. M. S., Sofi, P. A., Bhat, G. M., & Malik, A. R. (2015). Livelihood dependency of indigenous people on forest in Jharkhand. *India. Vegetos, 28*(3), 106–118.
Kumar, B. M. (2006). Agroforestry: The new old paradigm for Asian food security. *Journal of Tropical Agriculture, 44*(1–2), 1–14.
Lakhran, H., Kumar, S., & Bajiya, R. (2017). Crop diversification: An option for climate change resilience. *Trends in Biosciences, 10*(2), 516–518.
Leakey, R. R. B. (2010). *Should we be growing more trees on farms to enhance the sustainability of agriculture and increase resilience to climate change? A special report* (pp. 1–12). International Society of Tropical Foresters, USA.
Matocha, J., Schroth, G., Hills, T., & Hole, D. (2012). Integrating climate change adaptation and mitigation through agroforestry and ecosystem conservation. In P. K. R. Nair & D. Garrity (Eds.), *Agroforestry - The future of global land use. Advances in agroforestry* (pp. 105–126). Springer.
Mbow, C., Noordwijk, M., Prabhu, R., & Simons, T. (2014a). Knowledge gaps and research needs concerning agroforestry's contribution to sustainable development goals in Africa. *Current Opinion in Environment Sustainability, 6*, 162–170.
Mbow, C., Smith, P., Skole, D., Duguma, L., & Bustamante, M. (2014b). Achieving mitigation and adaptation to climate change through sustainable agroforestry practices in Africa. *Current Opinion in Environmental Sustainability, 6*, 8–14.
Mishra, V., Shah, R., & Garg, A. (2016). *Climate change in Madhya Pradesh: Indicators, impacts and adaptation*. Indian Institute of Management, Ahmedabad.
Mohanty, A. K., Kumar, G. A. K., Singh, B. B., & Meera, S. N. (2016). Developing multidimensional scale for effective measurement of rural leadership. *Indian Research Journal of Extension Education, 9*, 57–63.

Mustafa, M., Szalai, Z., Divéky-Ertsey, A., Gál, I., & Csambalik, L. (2022). Conceptualizing multiple stressors and their consequences in agroforestry systems. *Stress, 2*(3), 242–255.

National Research Centre for Agroforestry. (2007). *Vision-2025: NRCAF perspective plan*. Jhansi, India.

Ofori, D. A., Gyau, A., Dawson, I. K., Asaah, E., Tchoundjeu, Z., & Jamnadass, R. (2014). Developing more productive African agroforestry systems and improving food and nutritional security through tree domestication. *Current Opinion in Environment Sustainability, 6*, 123–127.

Olsson, L., Barbosa, H., Bhadwal, S., Cowie, A., Delusca, K., Flores-Renteria, D., Hermans, K., Jobbagy, E., Kurz, W., Li, D., Sonwa, D. J., & Stringer, L. (2019). Land degradation: IPCC special report on climate change, desertification, land 5 degradation, sustainable land management, food security, and 6 greenhouse gas fluxes in terrestrial ecosystems. In *IPCC special report on climate change, desertification, land 5 degradation, sustainable land management, food security, and 6 greenhouse gas fluxes in terrestrial ecosystems* (p. 1). Intergovernmental Panel on Climate Change (IPCC).

Rao, K. P. C., Verchot, L. V., & Laarman, J. (2007). Adaptation to climate change through sustainable management and development of agroforestry systems. *Journal of SAT Agricultural Research, 4*(1), 1–30.

Raj, A., Jhariya, M. K., Yadav, D. K., & Banerjee, A. (Eds.). (2020). *Climate change and agroforestry systems: adaptation and mitigation strategies*. CRC Press.

Ravindranath, N. H., Srivastava, N., Murthy, I. K., Malaviya, S., Munsi, M., & Sharma, N. (2012). Deforestation and forest degradation in India implications of REDD+. *Current Science, 102*(8), 1–9.

Ray, D. K., Ramankutty, N., Mueller, N. D., West, P. C., & Foley, J. A. (2012). Recent patterns of crop yield growth and stagnation. *Nature Communications, 3*, 1293.

Richard, L. C., & Munishi, P. K. T. (2013). Nzunda agroforestry as adaptation strategy under climate change in Mwanga District, Kilimanjaro, Tanzania. *International Journal of Environmental Protection, 3*(11), 29–38.

Riyadh, Z. A., Rahman, M. A., Saha, S. R., & Hossain, M. I. (2018). Soil properties under jackfruit based agroforestry systems in Madhupur tract of Narsingdi district. *J. Sylhet. Agric. Univ*, 5, 173–179.

Sahoo, G. R., Mishra, P. J., Dash, A. C., Wani, A. M., & Gupta, S. (2021). Agroforestry for adaptation, mitigation to climate change and livelihood security. In *Agroforestry - Prospective, strategies, and future aspects* (pp. 203–207). Taran Publication.

Sahoo, G. R., Mishra, P. J., Wani, A. M., Sharma, A., Mishra, D., Patra, D., Mishra, I., & Behera, M. (2023b). Impact of climate change on livelihood security and biodiversity – Issues and mitigation strategies. In *Climate change impacts on natural resources, ecosystems and agricultural systems* (pp. 1–27). Springer Nature.

Sahoo, G. R., Swamy, S. L., Wani, A. M., & Mishra, A. (2022). Agroforestry systems for carbon sequestration and food security: Implications for climate change mitigation. In *Soil health and environmental sustainability application of geospatial technology* (pp. 503–528). Springer Nature.

Sahoo, G. R., & Wani, A. M. (2019). Multifunctional agroforestry Systems in India for livelihoods. *Annals of Horticulture, 12*(2), 139–149.

Sahoo, G. R., & Wani, A. M. (2020). Effect of climate change on land degradation. *International Journal For Innovative Engineering and Management Research, 09*(12), 483–494.

Sahoo, G. R., Wani, A. M., Kishore, P., & Vijay, R. (2019). Biodiversity conservation and climate change approach. *International Archive of Applied Sciences and Technology, 10*(4), 01–09. https://doi.org/10.15515/iaast.0976-4828.10.4.19

Sahoo, G. R., Wani, A. M., Prusty, M., & Ray, M. (2023a). Effect of globalization and climate change on forest - A review. *Material Today Proceedings, 80*(3), 2060–2063.

Sahoo, G. R., Wani, A. M., & Satpathy, B. (2020a). Greening wastelands for environmental security through agroforestry. *International Journal of Advanced Research in Science and Technology, 7*, 2581–9429.

Sahoo, G. R., Wani, A. M., & Sharma, A. (2020b). Enhancing food security through agroforestry for sustainability – A review. *International Journal of Current Microbiology and Applied Sciences, Special Issue-11*, 2001–2020.

Sahoo GR, Wani AM (2022) Agroforestry in food security. *Recent Research in Agriculture for doubling of farmer's Income*, 73–78. ISBN: 978-81-946373-2-5

Schoeneberger, M., Bentrup, G., de Gooijer, H., Soolanayakanahally, R., Sauer, T., Brandle, J., Zhou, X., & Current, D. (2012). Branching out: Agroforestry as a climate change mitigation and adaptation tool for agriculture. *Journal of Soil and Water Conservation, 67*, 128A–136A.

Sharma, A., & Sahoo, G. R. (2021). Confronts and prospects for enhancing farmers livelihood protection. In *Sustainable agriculture and farmers welfare for rural development* (pp. 14–32).

Sinha, B., Vasisht, A., Omprakash, M. D., Ghosh, M., Devi, R. M., Patasaraiya, M. K., Yadava, R. N., Kinhal, G. A., & Bisaria, J. (2018). *Opportunities of renewable energy interventions in forest fringe villages of Madhya Pradesh*. Indian Institute of Forest Management.

Thomas, S. C., Halim, M. A., Gale, N. V., & Sujeeun, L. (2019). Biochar enhancement of facilitation effects in agroforestry: Early growth and physiological responses in a maize-leucaena model system. *Agroforestry Systems, 93*(6), 2213–2225.

Torquebiau, O. C., Detlefsen, E., Gauthier, G., & Buttoud, G. (2012). Improved policies for facilitating the adoption of agroforestry. In M. Kaonga (Ed.), *Agroforestry for biodiversity and ecosystems services: Science and practice* (pp. 113–128). IntechOpen.

Tscharntke, T., Clough, Y., Bhagwat, S. A., Buchori, D., Faust, H., Hertel, D., Holscher, D., Juhrbandt, J., Kessler, M., Perfecto, I., Scherber, S., Schroth, G., Veldkamp, E., & Wanger, T. C. (2011). Multifunctional shade-tree management in tropical agroforestry landscapes–A review. *Journal of Applied Ecology, 48*, 619–629.

Udawatta, R. P., Gantzer, C. J., & Jose, S. (2017). Agroforestry practices and soil ecosystem services. In M. A. Al-Kaisi & B. Lowery (Eds.), *Soil health and intensification of agroecosystems* (pp. 305–333). Academic Press.

Udawatta, R. P., Kremer, R. J., Adamson, B. W., & Anderson, S. H. (2008). Variations in soil aggregate stability and enzyme activities in a temperate agroforestry practice. *Applied soil ecology*, 39(2), 153–160. https://doi.org/10.1016/j.apsoil.2007.12.002

UNDP. (2012). *India: Greening rural development in India* (Vol. I, 61 p). United Nations Development Program, New Delhi.

Vermeulen, S. J., Campbell, B. M., & Ingram, J. S. I. (2012). Climate change and food systems. *Annual Review of Environment and Resources, 37*, 195–222.

World Bank. (2006). *Unlocking opportunities for forest dependent people in India*. The World Bank Agriculture and Rural Development Sector Unit, South Asia Region, New Delhi.

Zhao, C., Liu, B., Piao, S., Wang, X., Lobell, D. B., Huang, Y., Huang, M., Yao, Y., Bassu, S., Ciais, P., Durand, J. L., & Asseng, S. (2017). Temperature increase reduces global yields of major crops in four independent estimates. *Proceedings of the National Academy of Sciences, USA, 114*, 9326–9331.

Ensuring Irrigation Water Quality for Climate-Resilient Agriculture: Strategies and Solutions in Semi-Arid Regions

Shivani Garg and Pooja Arora

Abstract Climate change poses significant challenges to sustainable agriculture development in semi-arid regions, where water scarcity and temperature fluctuations are already pressing issues. Irrigation water quality plays a crucial role in maintaining soil health, crop productivity, and food security in these areas. The introduction highlights the importance of sustainable agriculture in semi-arid regions and the role of irrigation water quality in supporting this endeavor. The subsequent sections explain the impacts of climate change on water quality parameters, including temperature increase, altered precipitation patterns, salinization, and groundwater quality degradation. A comprehensive overview of strategies and solutions for maintaining and improving water quality in semi-arid areas will also be discussed. The strategies include water conservation techniques, such as efficient irrigation practices and rainwater harvesting, wastewater treatment and reuse, soil management practices to mitigate salinization and nutrient depletion and the integration of renewable energy sources into irrigation systems. The role of precision agriculture and supportive policies and institutional frameworks are also discussed. It aims to equip researchers, policymakers, and stakeholders with the knowledge and tools necessary to address the water-related challenges posed by climate change and ensure long-term food security and environmental sustainability.

Keywords Water quality · Sustainable agriculture · Water scarcity · Salinity · Water conservation · Food security

1 Introduction

Global agriculture faces unprecedented challenges due to climate change, particularly in semi-arid regions where water scarcity and soil degradation are pressing concerns (Smith et al., 2023). Low and erratic rainfall patterns characterize the

S. Garg · P. Arora (✉)
Institute of Environmental Studies, Kurukshetra University, Kurukshetra, Haryana, India

A. K. Singh et al. (eds.), *Sustainable Agriculture Management in Semi-Arid Climates*, https://doi.org/10.1007/978-3-031-94062-0_9

region as highly vulnerable to the detrimental effects of changing temperatures, precipitation patterns, and extreme weather events (Mohammed et al., 2020). The agricultural area in this region faces more problems with food security and crop productivity (Oommen et al., 2022). The most significant factor that influences agriculture sustainability is irrigation water quality (Li et al., 2023). Climate change intensifies the water scarcity problem and deteriorates the soil quality. The changes in the water source's physicochemical and biological parameters rendered them unsuitable for irrigation purposes (Ricart et al., 2019). Irrigation water with high salinity leads to soil salinization that inhibits plant growth, reduces crop yields, and affects food security. The presence of contaminants such as pesticides, pathogens, and heavy metals can pose risks to the health of crops, food, and humans (Dotaniya et al., 2023). Nutrient unavailability and degraded soil fertility can also enhance the problems of farmers in semi-arid regions and can also lead to problems of environmental degradation and low production (Visconti & de Paz, 2012). To address these challenges, a comprehensive approach is required that can integrate mitigation and adaptation strategies for semi-arid regions. Preserving water quality and mitigating water scarcity requires water conservation techniques such as rainwater harvesting, efficient irrigation systems, and advanced technologies such as precision agriculture and remote sensing (Gebreyess & Amare, 2019). Wastewater treatment and reuse of water offer a significant solution for augmenting water resources and reducing the pressure on freshwater supplies, provided that proper treatment techniques and monitoring measures are taken into account (Chen et al., 2017). Soil management practices such as nutrient management, salinity control, and the addition of organic amendments are required to maintain soil health, fertility, and crop productivity (Shi, 2022). Renewable energy sources such as solar and wind power can be integrated into irrigation systems to reduce carbon footprints in agriculture, and it is highly energy and water-use efficient (Falchetta et al., 2023). Precision agricultural practices and remote sensing techniques can provide valuable information for decision-making to the farmers for managing and scheduling irrigation practices (Pradipta et al., 2022). Crop selection and stress tolerance are important factors in adapting to climatic changes in semi-arid regions (Lee et al., 2023). For sustainable yields and food security, the identification and development of crop varieties that are resilient to heat, drought, and salinity are required. Genetic engineering and breeding techniques are modern strategies that are also helping in the development of these stress-tolerant varieties (Mir et al., 2022). This chapter explains the intricacies of ensuring irrigation water quality for climate-resilient agriculture in semi-arid regions. It also explains the impacts of climate change on water quality parameters, strategies and solutions for maintaining irrigation water quality, and the importance of precision agriculture and crop selection. The chapter aims to provide valuable information and recommendations for promoting sustainable agriculture development in these regions and enhancing food security and environmental sustainability.

2 Impact of Climate Change on Irrigation Water Quality

The climatic factors that affect semi-arid regions are water scarcity and temperature rise. The temperature rise leads to increased evapotranspiration which enhances the water loss from soil and water sources (Golla, 2021). This process made water scarcity a permanent challenge in semi-arid regions, straining the limited freshwater resources for irrigation. Water scarcity not only threatens crop yields but can also concentrate the contaminants and salts that deteriorate water quality (Torabi Haghighi et al., 2020). Another factor that is influenced by climate change in these regions is precipitation patterns. Semi-arid regions are expected to experience increased variability in rainfall distribution, with longer dry spells and more intense precipitation events. These events disrupt the natural replenishment of ground and surface water resources, affecting water availability for irrigation (Melki et al., 2017). Intense rainfall events lead to an increased risk of soil erosion that in turn increases the sedimentation and nutrient loading in water bodies, which degrade water quality and make it unsuitable for irrigation (Bhatti et al., 2021). The effect of climate change on water quality also occurs in the form of soil and groundwater resource salinization. Rising temperatures and low rainfall can lead to enhanced evaporation and reduce groundwater recharge, concentrating salts in the soil and groundwater (Etikala et al., 2021). This problem increases when seawater intrudes into coastal aquifers due to sea-level rise which further increases salinity levels. Salinization also affects crop growth and the long-term sustainability of agricultural systems in semi-arid regions (Dotaniya et al., 2023). Increased temperature and altered precipitation patterns can influence the mobility and transport of contaminants such as pesticides, salts, heavy metals, and pathogens into groundwater resources (Ibrahim & Al-Turki, 2020). Changes in land use patterns and agricultural practices in response to climate change may introduce new sources of contamination and compromise groundwater quality (Bornare et al., 2018).

3 Water Quality Parameters for Sustainable Irrigation

Water quality parameters play a crucial role in determining the suitability of water for sustainable irrigation practices, particularly in semi-arid regions where water resources are scarce and susceptible to degradation. These parameters are classified into physical, chemical, and biological, and each has importance for crop growth, soil health, and environmental sustainability.

3.1 Physical Parameters

Physical parameters comprise measurable and visible characteristics of water. The clarity of the water, which is measured by turbidity, is an essential parameter, as high turbidity can clog the irrigation equipment and reduce the effectiveness of water treatment processes (Liu et al., 2013). Water temperature is another factor, as extreme temperature can influence the solubility of dissolved gases and the metabolic processes of aquatic organisms (Li et al., 2024). Dissolved oxygen levels are vital for maintaining a healthy aquatic ecosystem and preventing the proliferation of anaerobic microorganisms that can produce harmful byproducts (Lu et al., 2022).

3.2 Chemical Parameters

These parameters involve the assessment of various substances dissolved or suspended in water. Salinity often measured by electrical conductivity or total dissolved solids is a key parameter as high salinity increases soil salinization and negatively impacts crop yield and growth (Etikala et al., 2021). Figure 1 outlines the physiological, biochemical, and morphological responses of salt stress on crop growth. Nutrient levels such as nitrogen, phosphorus, and potassium are essential for plant growth but can also contribute to eutrophication and algal blooms when present in high concentrations (Nieder et al., 2018). Cadmium, lead, and arsenic are some toxic heavy metal contaminants that accumulate in crops and harm human health and the environment (Xu et al., 2020).

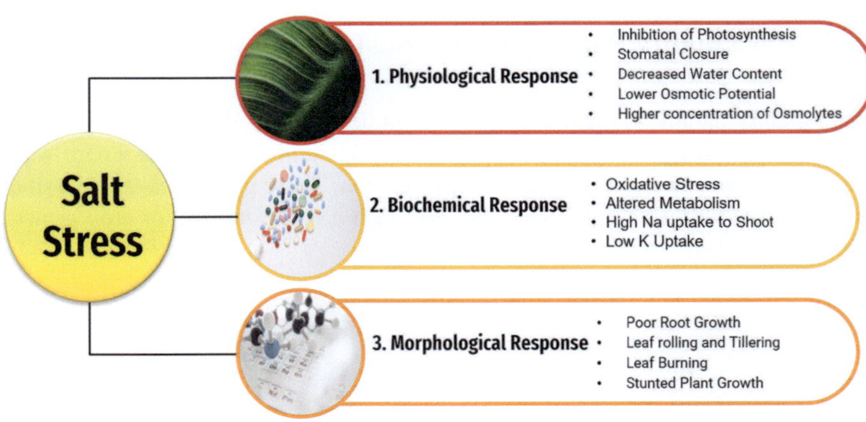

Fig. 1 Impacts of salt stress on crop growth

3.3 *Biological Parameters*

These are living organisms such as pathogens and algae present in water bodies. Pathogenic organisms such as viruses, bacteria, protozoa, and fungi can contaminate irrigation water and pose serious health risks to agricultural workers and consumers (Khalid et al., 2018). Algal blooms are often increased by excessive nutrient concentrations, and high temperatures can reduce dissolved oxygen, produce toxins, and clog irrigation systems (Xu et al., 2022). For sustainable irrigation practices, it is required to maintain these water quality parameters in semi-arid regions. Regular assessment and monitoring are important to identify potential problems and implement appropriate mitigation strategies.

4 Strategies for Maintaining and Improving Irrigation Water Quality

Sustainable irrigation practices in semi-arid regions are a complicated approach that comprises various strategies to improve and maintain water quality (Dotaniya et al., 2023). These strategies aim to optimize resource utilization and promote sustainable practices while ensuring long-term food security and environmental conservation. The following subsections explain these strategies for improving water quality in semi-arid regions.

4.1 *Water Conservation Strategies*

Water conservation is an important strategy in semi-arid regions to maximize efficiency and reduce wastage. Irrigation systems such as sprinkler and drip irrigation offer significant benefits in water efficiency (Shi, 2022). Table 1 compares the advantages and disadvantages of the different water conservation techniques that should be adopted in semi-arid regions. These systems precisely deliver water to the root zone, minimizing losses through evaporation, runoff, and deep percolation. Conserving water during irrigation reduces overconsumption and mitigates the risk of nutrient leaching and soil salinization (Hu, 2020). Rainwater harvesting techniques present a valuable solution for enhancing water supplies in semi-arid regions. Storage tanks and catchment systems facilitate the capture and storage of rainwater during wet periods, creating valuable alternative water sources for irrigation during dry spells (Kahinda et al., 2007). By harnessing this underutilized resource, farmers can reduce their support on groundwater or surface water sources, potentially reducing the strain on natural resources. By reducing water demand and diversifying water sources, these strategies help to mitigate the detrimental effects of water

Table 1 Advantages and disadvantages of different water conservation techniques

Technique	Advantages	Disadvantages	References
Drip irrigation	• High water use efficiency (up to 90%) • Precise water application near roots • Reduced weed growth • Suitable for saline water	• High initial cost • Potential clogging issues • Requires regular maintenance	Ojo et al. (2021)
Sprinkler irrigation	• Uniform water distribution • Suitable for various crop types • Can be automated • Cools crops and reduces heat stress	• Higher evaporation losses • Potential for wind drift • Risk of disease spread • Unsuitable for saline water	Priyan (2021)
Precision agriculture technologies	• Optimized water application based on real-time data • Improved water use efficiency • Reduced nutrient leaching • Automated and remote monitoring	• High initial investment costs • Requires technical expertise • Data management challenges • Potential system malfunctions	Hundal et al. (2023)

scarcity and climate change on water quality and promote sustainable agricultural practices (Sraïri, 2018).

4.2 Wastewater Treatment and Reuse

Treated water usage for agricultural irrigation presents a promising solution to enhance water resources and reduce the mounting stress on freshwater supplies in semi-arid regions (Smith et al., 2023). By harnessing this water source, farmers can mitigate the impact of climate change and water scarcity while reducing the demand for freshwater (Dong, 2021). The successful implementation of wastewater reuse requires a comprehensive approach that prioritizes public health and environmental protection. Before using wastewater for irrigation purposes, there is a need for rigorous treatment processes to meet stringent quality standards (Garrido-Cardenas et al., 2020). Membrane filtration, reverse osmosis, and disinfection are some advanced treatment technologies that are essential for removing a wide range of contaminants including heavy metals, emerging pollutants, and pathogens (Zhao et al., 2022). Continuous monitoring and quality control measures must be implemented to detect and mitigate any potential risks associated with the presence of residual contaminants (Isa et al., 2023).

4.3 Soil Management Practices

Effective soil management practices are essential to maintaining and enhancing irrigation water quality in semi-arid regions (Shi, 2022). Practical measures to control soil salinization are important, as excessive salt accumulation deteriorates the quality of soil and crop productivity and also degrades water quality by runoff and leaching (Sahab et al., 2021). The installation of subsurface drainage systems and the application of soil amendments like organic matter and gypsum can mitigate the effect of salinization and prevent further deterioration of soil and water resources (Singh, 2019). The excessive application of fertilizers can lead to nutrient leaching and subsequent eutrophication of water bodies, compromising their suitability for irrigation (Haque, 2021). Precision nutrient management techniques such as targeted fertilizer applications, soil testing and the addition of organic matter can help in maintaining optimal nutrient levels in the soil. This technique not only enhances efficient nutrient utilization by crops but also reduces the impacts of excess nutrients entering water systems, protecting aquatic ecosystems and water quality (Dotaniya et al., 2023). These approaches can effectively maintain and improve the quality of water in semi-arid regions. Mitigating soil degradation and controlling the leaching of contaminants contribute to the long-term sustainability of agricultural production while preserving the integrity of valuable water resources (Bornare et al., 2018).

4.4 Integration of Renewable Energy Sources in Irrigation Systems

Solar and wind power for powering irrigation systems presents a transformative opportunity to cultivate sustainable agricultural practices in semi-arid regions (Izquierdo & Blanchard, 2020). The integration of these clean energy technologies reduces carbon footprint and enhances overall energy efficiency (Tariq et al., 2021). By collecting abundant solar radiation and wind resources often found in these regions, farmers can minimize the dependency on fossil fuels and in turn, help mitigate greenhouse gas emissions that can contribute to the global efforts in combating climate change (Onah et al., 2021). Adopting renewable energy sources helps eliminate the costs associated with fossil fuel consumption (Sani et al., 2019). Solar and wind power have the potential to lower operational costs for farmers. Renewable energy sources can contribute to energy security and save farmers from fluctuations in conventional energy prices and supply disruptions (Kodirov et al., 2020).

4.5 Precision Agriculture and Decision Support Systems

In recent decades, RS (remote sensing) has emerged as a powerful tool for agricultural water management with significant advancements in its ability to monitor various physical parameters related to crop health and growth (Ahmed et al., 2022). By leveraging techniques such as geographic information systems (GIS), remote sensing, and sensor networks, farmers can precisely monitor nutrient levels, soil moisture, and crop conditions (Imbernón-Mulero et al., 2023). This data-driven approach enables informed decision-making, allowing for the required targeted application of water, nutrients, and other input, reducing wastage and minimizing the environmental footprints of agriculture practices (Pradipta et al., 2022).

4.6 Crop Selection and Breeding for Stress Tolerance

Semi-arid regions are characterized by harsh environmental conditions including high temperature, drought, and soil salinity which affect crop yield and growth (Navarro-Morillo et al., 2023). Farmers can identify these stress-tolerant crop varieties to increase their chances of achieving consistent and reliable yields even in challenging climatic conditions (Lamba et al., 2023). Through marker-assisted selection, traditional breeding techniques, and genetic engineering approaches, researchers can develop crop varieties better adapted to semi-arid regions'challenging conditions (Rai et al., 2023). These stress-tolerant varieties can help to maintain crop yields and ensure food security.

4.7 Policies and Institutional Framework

Effective policies and robust institutional frameworks are important for facilitating the adoption of sustainable practices and enhancing climate change mitigation efforts (Wichelns, 2004). Governments and policymakers play a crucial role in developing and implementing region-specific regulations and incentives that encourage the adoption of water-efficient technologies to promote smart-climate agriculture and support research and development in precision farming and crop breeding (Laamari et al., 2022). Policymakers must prioritize the formulation of comprehensive policies that provide a conducive environment for the integration of innovative solutions such as precision farming techniques, crop breeding programs, and efficient irrigation systems aimed at developing stress-tolerant varieties (Husaini, 2022). These policies should encompass a range of measures including subsidies, tax credits, financial incentives, and regulatory frameworks that facilitate the widespread adoption of these practices among farming communities (Pérez-Blanco et al., 2020). Robust institutional frameworks are essential for streamlining the implementation of these policies and ensuring coordination among various participants (Maleza & Nishimura, 2007). These frameworks should foster collaboration among research institutions, extension services, and farmers, facilitating knowledge-sharing, capacity building, and the dissemination of best practices (Nhemachena et al., 2017). They should prioritize investments in research and development initiatives focused on advancing precision agriculture technologies, developing drought-resistant crop varieties, and exploring sustainable water management strategies tailored to the unique challenges of semi-arid regions (Hu, 2020).

5 Successful Adaptation and Mitigation Strategies from Semi-Arid Regions

The study of ten arid river basins revealed that as economies develop, agricultural practices tend to shift toward more water-efficient and environmentally friendly approaches with greater emphasis on salinity control and sustainability of the overall system (Shi, 2022). The implementation of adaptation and mitigation strategies in semi-arid regions has yielded numerous success stories that serve as valuable case studies. In regions such as the Sahel in Africa, the adoption of water harvesting techniques, including Zai pits and contour bunds, has significantly improved soil moisture retention and crop yield (Vishwakarma et al., 2021). The integration of traditional water conservation methods like johads (earthen dams) with modern technologies like RS and precision agriculture has enabled more efficient water management and sustainable agricultural practices (Ali et al., 2019).

Sustainable agricultural practices including the development of drought-resistant crop varieties, agroforestry, and conservation agriculture have been adopted to enhance food security and soil health (Hu, 2020). Livestock management strategies such as rotational grazing and alternative fodder sources have been employed to sustain livestock production during dry periods (Ahmed et al., 2022). Ecosystem conservation and restoration efforts including afforestation and reforestation as well as soil and water conservation measures have been undertaken to mitigate desertification and improve ecosystem services (Amede & Whitbread, 2022).

6 Conclusion and Recommendation

Water scarcity, climate change, and soil degradation in semi-arid regions necessitate a comprehensive approach to ensure food security and sustainability of agriculture systems. This chapter has outlined the importance of maintaining and improving irrigation water quality through various strategies such as sustainable water conservation techniques, soil management practices, wastewater treatment and reuse, and integration of renewable energy sources into irrigation systems. These strategies not only contribute to preserving water resources but also help mitigate the deteriorating impact of climate change on irrigation water quality, soil health, and crop yield. This chapter also highlighted the role of crop selection and breeding of stress-tolerant varieties, precision agriculture, and decision support systems of governments and policymakers in enhancing the resilience of agriculture systems in semi-arid regions to changing climatic conditions and maintaining sustainable agriculture. Precision agriculture techniques and data-driven tools can optimize resource utilization, enhance overall productivity, and minimize environmental impacts. Effective implementation of these strategies requires supportive policies and institutional frameworks that prioritize water governance and smart-climate agriculture. Collaborative research efforts that integrate expertise from various fields including water resource management, soil science, climate science, agriculture, and renewable energy technologies need to be encouraged. Comprehensive training programs for farmers, workers, and local populations on sustainable agricultural practices, water conservation techniques, and the learning of new technologies such as remote sensing and GIS systems usage in agriculture and crop breeding programs should be executed. Effective knowledge transfer is essential for ensuring the successful implementation and long-term sustainability of these strategies.

References

Ahmed, M., Hayat, R., Ahmad, M., Ul-Hassan, M., Kheir, A. M., Ul-Hassan, F., et al. (2022). Impact of climate change on dryland agricultural systems: A review of current status, potentials, and further work need. *International Journal of Plant Production, 16*(3), 341–363.

Ali, S., Ma, X., Jia, Q., Ahmad, I., Ahmad, S., Sha, Z., et al. (2019). Supplemental irrigation strategy for improving grain filling, economic return, and production in winter wheat under the ridge and furrow rainwater harvesting system. *Agricultural Water Management, 226*, 105842.

Amede, T., & Whitbread, A. M. (2022). Restoring degraded landscapes and fragile food systems in sub-Saharan Africa: Synthesis of best practices. *Renewable Agriculture and Food Systems, 37*(S1), S1–S3.

Bhatti, M. T., Ashraf, M., & Anwar, A. A. (2021). Soil erosion and sediment load management strategies for sustainable irrigation in arid regions. *Sustainability, 13*(6), 3547.

Bornare, D., Nagarajan, R., & Barge, R. (2018). Improvement of supplementary irrigation water quality for rain-fed agriculture in the semi-arid region using magnetization techniques. *Journal of Water Resource and Protection, 10*(12), 1198.

Chen, L., Feng, Q., Li, C., Wei, Y., Zhao, Y., Feng, Y., et al. (2017). Impacts of aquaculture wastewater irrigation on soil microbial functional diversity and community structure in arid regions. *Scientific Reports, 7*(1), 11193.

Dong, Z. (2021). Study on irrigation pattern in irrigation area of Semi-arid region. *IOP Conference Series: Earth and Environmental Science, 769*(3), 032037.

Dotaniya, M. L., Meena, V. D., Saha, J. K., Dotaniya, C. K., Mahmoud, A. E. D., Meena, B. L., et al. (2023). Reuse of poor-quality water for sustainable crop production in the changing scenario of climate. *Environment, Development and Sustainability, 25*(8), 7345–7376.

Etikala, B., Adimalla, N., Madhav, S., Somagouni, S. G., & Keshava Kiran Kumar, P. L. (2021). Salinity problems in groundwater and management strategies in arid and semi-arid regions. In *Groundwater geochemistry: pollution and remediation methods* (pp. 42–56). Wiley.

Falchetta, G., Semeria, F., Tuninetti, M., Giordano, V., Pachauri, S., & Byers, E. (2023). Solar irrigation in sub-Saharan Africa: Economic feasibility and development potential. *Environmental Research Letters, 18*(9), 094044.

Garrido-Cardenas, J. A., Esteban-García, B., Agüera, A., Sánchez-Pérez, J. A., & Manzano-Agugliaro, F. (2020). Wastewater treatment by advanced oxidation process and their worldwide research trends. *International Journal of Environmental Research and Public Health, 17*(1), 170.

Gebreyess, B. F., & Amare, A. (2019). Water harvesting technologies in semi-arid and arid areas. *Journal of Degraded and Mining Lands Management, 7*(1), 1921.

Golla, B. (2021). Agricultural production system in arid and semi-arid regions. *International Journal of Agricultural Science and Food Technology, 7*(2), 234–244.

Haque, M. A. (2021). Management of saline soil using organic manure and gypsum fertilizer for growing sweet gourd in coastal region of Bangladesh. *Journal of the Bangladesh Agricultural University, 19*(1), 1–7.

Hu, R. (2020). Overview of water-saving irrigation methods in arid/semi-arid areas. *E3S Web of Conferences, 191*, 02001.

Hundal, G. S., Laux, C. M., Buckmaster, D., Sutton, M. J., & Langemeier, M. (2023). Exploring barriers to the adoption of internet of things-based precision agriculture practices. *Agriculture, 13*(1), 163.

Husaini, A. M. (2022). High-value pleiotropic genes for developing multiple stress-tolerant biofortified crops for 21st-century challenges. *Heredity, 128*(6), 460–472.

Ibrahim, H. M., & Al-Turki, A. M. (2020). Assessment of the environmental risk of pesticides leaching at the watershed scale under arid climatic conditions and low recharge rates. *Water, 12*(2), 418.

Imbernón-Mulero, A., Maestre-Valero, J. F., Martínez-Alvarez, V., García-García, F. J., Jódar-Conesa, F. J., & Gallego-Elvira, B. (2023). Evaluation of an autonomous smart system for optimal management of fertigation with variable sources of irrigation water. *Frontiers in Plant Science, 14*, 1149956.

Isa, M. H., Wong, L. P., & Mubarak, N. M. (2023). Advances in removal of toxic substances from wastewater. *Frontiers in Environmental Science, 11*, 1165882.

Izquierdo, J., & Blanchard, R. E. (2020). Solar desalination system design for irrigation/drinking water and electricity generation in desert or arid areas. *Journal of Sustainability Research, 2*(2), e200018.

Kahinda, J. M. M., Rockström, J., Taigbenu, A. E., & Dimes, J. (2007). Rainwater harvesting to enhance water productivity of rainfed agriculture in the semi-arid Zimbabwe. *Physics and Chemistry of the Earth, Parts A/B/C, 32*(15–18), 1068–1073.

Khalid, S., Shahid, M., Natasha, Bibi, I., Sarwar, T., Shah, A. H., & Niazi, N. K. (2018). A review of environmental contamination and health risk assessment of wastewater use for crop irrigation with a focus on low and high-income countries. *International Journal of Environmental Research and Public Health, 15*(5), 895.

Kodirov, D., Muratov, K., Tursunov, O., Ugwu, E. I., & Durmanov, A. (2020). The use of renewable energy sources in integrated energy supply systems for agriculture. *IOP Conference Series: Earth and Environmental Science, 614*(1), 012007.

Laamari, A., Laamrani, H., & Khoali, S. (2022). Smart water governance in Moroccan agriculture: New science and policy collaboration and partnership. *Open Journal of Social Sciences, 10*(9), 11–27.

Lamba, K., Kumar, M., Singh, V., Chaudhary, L., Sharma, R., Yashveer, S., & Dalal, M. S. (2023). Heat stress tolerance indices for identification of the heat tolerant wheat genotypes. *Scientific Reports, 13*(1), 10842.

Lee, T., Jang, W. S., Chun, B., Ahmad, M. J., Jung, Y., Kim, J., & Shin, Y. (2023). Development of irrigation schedule and management model for sustaining optimal crop production under agricultural drought. *Paddy and Water Environment, 21*(1), 31–45.

Li, S., Zhuang, Y., Liu, H., Wang, Z., Zhang, F., Lv, M., et al. (2023). Enhancing rice production sustainability and resilience via reactivating small water bodies for irrigation and drainage. *Nature Communications, 14*(1), 3794.

Li, W., Chen, X., Xu, S., Wang, T., Han, D., & Xiao, Y. (2024). Effects of storm runoff on the spatial–temporal variation and stratified water quality in Biliuhe Reservoir, a drinking water reservoir. *Environmental Science and Pollution Research, 31*, 19556–19574.

Liu, T., Xia, X., Liu, S., Mou, X., & Qiu, Y. (2013). Acceleration of denitrification in turbid rivers due to denitrification occurring on suspended sediment in oxic waters. *Environmental Science & Technology, 47*(9), 4053–4061.

Lu, Q., Feng, Y. Z., Liu, N. Q., Song, W., Huang, Y., Zhang, K., & Yu, Y. J. (2022). Effect of converting natural forests to plantations on the soil bacterial diversity in a subtropical mountainous area: Case study in southern China. *Applied Ecology and Environmental Research, 20*(4), 2831–2844.

Maleza, M. C. E., & Nishimura, Y. (2007). Participatory processes and outcomes: The case of national irrigation system management in Bohol, Philippines. *Irrigation and Drainage: The Journal of the International Commission on Irrigation and Drainage, 56*(1), 21–28.

Melki, A., Abdollahi, K., Fatahi, R., & Abida, H. (2017). Groundwater recharge estimation under semi arid climate: Case of Northern Gafsa watershed, Tunisia. *Journal of African Earth Sciences, 132*, 37–46.

Mir, R. R., Rustgi, S., Zhang, Y. M., & Xu, C. (2022). Multi-faceted approaches for breeding nutrient-dense, disease-resistant, and climate-resilient crop varieties for food and nutritional security. *Heredity, 128*(6), 387–390.

Mohammed, B., Salah, O., Driss, O., & Abdelghani, C. (2020). Global warming and groundwater from semi-arid areas: Essaouira region (Morocco) as an example. *SN Applied Sciences, 2*(7), 1245.

Navarro-Morillo, I., Pardo-Pina, S., Garcia-Sánchez, F., Ruiz, J. M., Laserna-Arcas, S., Plasencia, F., & Cámara-Zapata, J. M. (2023). Corn steep liquor application improves pepper (*Capsicum annum* L.) tolerance to salinity. *Horticulturae, 9*(7), 785.

Nhemachena, C., Matchaya, G., & Nhlengethwa, S. (2017). Strengthening mutual accountability and performance in agriculture in Southern Africa. *South African Journal of Science, 113*(5–6), 1–7.

Nieder, R., Benbi, D. K., Reichl, F. X., Nieder, R., Benbi, D. K., & Reichl, F. X. (2018). Reactive water-soluble forms of nitrogen and phosphorus and their impacts on environment and human health. In *Soil components and human health* (pp. 223–255). Springer.

Ojo, T. O., Baiyegunhi, L. J., Adetoro, A. A., & Ogundeji, A. A. (2021). Adoption of soil and water conservation technology and its effect on the productivity of smallholder rice farmers in Southwest Nigeria. *Heliyon, 7*(3), e06433.

Onah, B. N., Okafor, I. F., Oyigbo, D. N., & Nnate, K. N. (2021). Deployment of energy technologies for sustainable rural agricultural development in Nigeria. *IOP Conference Series: Earth and Environmental Science, 730*(1), 012039.

Oommen, M., Gopakumar, C. S., Varghese, J. J., & Shree, A. R. (2022). Climate variability and change in the spices and plantation cropping systems in Kerala state, India. *Indian Society for Plantation Crops, 50*(3), 169.

Pérez-Blanco, C. D., Hrast-Essenfelder, A., & Perry, C. (2020). Irrigation technology and water conservation: A review of the theory and evidence. *Review of Environmental Economics and Policy., 14*, 216.

Pradipta, A., Soupios, P., Kourgialas, N., Doula, M., Dokou, Z., Makkawi, M., et al. (2022). Remote sensing, geophysics, and modeling to support precision agriculture—Part 1: Soil applications. *Water, 14*(7), 1158.

Priyan, K. (2021). Issues and challenges of groundwater and surface water management in semi-arid regions. In *Groundwater resources development and planning in the semi-arid region* (pp. 1–17). Springer.

Rai, G. K., Khanday, D. M., Kumar, P., Magotra, I., Choudhary, S. M., Kosser, R., et al. (2023). Enhancing crop resilience to drought stress through CRISPR-Cas9 genome editing. *Plants, 12*(12), 2306.

Ricart, S., Olcina, J., & Rico, A. (2019). Introductory chapter: Addressing past claims and oncoming challenges for irrigation systems. In *Irrigation - Water productivity and operation, sustainability and climate change*. IntechOpen.

Sahab, S., Suhani, I., Srivastava, V., Chauhan, P. S., Singh, R. P., & Prasad, V. (2021). Potential risk assessment of soil salinity to agroecosystem sustainability: Current status and management strategies. *Science of the Total Environment, 764*, 144164.

Sani, G. D., Ibrahim, A., Sahabi, S., & Lailaba, B. B. (2019). Renewable energy: Environmental impacts and economic benefits for sustainable development. *International Journal of Engineering Research, 8*(08), 547–555.

Shi, H. (2022). To reduce soil salinity: The role of irrigation and water management in global arid regions across development phases. *arXiv preprint arXiv:2204.02029.*

Singh, A. (2019). Poor-drainage-induced salinization of agricultural lands: Management through structural measures. *Land Use Policy, 82*, 457–463.

Smith, M. D., Sikka, A., Dirwai, T. L., & Mabhaudhi, T. (2023). Research and innovation in agricultural water management for a water-secure world. *Irrigation and Drainage, 72*(5), 1245–1259.

Sraïri, M. T. (2018). Water uses in sustainable agriculture practices: Reconsidering the priorities in water scarce areas. *Advances in Plants and Agriculture Research, 8*(4), 333–334.

Tariq, G. H., Ashraf, M., & Hasnain, U. S. (2021). Solar technology in agriculture. *Technology in Agriculture, 387*, 536.

Torabi Haghighi, A., Abou Zaki, N., Rossi, P. M., Noori, R., Hekmatzadeh, A. A., Saremi, H., & Kløve, B. (2020). Unsustainability syndrome—From meteorological to agricultural drought in arid and semi-arid regions. *Water, 12*(3), 838.

Visconti, F., & de Paz, J. M. (2012). Soil, water and crop management for agricultural profitability and natural resources protection in salt-threatened irrigated lands. In M. Kumar (Ed.), *Problems, perspectives and challenges of agricultural water management* (pp. 293–310). IntechOpen.

Vishwakarma, A., Goswami, A., & Pradhan, B. (2021). Prioritization of sites for managed aquifer recharge in a semi-arid environment in western India using GIS-based multicriteria evaluation strategy. *Groundwater for Sustainable Development, 12*, 100501.

Wichelns, D. (2004). New policies are needed to encourage improvements in irrigation management. *Journal of Irrigation and Drainage Engineering, 130*(5), 366–372.

Xu, H., Liu, Y., Xu, X., Lan, H., Qi, W., Wang, D., et al. (2022). Spatiotemporal variation and risk assessment of phthalate acid esters (PAEs) in surface water of the Yangtze River Basin, China. *Science of the Total Environment, 836*, 155677.

Xu, J., Wang, X., Ding, Y., Mu, Q., Cai, H., Ma, C., & Saddique, Q. (2020). Effects of irrigation and nitrogen fertilization management on crop yields and long-term dynamic characteristics of water and nitrogen transport at deep soil depths. *Soil and Tillage Research, 198*, 104536.

Zhao, Y., Qiu, Y., Mamrol, N., Ren, L., Li, X., Shao, J., et al. (2022). Membrane bioreactors for hospital wastewater treatment: Recent advancements in membranes and processes. *Frontiers of Chemical Science and Engineering, 16*, 634–660.

Assessment of Soil Salinity and Environmental Parameters Variability Using Remote Sensing Indices and Landsat TM–OLI/TIRS Data: A Case Study of River Thamirabarani Basin, South India

S. Kaliraj, Hisham Mohamed, and C. Lakshumanan

Abstract The study investigates soil salinity and its spatiotemporal variability in the Thamirabarani River Basin (TRB) in Southern India to understand its relations with environmental parameters like soil moisture, land surface temperature, and potential evapotranspiration. The multivariate statistical indices were applied to analyze Landsat–OLI/TIRS, ETM+, and TM images for 1997, 2011, and 2022. The result shows the soil salinity index (SSI) value at the rate of 0.405–0.0207 in 1997, and it increased to 0.56 to −0.23 in 2011 while significantly decreasing to 0.26 to −0.44 in 2022, but higher salinity in soils spatially extended in a more extensive part of the basin. In the basin, SSI is directly influenced by land surface temperature (LST), soil moisture, potential evapotranspiration (PET), and rainfall factors (RF). It causes the spatial extent of salt-affected soils in various parts, typically related to the higher LST and PET but poor rainfall rates. An increase in soil salinity reduces the fertility rate and causes erosion and runoff. Besides, soil salinity reduces moisture content by decreasing water holding capacity, retention, and soil-water availability, affecting plant growth. Lower soil moisture can increase salt content in soils to a considerable extent. Overall, the study provides a primary database on spatiotemporal characteristics of soil salinity and its related properties. It has become a vital sustainable soil resource management source in the basin area.

Keywords Soil salinity · Soil moisture · Potential evapotranspiration · Remote sensing indices · GIS techniques · Thamirabarani River Basin

---

S. Kaliraj (✉)
National Centre for Earth Science Studies (NCESS), Ministry of Earth Sciences (MoES), Government of India, Thiruvananthapuram, Kerala, India
e-mail: s.kaliraj@ncess.gov.in

H. Mohamed · C. Lakshumanan
Department of Remote Sensing, Bharathidasan University, Tiruchirappalli, Tamil Nadu, India

A. K. Singh et al. (eds.), *Sustainable Agriculture Management in Semi-Arid Climates*, https://doi.org/10.1007/978-3-031-94062-0_10

1 Introduction

Soil is a vital resource for ecosystems that balance the biogeochemical cycle in the basin environment. In recent decades, soil sources have experienced adverse impacts due to the increase in salinity resulting from natural and anthropogenic factors that cause harmful effects on plant growth, agricultural productivity, and soil and water resource quality (Abbas et al., 2013; Allbed et al., 2017; Dehni & Lounis, 2012; Lalitha et al., 2021; Kaliraj et al. 2022). FAO (2010) report said that 3% of global soil reservoirs are impacted by salinity, especially in arid environments, and it directly affects plant growth and reduces agricultural productivity. Assessing and monitoring soil salinity are vital for managing agricultural productivity and socio-economic development. Assessing soil salinity through traditional methods depends on sampling methods and time-consuming processes. The recent development of remote sensing and GIS techniques provides valuable data sources to evaluate soil salinity at a spatiotemporal scale (Kaliraj et al., 2024a, 2024b). Multispectral images, i.e., Landsat OLI, ETM+, IRS-LISS3, LISS 4, and Sentinel 2-MSI., are widely used for assessment of soil salinity, EC, LST, and soil moisture, using remote sensing indices and other multivariate statistical indices (Elnaggar & Noller, 2020).

Remote sensing indices are widely used to measure soil salinity and other indices parameters like Normalized Difference Vegetation Index (NDVI). NDWI, SVI, Soil Moisture Index (SMI), SSI, and LST using multispectral images (Abbas & Khan 2007; Ivushkin, 2014; Abuelgasim & Ammad 2019; Kaliraj et al., 2024a, 2024b). Soil salinity directly relates to soil EC and is mainly influenced by LST, moisture, and soil physical properties (Azabdaftari & Sunar 2016; Stavi et al., 2021). Soils, with their salt content, influence the spectral properties of the image, and it becomes a pivotal point to measure salinity matter. In a soil layer, salt elements are high, reducing water infiltration and other chemical properties (Zewdu et al., 2017). FAO and UNESCO have reported that soil across various countries faces salt-affected issues due to irrigation schemes that influence secondary salinization and waterlogging. About ten million hectares of irrigated lands in various parts of the world are affected by salinity due primarily to secondary salinization and alkalinization activities (Szabolcs, 1991). Soil salinity is a severe issue in arid and semi-arid regions like Argentina, Egypt, India, Iraq, Pakistan, Syria, and Iran, drastically reducing agriculture productivity (Rhoades et al., 1990). Multispectral data like Landsat TM, ETM+, OLI, IRS-LISS3, LISS4, Sentinel2-MSI, MODIS-Terra, and other images are widely used for the measurement of soil salinity and its related parameters at a spatiotemporal scale, and the derived outcomes show a strong correlation with in-situ measured data. Soil salinity may be increased in various ways, including (1) mixing of salt matter from the parental rocky substance through the weathering process and climate impacts; (2) spatial distribution of salt matter from salt-affected sources by aeolian activities and rainfall-runoff; (3) accumulating salt contents from groundwater during uncontrolled irrigation and inadequate drainage practice, salt may be accumulated in topsoil layer as a result of the evaporation in the near water table areas, the areas occurred deeper water table may be increased soil salinity by enhancing irrigation activities (Douaoui et al. 2006; Allbed & Kumar 2013; Chen et al. 2017; Zhang & Niu 2019; Ghazali et al. 2020; Manickam et al. 2021; Gerardo & de Lima 2022).

The study area covers more extensive agricultural lands with various soil types that have recently faced salinity issues due to human-induced activities and farming practices. It also adversely impacts soil fertility and water quality in shallow water tables. Hence, the present study aims to assess the soil salinity and its related parameters at a spatiotemporal scale using time-series Landsat images in a GIS software environment. The study provides primary data sources for evaluating and mapping soil salinity with their spatial distribution for planning mitigation measures.

2 Study Area

The Thamirabarani River Basin (TRB) lies in the Tamil Nadu state of Southern India and comprises three administrative districts: Tirunelveli, Tenkasi, and Thoothukudi. Figure 1 shows the location of the study area. The geographical extent of the basin covers the longitude of 77°9′5″–78°9′23″E and the latitude of 8°27′7″–9°11′39″N, and the total areal extent is about 5565 km^2. The River Thamirabarani is a perennial river flowing eastward from the Western Ghats, along

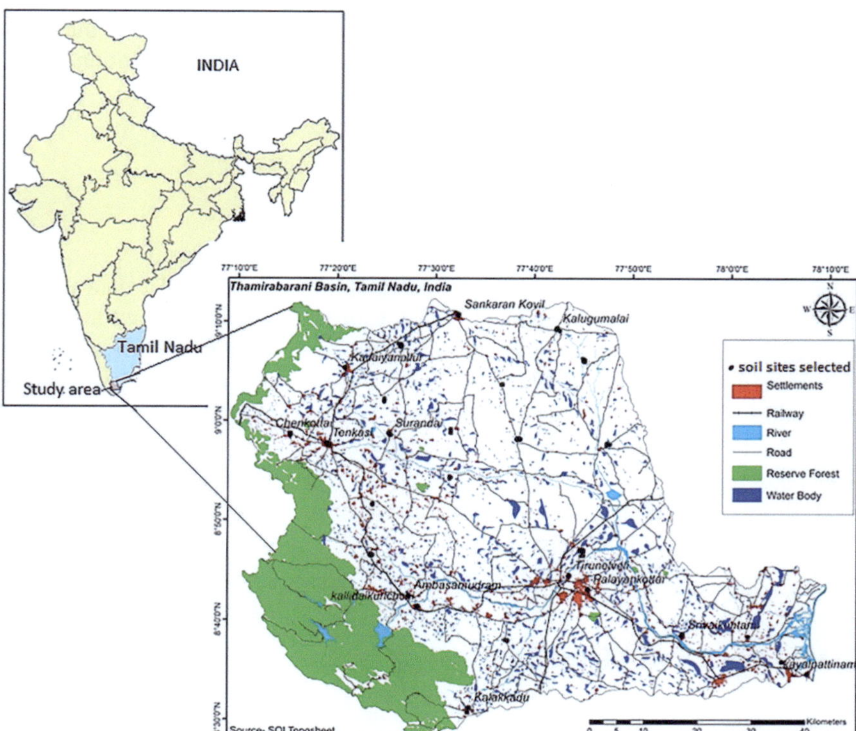

Fig. 1 Location map of the study area

a length of 128 km, and confluences with the Bay of Bengal Sea. The landscape of the TRB consists of upland hills, the middle parts of pediplains and alluvial plains, and the lower coastal plains in the eastern parts. The geological formations consist of crystalline rocks from the western part of the Archean age, and the coastal area comprises sedimentary rocks from the tertiary and quaternary ages. Almost 90% of the TRB comprises crystalline rocks like metamorphosed rocks, Gneisses, granites, basic and acidic intrusive rocks, and Charnockite. The basin covers various landforms, including dissected hills, salt pans, bajadas, dunes, quarries, and mining, deltaic plains, flood plains, and fluvio-marine plains. The TRB comprises various soil types, including loamy sand, clay loam, sandy loam, clay, and sandy clay loam, which spread across the basin at significant areal extents. Loamy sands are predominantly found in northern and northeastern parts, where isolated patches of red loamy, clayey, sandy clay, and clay loam soils are distributed. The agricultural lands cover 57% of the total area and cover almost all soil types in all parts except upland hills, and 15% of shrubs and scrubs are found in isolated patches across the basin. 6.9% of the area is covered with water bodies, including rivers, reservoirs, and ponds. 3.2% is occupied with build-ups, settlements, and other infrastructures. The TRB prevails in sub-tropical to semi-arid climate conditions with an optimum temperature of 23–32 °C. The basin receives an average rainfall of 985.77 mm per year, with higher rainfall found in the western parts, especially during the northeast monsoon, and the basin experiences higher rainfall variability, particularly in the lower basin areas. Notably, the monsoonal rainfall is only a significant irrigation source for central agricultural lands in the basin.

2.1 Landuse/Land Cover (LULC) Features

In the TRB, the LULC features are extracted from Landsat 9-OLI data using a supervised classification technique in ENVI 5.3 software. Figure 2 shows the spatial distribution of LULC features in the basin. About 16% of the forest cover is found in the Western Ghats and other upland hills, 57% of the agricultural lands are spread in the foothills, and middle parts of pediplains, flood plains, and alluvial plains, and 15% of shrubs and scrubs are found in isolated patches along the basin. The area of 6.9% is covered with water bodies, including rivers, reservoirs, channels, tanks, and ponds. Build-up settlements and other infrastructures occupy an area of 3.2%. The LULC details are given in Table 1.

3 Materials and Methods

In this analysis, the multi-temporal images of Landsat-9 OLI-2/TIRS, Landsat-7 ETM+, and Landsat-5 TM were used to estimate soil parameters like soil salinity, LST, moisture, erosivity, and PET using remote sensing indices and multivariate

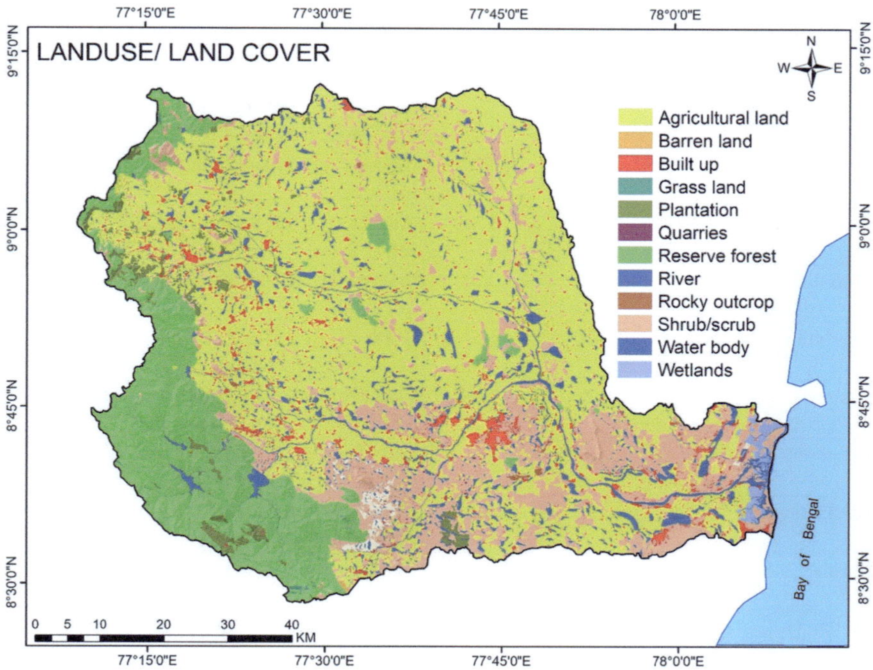

Fig. 2 Landuse/landcover (LULC) features of the study area

statistical indices in the GIS software environment. The SRTM-DEM (30 m) extracts parameters like slope and drainage patterns. The LULC features were extracted using the multispectral bands of Landsat images, and LST was estimated using TIR bands. The rainfall is estimated by the IMD (Indian Meteorological Department) for the years 1990–2020 using the GIS-IDW technique. Soil type is prepared using the soil map of Tamil Nadu Agriculture Department, and the geology and geomorphology layers were extracted using the GSI published resource map. Figure 3 shows the methodological flow chart used for this study.

3.1 Image Pre-processing

Multispectral bands of Landsat–OLI, ETM+, and TM images were systematically processed using ENVI software tools to enhance their spatial and spectral characteristics for soil salinity measurement. Notably, the SSI and other soil properties were estimated in the bare-soil areas after masking the non-soil features like water bodies, forest cover, and artificial features (Jiang et al. 2020; Ghada & Balázs 2021). The NDVI is estimated to demarcate the vegetative cover of the basin area. Firstly, suitable spectral bands are selected to execute the indices based on the

Table 1 The LULC features with its associated soil types in the TRB

LULC feature	Area (km^2)	Area (%)	LULC-associated soil types
Agriculture land	3209	57	Sandy loam, clay, clay loam, loamy, sandy clay, and loamy sand
Barren land	4	0.07	Coarse gravelly soil, sandy loam, and clay loam
Built up	183	3	Sandy clay, loamy sand, clay, clay loam, and sandy loam
Grassland	5	0.08	Sandy clay, lateritic soil, gravelly soil, and hilly soils
Plantation	90	1.5	Sandy clay, sand and gravel, loamy sand, clay loam, and sandy loam
Quarries	2	0.03	Coarse gravelly soil, stony waste, and clay soil
Reserve forest	890	16	Hilly soils, coarse gravelly soil, lateritic soil, sandy clay, and clay soil
Rocky outcrop	12	0.2	Clay, clay loam, loamy sand, sand and gravels, and sandy loam
Shrub/scrub	837	15	Loamy sand, coarse gravelly soil, lateritic soil, and clay
Wetlands	29	0.5	Sand silt clay, clay, and sandy clay
Waterbody (reservoir, lakes, and ponds)	305	5	–
River/channels	89	1.5	–

corresponding index equations, as different indices are constructed to analyze specific bands. Next, data pre-processing is conducted to correct for atmospheric interference, sensor artifacts, and geometric distortions. Once the data is refined, the indices are calculated using mathematical formulas that consider the reflectance values from the chosen spectral bands. For instance, the salinity index calculates the salinity over the soil by comparing the spectral bands. After index computation, the results are often classified into thematic maps or interpreted for actionable insights, aiding decision-making processes in agriculture, land management, and environmental assessments. Remote sensing indices are powerful tools that enable extracting valuable information from satellite or aerial imagery through a sequence of well-defined steps, ultimately contributing to informed and sustainable resource management.

3.2 Normalized Difference Vegetation Index (NDVI)

NDVI results from the band ratio used to quantify vegetation greenness and density and assess changes in plant health. NDVI uses NIR and red bands in its formula:

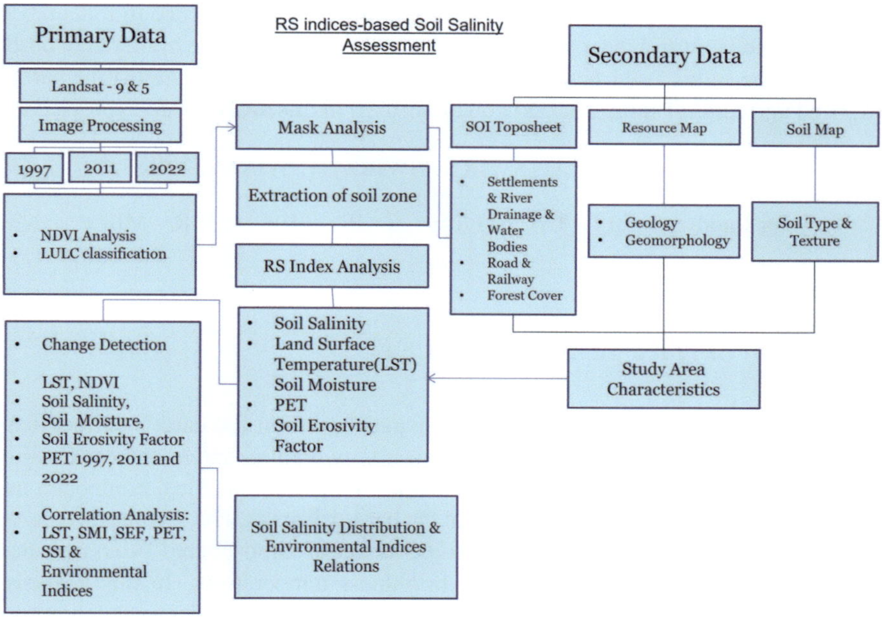

Fig. 3 The flowchart shows the methodology used in this study

$$\text{NDVI} = (\text{NIR} - \text{RED}) / (\text{NIR} + \text{RED}) \tag{1}$$

NDVI measures the ratio (−1 to +1) between R and NIR to reflect vegetation (chlorophyll). Wherein the lower NDVI (<0.1) indicates the non-vegetative features and the higher NDVI (>0.3) indicates the occurrence of vegetation. Thus, NDVI shows the spatial distribution of vegetative features that reflect their health conditions. At the same time, it is used to extract the bare soil zones from the basin by masking vegetative cover in the images.

3.3 Soil Salinity Index (SSI)

Soil salinity index (SSI) is a measure of the salt concentration in soils, and it is estimated using multispectral bands of Landsat OLI, ETM+, and TM images (30 m) for the years 1997, 2011, and 2022. Landsat images were collected from the USGS-Earth Explorer web portal, and the spectral bands of VNIR (Visible-Near infrared) and short-wave infrared (SWIR) were used to execute the SSI equation using map-algebra tool in ArcGIS software (Kaliraj et al., 2024a, 2024b). This analysis estimates SSI using the equations (2 and 3) adopted from the Fouad Al-Khaier (2003).

Thus, the bands of SWIR 1 & 2 showed a Landsat 9 OLI difference in reflectance pattern in the landuse-landcover feature, in which salinity may be higher than other features. The index used has an accurate detection for overall salinity in barely exposed agricultural soils. The SSI equations used in this analysis are expressed as:

$$\text{For Landsat 9} - \text{OLI} : (\text{SWIR1} - \text{SWIR2}) / (\text{SWIR1} + \text{SWIR2}) \quad (2)$$

$$\text{For Landsat 5 and 7} - \text{TM and ETM+} : (\text{NIR} - \text{MIR}) / (\text{NIR} + \text{MIR}) \quad (3)$$

3.4 *Land Surface Temperature (LST)*

The LST is measured using Landsat TIR bands based on the earth's surface features' thermal emissivity properties (brightness temperature) and in degrees Celsius or Fahrenheit. Soil properties, i.e., soil texture, moisture, chemical compositions, and salt matters, directly influence LST in the bare-soil areas. The TIR bands receive heat emissivity of EMR, which is quantified as top-of-atmosphere (TOA) radiance measurements and Top-of-Atmosphere brightness temperatures. In this analysis, the LST is at a higher rate in the salt-affected soils and indicates its spatial relations across the basin. LST is calculated using the Landsat TIS bands by executing the following equations,

1. TOA Radiance L⅄

$$= \text{Multi band Radiance} * \text{Qcal} + \text{Radiance Add band} \quad (4)$$

2. Brightness to Temperature

$$= \text{BT} = K2 / \ln((K1 / l⅄) + 1) - 272.15 \quad (5)$$

   K1 and K2 from metadata (for Band 10 & Band 11) Landsat-9 and (Band 6) Landsat-5
3. Cell Statistics
4. Vegetation proportion (e) by

$$(\text{NDVI} - \text{NDVImin} / \text{NDVImax} - \text{NDVImax})\text{\^{}}2 \quad (6)$$

5. Land Surface Temperature

$$\text{BT} / (1 + (⅄ * \text{BT} / \rho) * \ln(\text{e})) \quad (7)$$

3.5 Soil Moisture Index (SMI)

Soil moisture indicates the soil-water content that occurs in the soil layer, and it is spatially correlated with LST and NDVI at a site-specific scale. The SMI is estimated based on the proportion between current soil moisture and permanent wilting point to the field capacity and the residual soil moisture. The index values range from 0 to 1, with 0 indicating extreme dry conditions and 1 indicating extreme wet conditions. The SMI is estimated using eq. 8 and is expressed as,

$$\left(\mathrm{LST}_{\max} - \mathrm{LST}\right) / \left(\mathrm{LST}_{\max} - \mathrm{LST}_{\min}\right) \tag{8}$$

3.6 Rainfall Erosivity Factor (R)

The R factor indicates the holding capacity of soils against rainfall—volume, duration, and intensity in an area and time (Wischmeier & Smith, 1978). The R factor is an indicator of erosivity and is calculated using IMD-derived rainfall data. The R factor is one of the elements in the RUSLE estimation of soil erosion, and it quantifies the effect of raindrops on it (Renard 1997). This analysis calculates the R factor using decadal average rainfall data (1990–2022) of 22 rain gauges in the Thoothukudi and Tirunelveli districts. The GIS-based IDW method is used to interpolate the spatial distribution of rainfall in the basin. The R factor is calculated using the equation is expressed as,

$$R = \left(0.562 \times P\right) - 8.12 \tag{9}$$

where R refers to the decadal erosivity factor (MJcm/ha/year), and P refers to the decadal average rainfall (cm).

3.7 Potential Evapotranspiration (PET)

The PET indicates the spatial relation of LST, soil moisture (soil-water content), and NDVI in the area. It measures the ability of the atmosphere to remove water through evapotranspiration (conversion of water into gaseous form) processes. The PET is estimated using the Global Potential Evapotranspiration (Global-PET) dataset collected from Global Data Assimilation System (GDAS) analysis fields, and it provides high-resolution grid data on evapotranspiration and rainfall deficit. While PET increases in the basin during summer and decreases in the monsoons. Higher PET was observed from April to July, possibly due to the very high temperatures in these months.

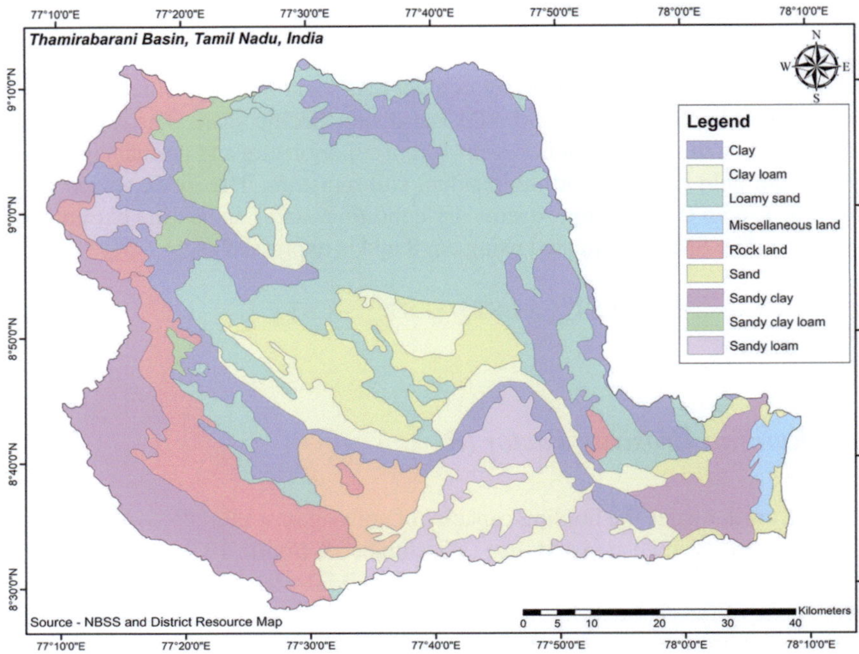

Fig. 4 The soil type of TRB and its spatial distribution

4 Results and Discussion

The TRB consists of various soil types: clay, clay loam, loamy sand, miscellaneous sand, sand, sandy loam, and sandy clay loam. Figure 4 shows the soil types distributed in the TRB. The loamy sand soils spread over the middle part of pediplains and are associated with sandy clay and clay loam soils. Clay soils are found in the northeastern parts, and clay loam is primarily found along the river, minor tributaries, and depositional side landforms. Sandy loam is identified south of the basin along a tributary, which later connects with the river channel. The west side of the basin consists of loamy sand, and the coastal belt in the eastern part consists of sand and sandy clay, majorly due to its fluvio-marine activities.

4.1 Assessment of Soil Salinity Index

Figure 5a–c shows the oil salinity index maps with their spatial distribution in 1997, 2011, and 2022. The salinity is high in flood plain and water bodies where the fluvial action must deposit it; salinity has different types of origin and may be accumulated in various ways, such as over evapotranspiration than precipitation rate, the

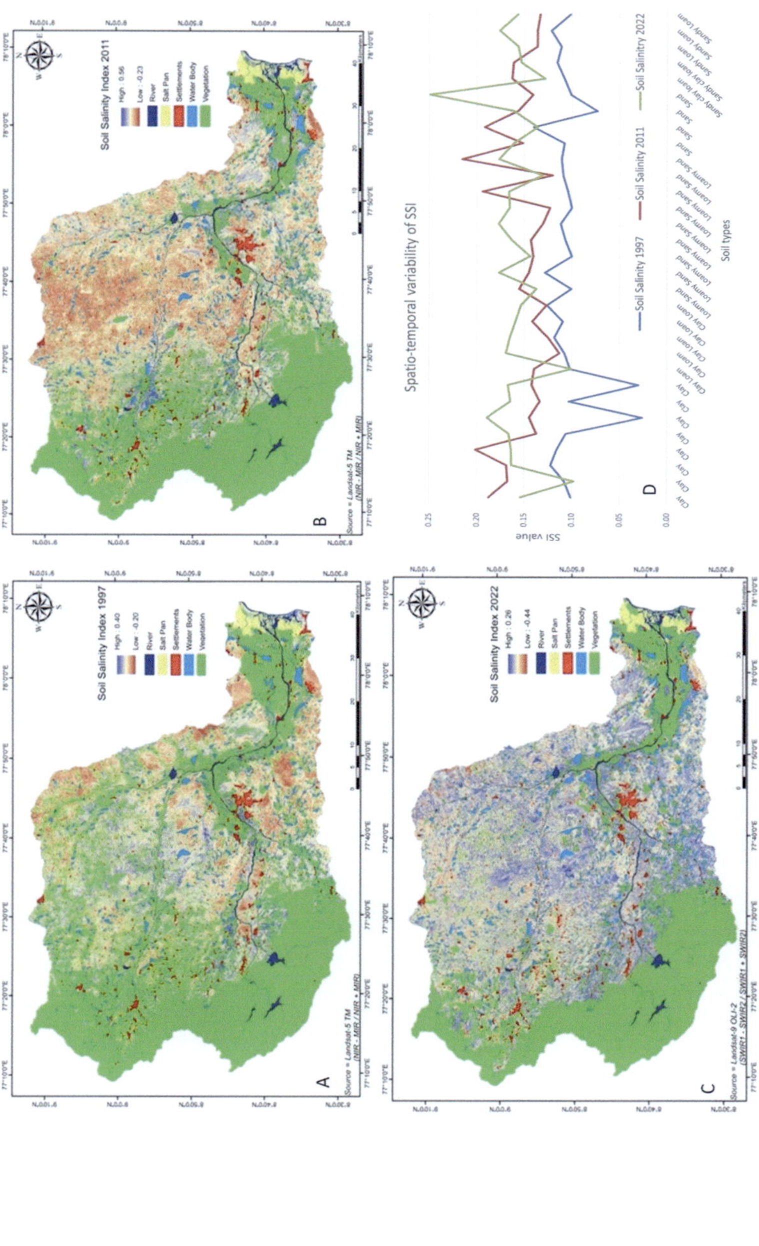

Fig. 5 Spatial distribution of SSI in 1997 (**a**), 2011 (**b**), 2022 (**c**), and SSI spatiotemporal changes in TRB soils (**d**)

addition of fertilizers for agricultural use, erosion, and deposition process, weathering of saline rocks in the hilly regions. Figure 5d illustrates the spatiotemporal changes of SSI in the different soil types of the basin.

Table 2 Soil Salinity Index—spatiotemporal changes in different soil types

Site no.	Soil type	Soil salinity 1997	Soil salinity 2011	Soil salinity 2022
1	Clay	0.101	0.188	0.154
2	Clay	0.108	0.167	0.098
3	Clay	0.122	0.168	0.163
4	Clay	0.116	0.201	0.163
5	Clay	0.106	0.137	0.166
6	Clay	0.026	0.145	0.189
7	Clay	0.103	0.133	0.164
8	Clay	0.030	0.142	0.167
9	Clay loam	0.102	0.140	0.100
10	Clay loam	0.108	0.113	0.169
11	Clay loam	0.118	0.125	0.165
12	Clay loam	0.109	0.141	0.159
13	Clay loam	0.126	0.126	0.153
14	Loamy sand	0.101	0.155	0.137
15	Loamy sand	0.128	0.145	0.176
16	Loamy sand	0.100	0.140	0.144
17	Loamy sand	0.107	0.143	0.163
18	Loamy sand	0.112	0.130	0.177
19	Loamy sand	0.100	0.123	0.165
20	Loamy sand	0.104	0.194	0.166
21	Loamy sand	0.109	0.120	0.131
22	Loamy sand	0.111	0.216	0.177
23	Sand	0.108	0.152	0.161
24	Sand	0.140	0.192	0.136
25	Sand	0.073	0.156	0.161
26	Sand	0.101	0.141	0.249
27	Sandy clay loam	0.108	0.163	0.128
28	Sandy clay loam	0.115	0.161	0.153
29	Sandy loam	0.111	0.136	0.156
30	Sandy loam	0.122	0.136	0.176
31	Sandy loam	0.102	0.133	0.157

4.2 *Spatiotemporal Changes of Soil Salinity Index*

The spatiotemporal changes of SSI are estimated using the output maps of SSI for the years 1997, 2011, and 2022. From these data, spatiotemporal variability of the SSI was extracted from 31 sites with different soil types in the TRB. Table 2 shows the SSI and its characteristics subject to soil types in site-specific areas. Figure 5d indicates a line graph that shows attribute values of SSI with its corresponding soil types across the basin area. It indicates the gradual increase in soil salinity index values during these periods.

4.3 *Assessment of Land Surface Temperature (LST)*

The LST is observed with a gradual increase in temperature of 5 °C; this could be due to a reduction in precipitation over 30 years and global temperature rise, salinity, anthropogenic activities, or global temperature rise (greenhouse effect). In this study area, the LST is at a lower rate in the wetland area water bodies due to higher soil moisture. Figure 6a shows the spatial distribution of LST during the years 1997, 2011, and 2022; Fig. 6d illustrates the spatiotemporal changes of LST in the different soil types of the basin. Table 3 shows the LST and its characteristics subject to soil types in site-specific areas.

4.4 *Assessment of Soil Moisture Index (SMI)*

The SMI depends on the LST and NDVI values spread over the TRB. The resulting map shows that the higher SMI is estimated along the river channels associated with loamy, clay loam, and sandy clay loam soils. In contrast, the lower SMI is found in larger areas of clay soils, mainly in the northern and northeastern parts. The LULC features found that the saline soils have low porosity due to filling fine saline particles in the pore spaces, low infiltration and storage capacity, and high runoff and erosion. The index value falls between 0 and 1, where an SMI value of 1 indicates higher moisture, while zero is lower moisture (Fig. 7).

4.5 *Assessment of Rainfall Erosivity (R)*

The R factor is estimated using decadal average rainfall data to show its erosivity characteristics at the site-specific scale. R factors indicate the rate of soil erosion, which is directly influenced by soil salinity, LST, moisture, and rainfall (Jain et al., 2001; Dabral et al., 2008). Figure 8a–c shows the spatial distribution of erosivity characteristics (*R*-factor) between 1997, 2011, and 2022. Figure 8d illustrates the

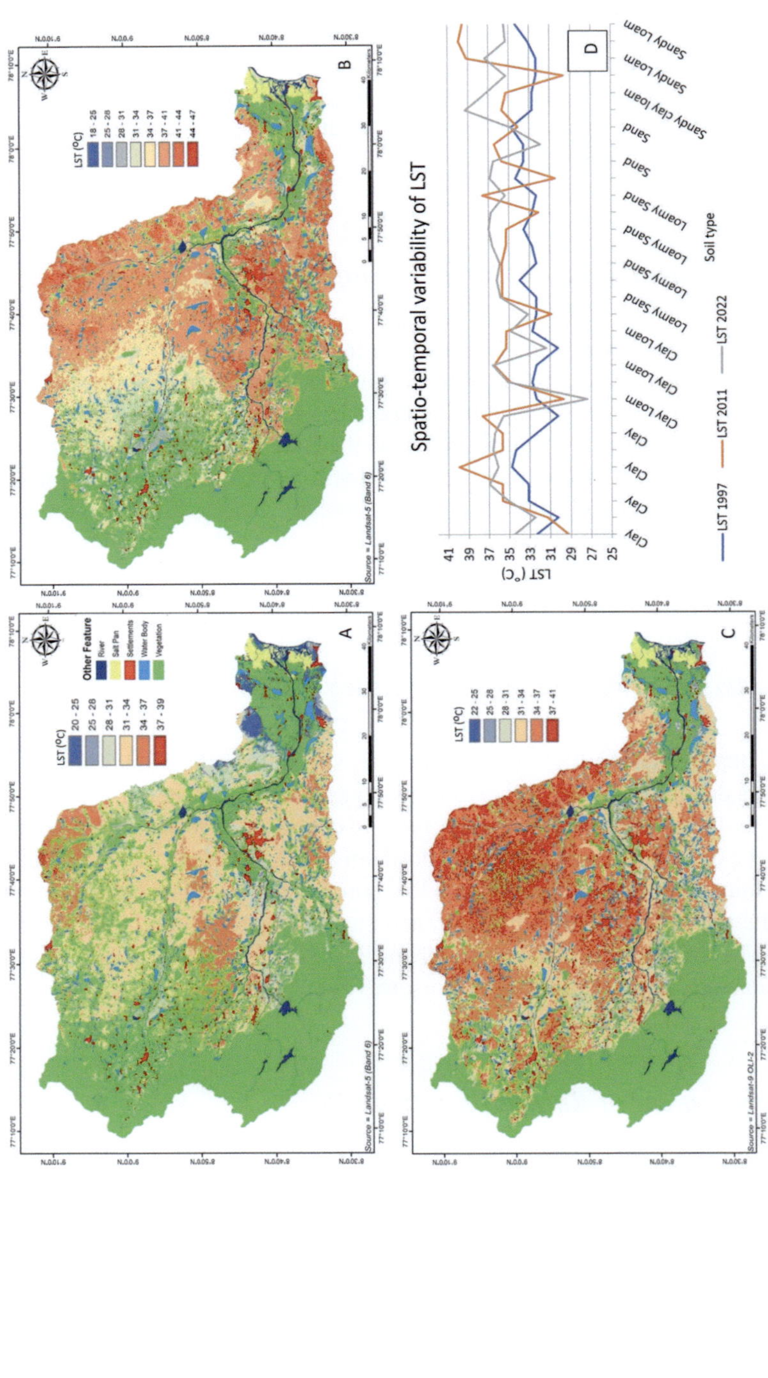

Fig. 6 Spatial distribution of LST in 1997 **(a)**, 2011 **(b)**, 2022 **(c)**, and LST spatiotemporal changes in TRB soils **(d)**

Table 3 Land surface temperature values collected in different soil types

Site no.	Soil type	LST 1997 (°C)	LST 2011 (°C)	LST 2022 (°C)
1	Clay	32.332	29.288	34.493
2	Clay	30.264	31.316	32.506
3	Clay	33.147	35.683	35.105
4	Clay	33.148	35.681	36.955
5	Clay	34.768	39.930	36.074
6	Clay	34.371	35.691	36.598
7	Clay	32.329	35.694	36.402
8	Clay	30.268	37.636	35.585
9	Clay loam	32.330	29.705	27.394
10	Clay loam	32.741	34.910	35.100
11	Clay loam	32.330	36.471	36.629
12	Clay loam	30.265	35.311	31.398
13	Clay loam	32.741	35.296	34.865
14	Loamy sand	32.329	30.916	33.159
15	Loamy sand	32.330	35.681	35.827
16	Loamy sand	33.961	35.682	36.210
17	Loamy sand	32.331	35.699	35.911
18	Loamy sand	32.740	35.300	36.625
19	Loamy sand	33.557	35.287	36.987
20	Loamy sand	32.741	32.124	36.779
21	Loamy sand	32.330	37.639	35.385
22	Loamy sand	34.369	30.508	36.864
23	Sand	33.558	35.699	36.540
24	Sand	33.557	36.460	31.924
25	Sand	34.774	34.113	34.866
26	Sand	32.742	35.693	39.289
27	Sandy clay loam	32.738	35.289	37.197
28	Sandy clay loam	32.331	29.693	35.297
29	Sandy loam	32.331	39.174	37.348
30	Sandy loam	33.149	39.933	35.304
31	Sandy loam	34.370	39.557	35.464

spatiotemporal changes of LST in the different soil types of the basin. The *R* factor is at a higher rate in areas having higher SSI and lower SMI, and the lowest value influences the semi-arid climatic conditions of the northeastern part. Table 3 shows the erosivity (*R*-factor) and its characteristics subject to soil types in site-specific areas.

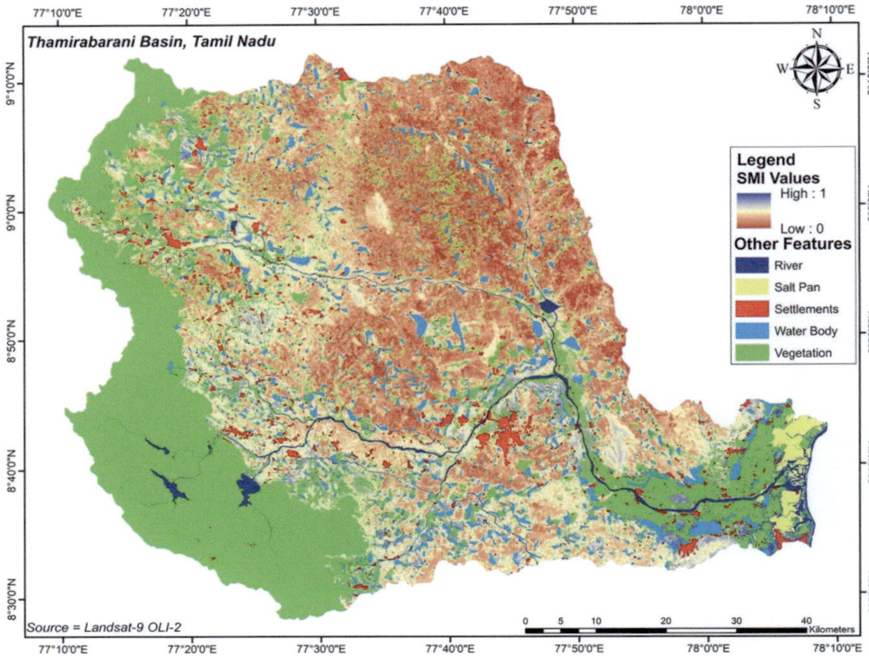

Fig. 7 Spatial distribution of Soil Moisture Index in 2022

4.6 Assessment of Potential Evapotranspiration (PET)

The potential evapotranspiration (PET) rate varies in soil types based on soil moisture, LST, and NDVI. A higher PET rate is observed in clay, loamy sand, and sandy loam soils in middle and northeastern pediplains and coastal areas. However, the uplands and foothills were found to lower PET rates. Figure 9 shows the spatial distribution of potential evapotranspiration (PET) in the basin. The temperatures increased with LULC, and soil types typically accelerated evaporation from the soil's surface and plant transpiration. Because there is more energy available for water vaporization in regions with high solar radiation, such as those with low cloud cover, there is a higher potential for extraterrestrial life. Spatiotemporal changes of R-factor (erosivity) rate in different soils in the TRB are given in Table 4.

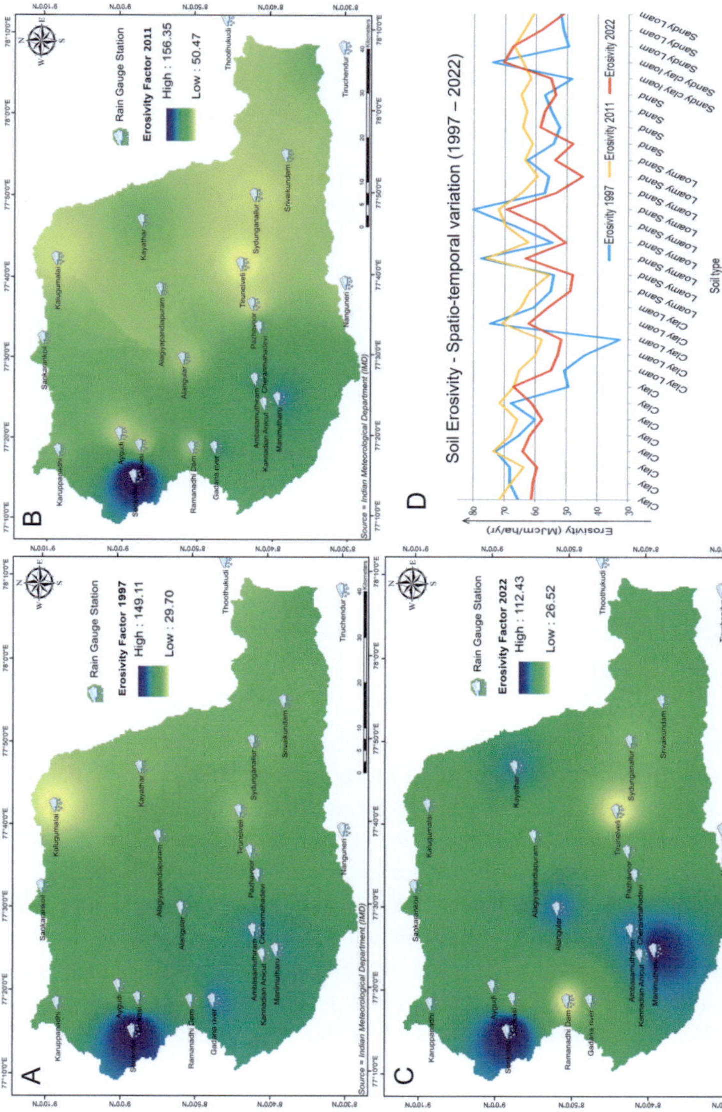

Fig. 8 Spatial distribution of erosivity in 1997 (**a**), 2011 (**b**), 2022 (**c**), and R Factor spatiotemporal changes in TRB soils (**d**)

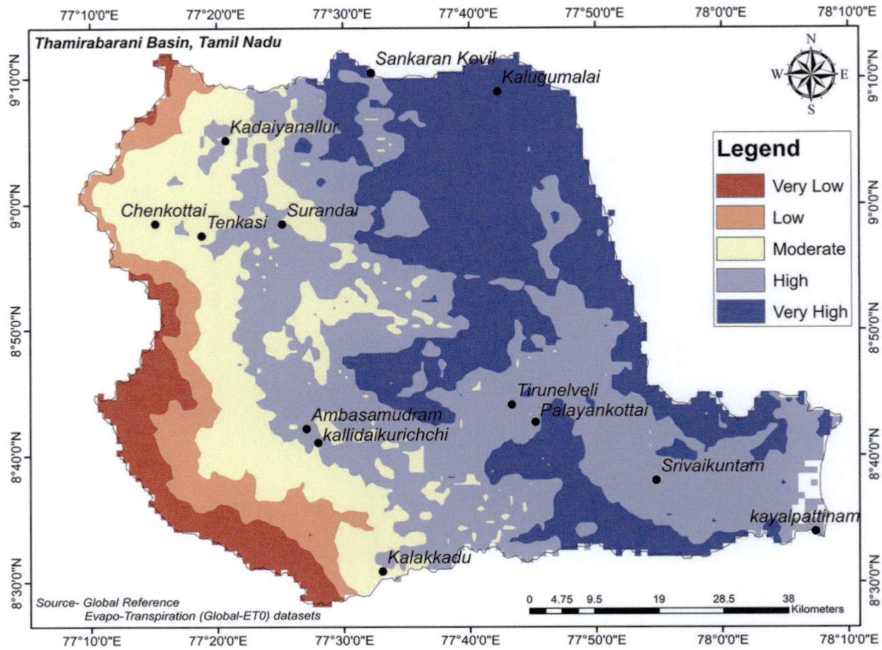

Fig. 9 Spatial distribution of potential evapotranspiration in the study area

4.7 Spatial Relation of Potential Evapotranspiration (PET) and Soil Salinity Index (SSI)

The combined process of evaporation from the soil surface and transpiration from plants, known as evapotranspiration (ET) potential, can be affected by the salinity of the soil. Salinity levels in the soil can also change due to variations in evapotranspiration. High evapotranspiration rates can encourage water to move upward, which can cause salt to build up near the soil surface and raise soil salinity (Sahbeni 2021; Yu et al. 2018; Nguyen et al. 2020; Manickam et al. 2021). On the other hand, low evapotranspiration rates, brought on by insufficient water supply or plant water stress, can restrict salt leaching, potentially escalating soil salinity problems. Thus, Saline soils may have a decreased capacity to hold water, resulting in lower soil moisture levels and a possible impact on evaporation rates (Khan et al. 2005; Zheng et al. 2009).

4.8 Spatial Relation of SSI, LST, SMI, and R Factor

Spatial relation among the Soil Salinity Index (SSI), land surface temperature (LST), Soil Moisture Index (SMI), and R-factor (Erosivity) is explored by comparative analysis of 31 site values from their derived outputs for the years 1997, 2011, and 2022.

Table 4 Spatiotemporal changes of R-Factor (erosivity) rate in different soils in the TRB

Site no.	Soil type	Erosivity 1997	Erosivity 2011	Erosivity 2022
1	Clay	65.65	71.52	61.37
2	Clay	68.21	67.66	60.81
3	Clay	68.22	63.70	59.44
4	Clay	72.87	73.53	64.13
5	Clay	65.43	68.10	61.77
6	Clay	60.07	65.88	57.78
7	Clay	67.93	71.51	61.27
8	Clay	49.39	66.85	67.14
9	Clay loam	50.72	65.25	54.97
10	Clay loam	44.09	59.98	52.94
11	Clay loam	32.71	57.92	51.67
12	Clay loam	74.65	70.68	62.24
13	Clay loam	60.59	64.73	55.96
14	Loamy sand	55.19	61.95	48.91
15	Loamy sand	54.12	55.68	48.18
16	Loamy sand	77.57	76.01	63.25
17	Loamy sand	54.64	62.46	50.43
18	Loamy sand	66.92	69.77	57.59
19	Loamy sand	80.23	72.05	69.36
20	Loamy sand	57.22	64.60	53.52
21	Loamy sand	55.89	59.45	44.90
22	Loamy sand	62.78	63.60	53.85
23	Sand	54.40	61.14	48.06
24	Sand	52.37	64.34	58.58
25	Sand	55.11	62.92	57.18
26	Sand	57.12	64.79	53.67
27	Sandy clay loam	48.52	61.71	55.24
28	Sandy clay loam	74.07	64.32	70.41
29	Sandy loam	49.62	67.12	67.41
30	Sandy loam	51.20	63.70	58.36
31	Sandy loam	52.08	60.76	51.16

The sample points were collected from site-specific soils and correlated with the estimated SSI, SMI, LST, and erosivity. Figures 10 and 11 show the spatial relation of SSI, SMI, LST, and erosivity in different soil types of the basin. The attribute values of these parameters in 31 sites of the different soil types were compared to estimate the spatial relations of SSI, SMI, LST, and erosivity characteristics. The attribute data were plotted as line graphs A, B, and C that denote its spatial variability in 1997, 2011, and 2022, respectively. The salinity of the soil can affect how quickly it evaporates and how much water is present in the soil, both of which affect soil temperature (Gao et al. 2018). Salinity levels that are too high can speed up soil surface evaporation. The energy-intensive evaporation process can cause the soil to cool. The PET

Fig. 10 The spatial relation of SSI, SMI, LST, and erosivity in different soil types in 1997 (**a**) and 2011 (**b**)

rate decreased in saline soils with less water availability, leading to higher soil temperatures. In comparison, soil salinity and temperature showed complexity in spatial relations due to the influence of various dynamic factors like rainfall, crop management practices, and human-induced activities (Zhu 2001; Al-Khakani 2018).

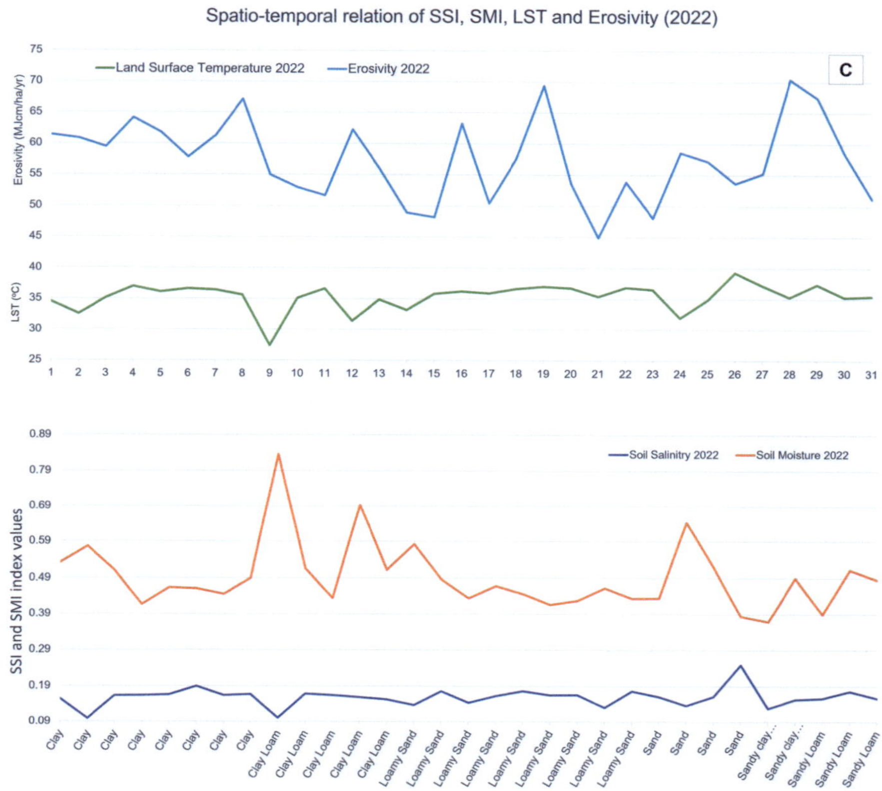

Fig. 11 The spatial relation of SSI, SMI, LST, and erosivity with the basin's different soil types in 2022

Increasing salt content in different soils at various sites significantly alter soil properties, moisture, and permeability, and excessive salt buildup can lead to compacted soils with smaller pore spaces. Meanwhile, the lower soil moisture content resulting from poor infiltration and surface runoff causes soil erosion in the basin area.

5 Discussion

This study describes the spatiotemporal relations among soil salinity (SSI) maps, land surface temperature (LST), and erosivity in various sites of different soil types for the years 1997, 2011, and 2022. Soil salinity significantly impacts soil health and agricultural productivity. High salinity levels can hinder plant growth and degrade soil. The dataset reveals SSI trends, showing an increase at site one from 0.101 in 1997 to 0.188 in 2011, then a slight decrease to 0.154 in 2022. At site 4, SSI rose from 0.116 in 1997 to 0.201 in 2011 and stabilized around 0.163 in 2022. Various factors, such as soil moisture, vegetation cover, and climate change, influence LST, which is crucial in understanding local climatic conditions and ecosystem stress (Wang et al. 2020). LST values varied considerably across sites and over the years. Site 1's LST decreased from 32.332 in 1997 to 29.288 in 2011 and then increased to 34.493 in 2022. Similarly, at site 4, LST consistently increased from 33.148 in 1997 to 36.955 in 2022.

Erosivity measures the potential for soil erosion due to rainfall and runoff, which is vital in assessing soil conservation and land management practices. Erosivity values generally exhibited a decreasing trend from 1997 to 2022. Site 1's erosivity in the basin decreased from 65.65 in 1997 to 61.37 in 2022, peaking in 2011 at 71.52.

Moreover, site 4's erosivity increased from 65.43 in 1997 to 73.53 in 2011, then decreased to 64.13 in 2022. The spatial correlation between soil salinity, LST, and erosivity highlights that higher soil salinity affects soil moisture content, influencing LST. Saline soils retain less moisture, potentially leading to higher surface temperatures. Increased soil salinity can lead to soil structure breakdown, making it more susceptible to erosion. However, erosivity affects soil nutrients and plant growth through the availability of sufficient rainfall. Significant changes in these parameters lead to adverse environmental impacts by increasing soil salinity and reducing agricultural productivity in the basin (Kholdorov et al. 2022; Wang et al. 2020). Effective management strategies, like improved irrigation practices and soil amendments, are essential to mitigate these effects. Monitoring soil salinity and adopting sustainable practices for managing soil and land resources is necessary. Maintaining soil health through organic matter addition and reduced tillage can also help mitigate the negative impacts of salinity and erosion. It is noted that the spatial variability of LST, soil moisture, and PET in the basin is mainly due to climate change, rainfall variations, and human-induced activities (Kholdorov et al. 2022). Wherein the LST change and its spatial trends influence soil moisture, salinity, and physical properties, they significantly impact local agricultural productivity and ecosystems. Overall, the study's findings provide valuable insights into soil salinity, LST, and erosivity trends across different sites and soil types in the TRB, and the outputs show their spatial patterns to address monitoring and adaptive measures.

6 Conclusion

This study explored the spatiotemporal changes of SSI, LST, PET, soil moisture, and erosivity for 1997, 2011, and 2022 using Landsat ETM+. OLI and Sentinel 2A–MSI multispectral images. It is observed that the soil salinity and temperature were increased in different soil types in the various parts of the basin. The relation between SSI and LST indicates that soils with higher salinity mainly exhibit higher LST due to altering soil properties and increased PET rates. The spatial relation of SSI and SMI indicates the indirect proportional relationship between soil salinity and moisture. Salinity affects water availability and retention, reducing soil moisture and impacting plant growth. Meanwhile, the area found in lower moisture levels can accumulate salt and increase soil salinity. The spatial relation of SSI and *R*-factor shows a strong correlation between higher SSI and erosion rate, which is mainly distributed in soils with lower SMI. Saline soils often have poor structure and permeability, leading to reduced water infiltration and increased surface runoff, which can enhance erosion. Additionally, vegetation cover may be compromised, leaving soil vulnerable to erosion. Overall, the TRB faces drastic changes in soil salinity and its physical properties due to the impacts of these interconnected factors. Hence, this study provides the primary database to implement effective management strategies to mitigate and measure the impacts of soil salinity, promote soil health, and ensure sustainable management practices.

Acknowledgments The corresponding author thanks the Director, National Centre for Earth Science Studies (NCESS), Ministry of Earth Sciences (MoES), Government of India, Thiruvananthapuram, for continuous support and encouragement to complete this successful research work. The corresponding author thanks Dr. K. Maya, Scientist–G, Group Head, BGG, NCESS, for continuous support. The authors thank the anonymous reviewers for their insightful comments and suggestions. The authors acknowledge the IMD, GSI, SOI, USGS–Earth Explorer, and EU–Copernicus Data Space Ecosystem for providing the necessary data sources.

References

Abbas, A., & Khan, S. (2007). *Using remote sensing technology to appraise irrigated soil salinity* (pp. 2632–2638).

Abbas, A., Khan, S., Hussain, N., Hanjra, M. A., & Akbar, S. (2013). *Characterizing soil salinity in irrigated agriculture using a remote sensing approach* (Vol. 55–57, pp. 43–52). Physics and Chemistry of the Earth, Parts A/B/C.

Abuelgasim, A., & Ammad, R. (2019). Mapping soil salinity in arid and semi-arid regions using Landsat 8 OLI satellite data. *Remote Sensing Applications: Society and Environment, 13*, 415–425.

Fouad Al-Khaier, Soil salinity detection using satellite remote sensing, , 2003.

Al-Khakani, E. T. (2018). Using Landsat 8 OLI data to predict and map soil salinity for part of An-Najaf governorate. Research Gate.

Allbed, A., & Kumar, L. (2013). Soil salinity mapping and monitoring in arid and semi-arid regions using remote sensing technology: A review. *Advanced Remote Sensing, 2*, 373–385.

Allbed, A., Kumar, L., & Sinha, P. (2017). Soil salinity and vegetation cover change detection from multi-temporal remotely sensed imagery in Al Hassa Oasis in Saudi Arabia. *Geocarto International, 33*, 1–17.

Azabdaftari, A., & Sunar, F. (2016). Soil salinity mapping using multitemporal landsat data. *The International Archives of the Photogrammetry, Remote Sensing and Spatial Information Sciences, XLI-B7*, 3–9.

Chen, Z., Zhang, Y., & Liu, X. (2017). Mapping and analyzing soil salinity using remote sensing and geostatistics in an arid area of China. *Journal of Arid Land, 9*(5), 639–652.

Dabral, P. P., Baithuri, N., & Pandey, A. (2008). Soil erosion assessment in a hilly catchment of North Eastern India using USLE, GIS and remote sensing. *Water Resources Management*, 22, 1783–1798. https://doi.org/10.1007/s11269-008-9253-9

Dehni, A., & Lounis, M. (2012). Remote sensing techniques for salt affected soil mapping: Application to the Oran region of Algeria. *Procedia Engineering, 33*, 188–198.

Douaoui, A. E. K., Nicolas, H., & Walter, C. (2006). Detecting salinity hazards within a semi-arid context by means of combining soil and remote-sensing data. *Geoderma, 134*, 217–230. https://doi.org/10.1016/j.geoderma.2005.10.009

Elnaggar, A. A., & Noller, J. S. (2020). Application of remote-sensing data and decision-tree analysis to mapping salt-affected soils over large areas. *Remote Sensing, 2*(1), 151–165. https://doi.org/10.3390/rs2010151

Gao, J., Zhang, Y., Huang, J., & Zhang, X. (2018). Spatial variability and prediction of soil salinity using geostatistics and remote sensing in an arid area of China. *Geoderma, 319*, 1–11.

Gerardo, R., & de Lima, I. P. (2022). Sentinel-2 satellite imagery-based assessment of soil salinity in irrigated Rice fields in Portugal. *Agriculture, 12*, 1490.

Ghada, S., & Balázs, S. (2021). *Estimation of soil salinity using sentinel-1 sar and landsat-8 oli data combined with machine learning-based models*. Retrieved from https://www.researchgate.net/publication/359717005.

Ghazali, M. F., et al. (2020). Generating soil salinity, soil moisture, soil pH from satellite imagery and its analysis. *Information Processing in Agriculture, 7*, 294–306.

Konstantin Ivushkin, Soil salinity assessment using remote sensing and GIS techniques in Syrdarya province of Uzbekistan, , 2014.

Ivushkin, K. (2014). Soil salinity assessment using Remote Sensing and GIS techniques in Syrdarya province of Uzbekistan. *Innovations for sustainability and food security in arid and semiarid lands, 50*

Jain, S.K., Kumar, S., Varghese, J. (2001) Estimation of soil erosion for a Himalayan watershed using GIS technique. Water Resources Management, 15, pp. 41–54

Jiang, X., Wang, Z., Lu, C., & Liu, H. (2020). Mapping soil salinity in an arid area using machine learning algorithms and remote sensing data. *Ecological Indicators, 116*, 106498.

Kaliraj, S., Chandrasekar, N., Parmar, M., Rejith, R. G., Dharumarajan, S., Lalitha, M., & Chandramohan, K. (2024b). Assessment of land degradation vulnerability in the semi-arid region of southern India using GIS-based MEDALUS approach. In S. Dharumarajan et al. (Eds.), *Remote sensing of soils: Mapping, monitoring, and measurement*. Elsevier. isbn:978-0-44-318773-5. https://doi.org/10.1016/B978-0-443-18773-5.00038-7

Kaliraj, S., Srinivas, R., Palanivel, K., Lakshumanan, C., & Chandrasekar, N. (2024a). Remote sensing indices based soil properties measurement—A case study of the Thamirabarani River Basin, South India. In S. Dharumarajan et al. (Eds.), *Remote sensing of soils: Mapping, monitoring, and measurement*. Elsevier. isbn:978-0-44-318773-5. https://doi.org/10.1016/B978-0-443-18773-5.00014-4

Kaliraj, S., et al. (2022). *Assessment of desertification and land degradation vulnerability in humid tropics and sub-tropical regions of India using remote sensing and GIS techniques*. Springer.

Khan, N. M., Rastoskuev, V. V., Sato, Y., & Shiozawa, S. (2005). Assessment of hydrosaline land degradation by using a simple approach of remote sensing indicators. *Agricultural Water Management, 77*(1–3), 96–109. https://doi.org/10.1016/j.agwat.2004.09.038

Kholdorov, S., et al. (2022). Soil salinity assessment research using remote sensing techniques: A particular focus on recent research. *IOP Conference Series: Earth and Environmental Science, 1068*, 012037.

Lalitha, M., Subramanian, D., Khandal, S., & Hegde, R. (2021). Modeling and mapping of salt-affected soils through spectral indices in Inland Plains of semi-arid agro-ecological region. *Journal of the Indian Society of Remote Sensing, 49*, 1475–1481. https://doi.org/10.1007/s12524-021-01321-w

Manickam, L., et al. (2021). Assessment of salt affected soils through spectral indices in inland plains of semi-arid agro-ecological region. *Journal of the Indian Society of Remote Sensing, 49*, 1475–1148.

Nguyen, K.-A., et al. (2020). Soil salinity assessment by using near-infrared channel and vegetation soil salinity index derived from Landsat 8 OLI data: A case study in the Tra Vinh Province, Mekong Delta, Vietnam. *Progress in Earth and Planetary Science, 7*, 1.

Renard, K.G., Foster, G.R., Weesies, G.A., McCool, D.K., Yoder, D.C. (1997). Predicting soil erosion by water: a guide to conservation planning with the revised universal soil loss equation (RUSLE). (USDA. Agri), U.S. Gov. Print Office, Washington DC.

Rhoades, J. D., Shouse, P. J., Alves, W. J., Manteghi, N. A., & Lesch, S. M. (1990). Determining soil salinity from soil electrical conductivity using different models and estimates. *Soil Science Society of America Journal*, 54(1), 46–54. https://doi.org/10.2136/sssaj1990.03615995005400010007x

Sahbeni, G. (2021). A PLSR model to predict soil salinity using Sentinel-2 MSI data. *Open Geosciences, 13*, 977–987.

Stavi, I., Thevs, N., & Priori, S. (2021). Soil salinity and sodicity in drylands: A review of causes, effects, monitoring, and restoration measures. *Frontiers in Environmental Science, 9*, 712831.

Szabolcs, I. (1991). Salinization Potential of European Soils. In: Brouwer, F.M., Thomas, A.J., Chadwick, M.J. (eds) Land Use Changes in Europe. The GeoJournal Library, vol 18. Springer, Dordrecht. https://doi.org/10.1007/978-94-011-3290-9_13

Wang, Q., Zhang, Y., & Zhang, X. (2020). Mapping soil salinity using remote sensing and a random forest algorithm in an arid area of China. *Catena, 194*, 104–704.

Wischmeier, W.H., Smith, D.D. (1978) Predicting rainfall erosion losses. A guide to conservation planning. *The USDA Agricultural Handbook* No. 537, Maryland.

Yu, H., et al. (2018). Mapping soil salinity/Sodicity by using Landsat OLI imagery and PLSR algorithm over Semiarid West Jilin Province, China. *Sensors (Basel), 18*(4), 1048.

Zewdu, S., Suryabhagavan, K. V., & Balakrishnan, M. (2017). Geospatial approach for soil salinity mapping in sego irrigation farm, South Ethiopia. *Journal of the Saudi Society of Agricultural Sciences, 16*, 16–24.

Zhang, Y., & Niu, Y. (2019). Spatial distribution of soil salinity and its relationship with soil properties and land use in arid areas of northwestern China. *Land Degradation & Development, 30*(6), 746–756.

Zheng, Z., Zhang, F., Ma, F., Chai, X., Zhu, Z., Shi, J., & Zhang, S. (2009). Spatiotemporal changes in soil salinity in a drip-irrigated field. *Geoderma, 149*, 243–248.

Zhu, J. K. (2001). Plant salt tolerance. *Trends in Plant Science, 6*(2), 66–71.

Landsat 9 OLI/TIRS Image-Derived Soil Properties as a Tool to Evaluate Soil Quality Zone Using GIS-Based AHP Technique: A Case Study of Thamirabarani River Basin, Southern India

S. Kaliraj, N. Kiruthika, E. Vairaveni, and K. Palanivel

Abstract This study is used to assess the soil quality in the Thamirabarani river basin in South India using remote sensing (RS) and geographic information system (GIS) techniques. In this study, 13 parameters contributing to soil quality were analyzed to demarcate the soil quality zone in the basin using the remote sensing indices and in-situ sampling data. The parameters include soil organic matter, organic carbon, clay minerals, soil iron oxide, soil carbonate, soil salinity, soil moisture, soil erodibility, land use/land cover (LULC), soil texture, land surface temperature (LST), Normalized Difference Moisture Index (NDMI), and slope. They are derived from multiple data sources, including Landsat 9 Operational Land Imager (OLI) and ASTER DEM satellite images, in-situ soil sampling data, Survey of India—topographical maps, GSI—resource maps, IMD—rainfall data, and other collateral data sets. The GIS-based Analytical Hierarchy Process (AHP) technique analyses soil quality zonation at a site-specific scale. The soil quality zone (SQZ) map is classified into five soil quality zones (Very good, good, moderate, poor, and very poor) to represent the spatial characteristics of soil quality in the basin, wherein low values (soil quality index (SQI) = 36–44, and 45–51) showed the area with inferior and poor soil quality that covers 1% and 10% respectively. The moderate soil quality zone (SQI = 52–59) spread over the area of 46%. However, the soil quality zones showed good (SQI = 60–66) and excellent (SQI = 67–74) categories found in the areal coverage of 41% and 2%, respectively, mainly associated with riverbeds and irrigation lands in the basin. The results indicated that moderate to good soil quality zones suit agricultural and plantation practices, while impoverished zones require

---

S. Kaliraj (✉)
National Centre for Earth Science Studies (NCESS), Ministry of Earth Sciences (MoES), Government of India, Thiruvananthapuram, Kerala, India
e-mail: s.kaliraj@ncess.gov.in

N. Kiruthika · E. Vairaveni · K. Palanivel
Department of Remote Sensing, Bharathidasan University, Tiruchirappalli, Tamil Nadu, India

A. K. Singh et al. (eds.), *Sustainable Agriculture Management in Semi-Arid Climates*, https://doi.org/10.1007/978-3-031-94062-0_11

targeted soil management strategies. However, a few limitations include the complexity of soil resource assessment and the need for periodic monitoring and field-based validation to ensure the accuracy and sustainability of soil quality over time. This study underscores the importance of integrating remote sensing and GIS techniques for soil quality assessment, providing critical insights for sustainable land management and environmental conservation. This approach provides a significant methodological advancement for sustainable soil resource management and can aid rural development.

Keywords Soil quality · SQI index · Soil properties · AHP model · Remote sensing · GIS · Thamirabarani basin · South India

1 Introduction

Soil is a vital livelihood source that balances the equilibrium in the environmental ecosystems. It acts as a medium for transferring minerals and nutrients, water purification, carbon sequestration, and biodiversity conservation while aiding in groundwater recharge (de Paul Obade & Lal, 2013; Mahmoudabadi et al., 2017). Soil quality can be considered as the availability of minerals and nutrients to plant growth in changing environmental conditions (Kaliraj et al., 2017; Else et al., 2018; Dharumarajan et al., 2023). Soil chemical properties include nutrient content, pH, organic matter content, and pollutants. In contrast, biological properties include the diversity and abundance of microorganisms, such as bacteria, fungi, and other soil organisms, which are also crucial for soil quality. Hence, soil quality measurement and monitoring are necessary for soil conservation and sustainable management in the basin environment (Else et al., 2018; Thaler et al., 2019; Abdellatif et al., 2021; Katebikord et al., 2022; Chandramohan et al., 2023). However, soil resources fall under critical impacts due to erosion, salinization, pollution degradation, etc. Soil quality in any place is determined by the soil's physical, chemical, and biological properties that affect its ability to perform these functions (Aksoy et al., 2009; Mehrez et al., 2011; Sahu et al., 2015; Vani, 2017; Kaliraj et al., 2023). USDA (2006) has stated that soil quality is one of the critical parameters in preparation for sustainable management of soil resources and conservation activities. Soil quality depends on the availability of nutrients such as nitrogen, phosphorus, and potassium, which are also variable across different regions. The soil quality of an area is mainly affected by deforestation, improper land use practices, heavy metals pollution, mixing pesticides, and other toxic substances (Anderson et al., 2009). Moreover, soil erosion causes the removal of topsoil due to water and wind erosion, affecting soil quality and crop productivity (Simon et al., 2018). Maintaining and improving soil quality is crucial for promoting sustainable agriculture, protecting the environment, and maintaining ecosystem services (Manchanda et al., 2002; Lakhankar et al., 2009; Mokarram et al., 2015; Suresh, 2016; Taloor et al., 2021). Mapping soils over a large area using conventional methods could be tedious and time-consuming(Shabou et al., 2015; Strashok et al., 2022). Integrated remote sensing and GIS techniques provide multiple data sources to estimate soil quality and spatial distribution. The conventional measurement of soil quality is a complex task that requires the integration of different indicators from physical, chemical, and biological properties (Raza et al., 2018; Fellipe et al., 2021). The GIS-based AHP technique analyzes multiple parameters to assess soil quality in the area. Moreover, the multi-temporal remotely sensed data provides time-series images for continuous monitoring of soil sources at a high spatial resolution and the ability to sense over a large area (Karlen et al. 2008; Shankar 2017).

The analytic hierarchy process (AHP) is a decision-making process tool used to analyze multi-criteria based on assigned weights and rating of the factors through pair-wise comparison (Saaty, 1980), and it is updated to evaluate the consistency of the result and to reduce discrimination in the decision-making process (Saaty, 1990). Mapping and assessment of soil quality play a pivotal role in soil-land-water resources management, global food security, and ecosystem sustainability. Further, the GIS-based AHP has emerged as a powerful tool for soil quality zone mapping, offering a methodical and spatially explicit approach to assessing soil conditions on a larger scale. The GIS-based analytic hierarchy process (AHP) technique is widely used for soil quality, wherein the parameters are systematically analyzed based on the degree of magnitude of soil quality characteristics (Aksoy et al., 2009; Muralitharan & Palanivel, 2015; Kaliraj et al., 2022). Many researchers have used the GIS-based AHP techniques in soil quality studies, and they have proven the reliability of this model, particularly effective for soil quality mapping, sustainable land use planning, and agricultural management (Prasad et al., 2017; Liu et al., 2022). The GIS techniques involve soil quality mapping by integrating various spatial and attribute data sources, including remote sensing derived layers, in-situ sampling data, field surveys, and tabulated data (e.g., soil databases). Using GIS, the user can generate soil quality maps by overlaying and analyzing multiple geo-environmental parameters such as soil texture, nutrients, moisture, LST, soil pH, organic matter, salinity, and erosion risk, and are often combined to identify the soil quality zone using AHP technique (Kaliraj et al., 2013).

The GIS-based AHP technique aids in structuring complex problems into a hierarchy, making it easier to evaluate multiple criteria simultaneously (Asfaw et al. 2018; Babaeian et al. 2019), and helps prioritize and assign weights to various criteria based on their relative importance to delineate a potential soil quality zone. In this context, the AHP technique is used to weigh the significance of various soil quality indicators, such as soil texture, moisture, organic content, and pH, systematically and consistently for mapping the soil quality zone (Karale et al., 1983; Huete, 1988; Paul et al. 2014; Mandal, 2016; Rasul et al., 2018; Nejad & Zoratipour, 2019). Many researchers have successfully applied GIS-based AHP techniques for soil quality zone mapping across different geographical regions. Koushika et al. (2024) used GIS-based AHP to map soil fertility zones in an agricultural region in India by analyzing various parameters such as organic carbon, available nitrogen, and phosphorus, and the output is derived from the soil fertility on a zone-by-zone basis, improving crop yields while reducing fertilizer usage. GIS-AHP technique is applied for land degradation assessment in arid and semi-arid regions by prioritizing the soil indicators inducing land degradation activities, and it maps risk areas, providing solutions for land management and rehabilitation efforts (Rakshit et al., 2021; Kaliraj et al., 2024b). The GIS-based AHP is widely used for mapping soil erosion risk zones by calculating a pair-wise matrix of parameters like slope, rainfall, vegetation, and soil texture, in which the AHP assigns weights and ratings of these parameters based on their relative importance. GIS facilitates the spatial analysis and visualization of erosion-prone areas, and the resultant maps enable policymakers to make conservation efforts in successful ways (Kaliraj et al. 2024a; Nguyen et al., 2021).

This study investigates soil quality zones based on analysis of 13 parameters, including soil organic matter, organic carbon, clay mineral, soil iron oxide, soil carbonate, soil salinity, soil moisture, soil erodibility (K-Factor), LULC, soil texture, LST, NDMI, and slope obtained using both primary data using Landsat—9 and secondary data running on a GIS-based AHP model to get a view on the soil quality of South India's perennial river Thamirabarani basin. The outcome provides a primary database on soil quality characteristics at a site-specific scale and becomes vital for sustainable soil resource management and conservation activities.

2 Study Area

The Thamirabarani river basin is in southern Tamil Nadu, India. The geographical extent of the basin covers an area of around 5565 sq. km with a longitude of 77°9′5″E-78°9′23″E and a latitude of 8°27′7″N-9°11′39″N. The Thamirabarani River originates from the Western Ghats and flows eastwards along the length of 128 km, confluences with the Bay of Bengal Sea. Figure 1 shows the geographical location of the study area. The TRB covers the uplands of the Western Ghatson on the western side, pediplains and alluvial plains in the middle parts, and coastal plains in the eastern parts with structural hills, pediplains, fluvial plains, and coastal plains. Geological formations consist of crystalline rocks of the Archean age in the western part and sedimentary rocks from the tertiary and quaternary ages in the middle and eastern parts. Metamorphosed rocks, charnockite, gneisses, granites, and primary and acidic intrusive rocks have formed the basin. The TRB is found in different soil types, including loamy sand, clay, clay loam, sandy clay loam, and sandy loam, and is spatially distributed in various parts with larger areal extend. Loamy sand is spread over a larger area, mainly in northern and north-eastern parts, and is associated with isolated patches of red loamy, clayey, sandy clay, and clay loam soils (Kaliraj et al., 2024b). The TRB consists of 57% of the agricultural lands, found in almost all soil types except upland hills, and 15% of shrubs and scrubs are found in isolated patches across the basin. 6.9% of the area is covered with water bodies, including rivers, reservoirs, channels, tanks, and ponds. 3.2% is occupied with build-ups, settlements, and other infrastructures. The basin prevails in subtropical to semi-arid climate conditions with an annual average rainfall of 985.77 mm, and the optimum temperature range is at 23–32 °C. Higher rainfall is

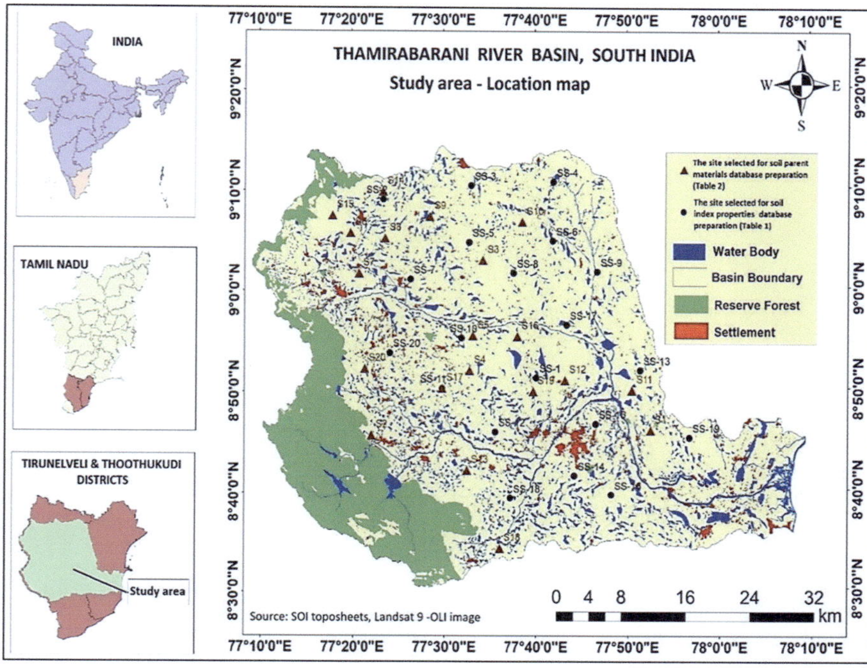

Fig. 1 Geographical location of the study area

found during the northeast monsoon season, especially in the western parts, and other parts have noted drastic rainfall variability, particularly in the lower basin areas.

3 Materials and Methods

The use of remote sensing indices to estimate soil properties provides a valuable quantitative database for soil and land resource management (Kalinowski et al., 2004; Gholizadeh et al., 2018). In this study, remote sensing indices are employed to assess the relationship between soil and land characteristics for various soil types. These indices play a crucial role in the development of soil and land resources, as well as in agriculture, food production, and environmental-ecosystem conservation and management at increasingly detailed scales (Fig. 2). Figure 3 illustrates the methodological workflow of SQZ mapping. The study utilizes Level-2 and Collection-2 products of Landsat 9 OLI (Operational Land Imager) images to analyze the spatial characteristics of different soil types using indices such as Normalized Difference Vegetation Index (NDVI), SAVI, Bare Soil Index (BSI), NDMI, SOCI, clay mineral index, Ferrous mineral index, Carbonate index, Iron Oxide Index, and Soil Salinity Index, along with GIS techniques in the Thamirabarani river basin study area (Kaliraj et al., 2023). Additionally, a GIS-based AHP model

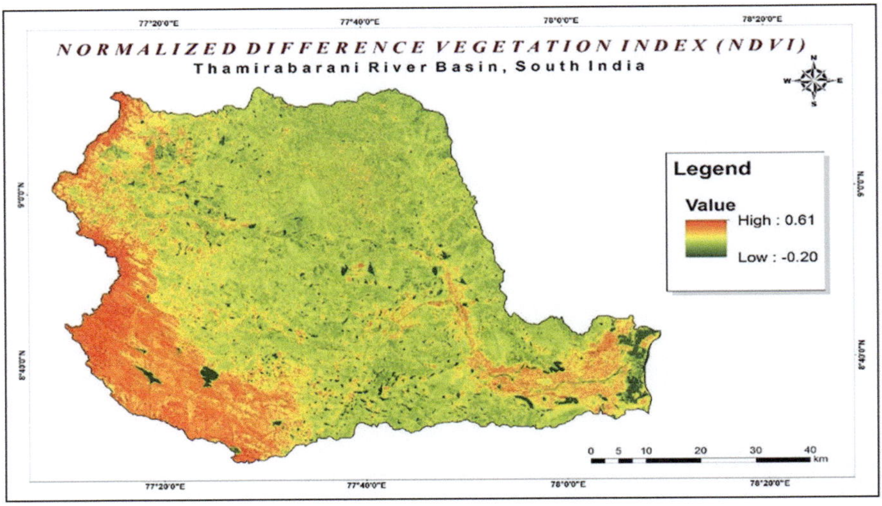

Fig. 2 Spatial characteristics of NDVI in the basin

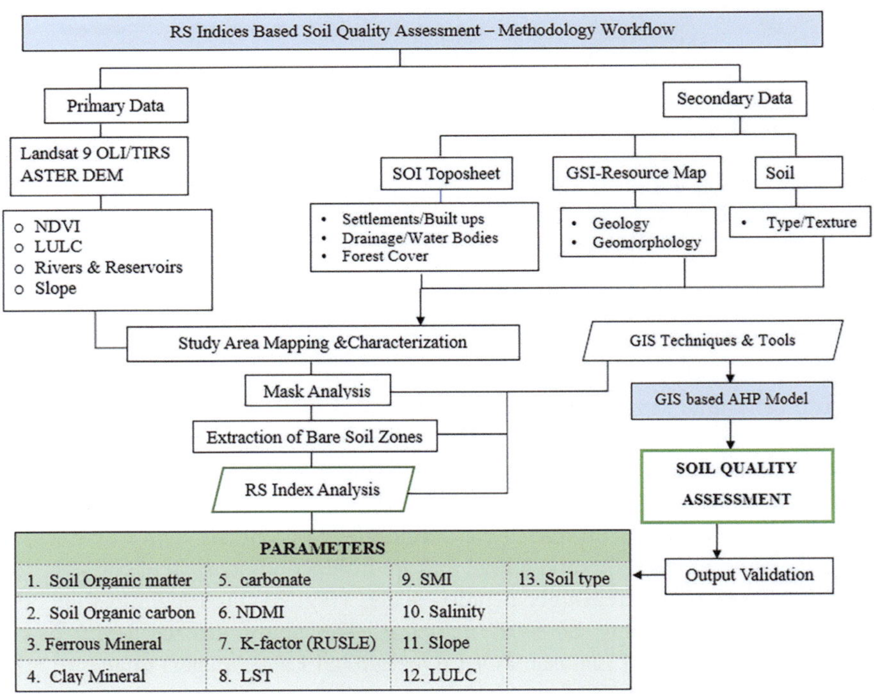

Fig. 3 The flowchart showing the methodology used in the study

is employed to assess soil quality by prioritizing each of the mentioned indices and producing a composite index that aids in understanding soil quality in the study area.

3.1 Preparation of Thematic Layers

The study uses both primary and secondary data to assess soil quality. Various data such as NDVI, soil organic matter, organic carbon, clay mineral, ferrous mineral, carbonate, soil salinity, soil erodibility, LULC, LST, and NDMI were derived using Landsat 9 Thermal Infrared Sensor (TIRS)/OLI & ASTER DEM satellite imagery. These derived index products can be used to analyze soil properties in the study area, directly influencing soil quality. Copernicus provides soil moisture-gridded data for 2023, which were used to estimate soil moisture globally with the help of a large set of satellite sensors. Secondary datasets were used to prepare maps such as location maps, soil texture, slope, and other thematic layers.

The study area layer was created using GIS techniques from the map of the River Basin Atlas of India published by the Central Water Commission (CWC) and ISRO in 2012. The NDVI layer was generated for April 2022 and has a range of values from −0.20 to 0.61. The NDVI was then reclassified into two categories; values less than 0.3 were classified as exposed soil zones, while areas with values greater than 0.3 have vegetative cover in the basin. The reclassified NDVI was vectorized, and the values greater than 0.3 were removed. Furthermore, LULC features like settlements, water bodies, and reserved forests were masked for analysis using the masking tool in GIS. The soil exposed zones (<0.3) are ideal for soil quality assessment as vegetation might hinder deriving the remote sensing index products.

Landsat 9 OLI multispectral images with four visible spectral bands (0.43–0.67 μm), one near-infrared band (0.85–0.88 μm), and three shortwave-infrared bands (1.57–2.29 μm) at 30 m spatial resolution are used for monitoring and assessing soil and land properties. The remote sensing index analysis is performed on the extracted bare soil using Landsat—9 bands to estimate soil properties such as organic matter, organic carbon, clay mineral, ferrous mineral, carbonate matter, soil salinity, LST, and NDMI. Other spatial data sources are used to create thematic layers of LULC, soil texture, slope gradient, K-factor, and soil moisture. Finally, the GIS-based AHP technique is applied to the 13 parameters to estimate the soil quality zone at a site-specific scale.

3.2 GIS-Based AHP Technique for Soil Quality Zone (SQZ) Mapping

The Analytical Hierarchy Process (AHP) is a decision-making tool developed by Professor Thomas L. Saaty in 1970 to handle multi-criteria decision-making problems. The AHP method involves breaking down complex decision problems into a hierarchy based on the importance of different criteria (Saaty, 2005). The first step in using AHP is to create a hierarchy that identifies the parameters or criteria for the study's goal. These identified criteria are then compared against each other using a pair-wise comparison matrix (Pij), which measures the relative preference among the factors. A scale for comparison (Saaty, 1977; Saaty & Vargas, 1991) with values from 1 to 9 is used to express the importance of one criterion over another. The pair-wise comparison matrix columns are summed to calculate the scale weights (SW). Next, to create a normalized pair-wise matrix, each matrix element is divided by its respective scale weights. The row average is then calculated to find the Geometric mean (GM). The pair-wise matrix values calculated from the parameter's assigned values should be satisfied with the consistency ratio (Saaty's CR value <0.1), which is calculated using eq. 1.

$$\text{CR} = \text{CI}/\text{RI}(< 0.1) \tag{1}$$

where CI is the consistency index, RI is random inconsistency.

The CR value is calculated by multiplication of the parameter's pair-wise matrix value and the geometric mean (P_{ij}*GM). The sum of each row is found and divided by its geometric mean. The λ_{max} denotes the average of the sum values, and the consistency index (CI) value is calculated from it.

$$\text{CI} = \lambda_{\text{max}} - n / n - 1 \tag{2}$$

where λ_{max} is the highest eigenvalue of a matrix of P_{ij}; n is the number of parameters.

In this analysis, the random inconsistency (RI) value is adopted from the standard RI table proposed (by Saaty and Vargas in 1991). The CR is calculated by equation (1), which, according to Saaty's consistency ratio condition, should be less than 0.1. If not, the decision maker must revise the decision so that the consistency ratio is acceptable (<10%). Then, the matrix is acceptable.

4 AHP Analysis of Soil Quality Zone Mapping

The AHP technique was introduced by Saaty in 1970, and this was adopted to analyze the soil quality zones in the study area using GIS software. The process of AHP is explained in Sect. 3.2. A scale of preference ranging from 1 to 5 is used in the pair-wise comparison matrix table proposed by Saaty and Vargas in 1991. Based on this, the relative importance of the 13 parameters was determined to indicate their potential impact on soil quality. The AHP method allows for evaluating the geometric mean and assigning a normalized weight to different parameters to finalize the decision-making process. The consistency of the matrix must meet Saaty's consistency condition as defined by equation (1) (Tables 1 and 2).

The consistency index (CI) value is calculated based on the equation (2), and it is given in the Table 3. The eq. CI = $\lambda_{max} - n/n - 1$ is assigned to the values of CI = 13.916 – 13/13 – 1, and the calculated CI value is 0.0763.

The average Random inconsistency (RI) value for 13 parameters is 1.56, is given in Table 4, and is substituted in equation (1) to calculate the consistency ratio (CR), and the CR value is 0.049. The eq. CR = CI/RI is assigned to the values of CR = 0.0763/1.56, and the calculated CR value is 0.049 (and it is <0.1). The consistency ratio value is 0.049, which lies below 0.1. Hence, the pair-wise comparison matrix has passed the consistency test, and the assigned weights of the AHP parameters were highly accepted.

The normalized weight of the AHP parameter is calculated by dividing the assigned weights by the geometric mean. The assigned weights for each feature class of the parameter are listed in Table 5. These weights range from 1 (low) to 5 (high) based on the feature class's impact on soil quality. For instance, areas with steep slopes, which contribute to soil erosion, are assigned a lower value (1), while areas with fertile soils and favorable conditions are assigned a higher weight (5). Soil quality index is calculated using normalized weights of the AHP parameters using the map algebra tool in ArcGIS software.

Table 1 Assignment of scale weight for AHP parameters

Table 1. Pairwise matrix for calculating the scale weight (Pij)

Parameter	Soil Class	SOM	Soil moisture	Soil clay	SOC	Soil Iron oxide	Soil carbonate	NDMI	Soil salinity	K-factor	LST	LU/LC	Slope
Soil Class	1	1	1	2	2	2	3	3	4	4	4	5	5
SOM	1	1	1	1	1	2	2	4	1	1	3	2	5
Soil moisture	1	1	1	2	2	2	2	1	2	1	1	2	3
Soil clay	0.5	1	0.5	1	1	1	2	3	2	3	2	1	4
SOC	0.5	0.5	0.5	1	1	1	2	2	1	1	2	2	3
Soil Iron oxide	0.5	0.5	0.5	1	1	1	2	3	2	2	2	3	4
Soil carbonate	0.33	0.5	0.5	0.5	0.5	0.5	1	1	2	2	2	3	3
NDMI	0.33	0.25	1	0.33	0.5	0.33	1	1	2	2	2	2	3
Soil salinity	0.25	1	0.5	0.5	1	0.5	0.5	0.5	1	1	1	2	3
K-factor	0.25	1	1	0.33	1	0.5	0.5	0.5	1	1	1	2	2
LST	0.25	0.33	1	0.5	0.5	0.5	0.5	0.5	1	1	1	2	2
LU/LC	0.2	0.5	0.5	1	0.5	0.33	0.33	0.5	0.5	0.5	0.5	1	1
Slope	0.2	0.2	0.33	0.25	0.33	0.25	0.33	0.33	0.33	0.5	0.5	1	1
Scale weight (SW)	6.31	8.78	9.33	11.41	12.33	11.91	17.16	20.33	19.83	20	22	28	39

Note: SOM - soil organic matter; SOC - soil organic carbon; NDMI - normalized difference moisture index; K-factor - soil erodibility; LST - land surface temperature; LULC - landuse and land cover

Table 2 Pair-wise matrix of AHP parameters for calculating the geometric mean

Table 2. Pair-wise matrix for estimating the geometric mean														
Parameter	Soil texture	SOM	Soil moisture	Soil clay	SOC	Soil ferrous	Soil carbonate	NDMI	Soil salinity	K-factor	LST	LU/LC	Slope	Geometric mean (G = Pij / SW)
Soil texture	0.158	0.114	0.107	0.175	0.162	0.168	0.175	0.148	0.202	0.200	0.182	0.179	0.128	0.161
SOM	0.158	0.114	0.107	0.088	0.081	0.168	0.117	0.197	0.050	0.050	0.136	0.071	0.128	0.113
Soil moisture	0.158	0.114	0.107	0.175	0.162	0.168	0.117	0.049	0.101	0.050	0.045	0.071	0.077	0.107
Soil clay	0.079	0.114	0.054	0.088	0.081	0.084	0.117	0.148	0.101	0.150	0.091	0.036	0.103	0.096
SOC	0.079	0.057	0.054	0.088	0.081	0.084	0.117	0.098	0.050	0.050	0.091	0.071	0.077	0.077
Soil ferrous	0.079	0.057	0.054	0.088	0.081	0.084	0.117	0.148	0.101	0.100	0.091	0.107	0.103	0.093
Soil carbonate	0.052	0.057	0.054	0.044	0.041	0.042	0.058	0.049	0.101	0.100	0.091	0.107	0.077	0.067
NDMI	0.052	0.028	0.107	0.029	0.041	0.028	0.058	0.049	0.101	0.100	0.091	0.071	0.077	0.064
Soil salinity	0.040	0.114	0.054	0.044	0.081	0.042	0.029	0.025	0.050	0.050	0.045	0.071	0.077	0.056
K-factor	0.040	0.114	0.107	0.029	0.081	0.042	0.029	0.025	0.050	0.050	0.045	0.071	0.051	0.057
LST	0.040	0.038	0.107	0.044	0.041	0.042	0.029	0.025	0.050	0.050	0.045	0.071	0.051	0.049
LU/LC	0.032	0.057	0.054	0.088	0.041	0.028	0.019	0.025	0.025	0.025	0.023	0.036	0.026	0.037
Slope	0.032	0.023	0.035	0.022	0.027	0.021	0.019	0.016	0.017	0.025	0.023	0.036	0.026	0.025
														1.000

5 Results and Discussion

The spatial characteristics of the Thamirabarani River basin are defined using Landsat 9 OLI/TIRS images. The soil quality zone of the basin is determined by analyzing 13 parameters using the GIS-based AHP technique. The spatial characteristics of these parameters are explained with associated maps and quantitative details.

Table 3 Consistency ratio matrix of AHP parameters

Table 3. Consistency Ratio matrix (Pij*GM)															
Parameter	Soil texture	SOM	Soil moisture	Soil clay	SOC	Soil ferrous	Soil carbonate	NDMI	Soil salinity	K-factor	LST	LU/LC	Slope	Sum	λmax = Average of (sum/GM)
Soil texture	0.161	0.113	0.107	0.192	0.154	0.186	0.201	0.192	0.224	0.232	0.196	0.185	0.125	2.268	14.056
SOM	0.161	0.113	0.107	0.096	0.077	0.186	0.134	0.256	0.056	0.058	0.147	0.074	0.125	1.590	14.100
Soil moisture	0.161	0.113	0.107	0.192	0.154	0.186	0.134	0.064	0.112	0.058	0.049	0.074	0.075	1.479	13.779
Soil clay	0.081	0.113	0.054	0.096	0.077	0.093	0.134	0.192	0.112	0.174	0.098	0.037	0.100	1.360	14.217
SOC	0.081	0.057	0.054	0.096	0.077	0.093	0.134	0.128	0.056	0.058	0.098	0.074	0.075	1.080	14.074
Soil ferrous	0.081	0.057	0.054	0.096	0.077	0.093	0.134	0.192	0.112	0.116	0.098	0.111	0.100	1.320	14.199
Soil carbonate	0.053	0.057	0.054	0.048	0.039	0.047	0.067	0.064	0.112	0.116	0.098	0.111	0.075	0.939	13.993
NDMI	0.053	0.028	0.107	0.032	0.039	0.031	0.067	0.064	0.112	0.116	0.098	0.074	0.075	0.895	13.976
Soil salinity	0.040	0.113	0.054	0.048	0.077	0.047	0.034	0.032	0.056	0.058	0.049	0.074	0.075	0.756	13.608
K-factor	0.040	0.113	0.107	0.032	0.077	0.047	0.034	0.032	0.056	0.058	0.049	0.074	0.050	0.768	13.582
LST	0.040	0.037	0.107	0.048	0.039	0.047	0.034	0.032	0.056	0.058	0.049	0.074	0.050	0.670	13.759
LU/LC	0.032	0.057	0.054	0.096	0.039	0.031	0.022	0.032	0.028	0.029	0.025	0.037	0.025	0.505	13.785
Slope	0.032	0.023	0.035	0.024	0.025	0.023	0.022	0.021	0.018	0.029	0.025	0.037	0.025	0.340	13.782
														λmax	13.916

Table 4 Consistency ratio (CR) value of cumulated AHP parameters

Consistency index	0.0763
Average random index	1.56
Consistency ratio	0.049 (<0.1)

5.1 Soil Types and Soil Texture

In Fig. 4, various types of soils present in the Thamirabarani basin are illustrated. The predominant soils in the area include clay, clay loam, loamy sand, random sand, sand, sandy loam, and sandy clay loam. In the northern region of the basin, most of the land consists of loamy sand, followed by clay loam and sandy clay. The soil layer comprises minerals, liquids, gases, and organic materials. Soil nutrients vary according to geo-environmental factors, directly influencing soil quality at specific locations.

5.2 Soil Organic Matter (SOM)

The organic matter in soils indicates the richness of soil quality, and it depends on the availability of soil carbon matter. In this analysis, the organic matter for different soil types is assessed by estimation of soil organic carbon (SOC) using the following equation, which is expressed as,

$$SOM = SOC^{*}1.72 \tag{3}$$

The study evaluates soil organic carbon (SOC), which accounts for 58% of soil organic matter, in addition to nitrogen, hydrogen, potassium, and other elements. The constant 1.72 used in the equation is sourced from Howard & Howard (1990). Figure 5 interprets the values of soil organic matter (SOM) for different soil textures, and land use/land cover (LULC) features in the Thamirabarani basin. High SOM values are observed in loams, sandy loams, and sandy clay loam soil textures, as well as in reserve forests, agricultural land, and land with scrub among the LULC classes. Moderate values are linked to sand, land with scrub, and agricultural land, while low values are present in clay, clay loam, and land with scrub and barren land among the LULC classes.

5.3 Soil Organic Carbon (SOC)

The soil organic carbon (SOC) indicates the amount of carbon present in the soil layers. It is determined using remote sensing indices such as NDVI (Normalized Difference Vegetation Index) and BSI (Bare Soil Index) derived from LANDSAT 9

Table 5 Calculation of normalized weights of AHP parameters

S. no.	Parameter	Feature class	Assigned weight (AW)	Geometric mean (G)	Normalised weight (*N* = AW/G)
1	Soil texture	Loamy sand	5	0.161	31.06
		Clay loam	4		24.84
		Sandy clay loam	3		18.63
		Sandy loam, clay	2		12.42
		Sand, other features	1		6.21
2	Soil organic matter (SOM) (%)	0.79–0.49	5	0.113	44.24
		0.49–0.43	4		35.39
		0.43–0.39	3		26.54
		0.39–0.32	2		17.69
		0.32 to −0.01	1		8.84
3	Soil moisture (%)	67.1–62.7	5	0.107	46.72
		62.7–59.9	4		37.38
		59.9–57.4	3		28.03
		57.4–53.1	2		18.69
		53.1–47.6	1		9.34
4	Soil clay (g. kg^{-1})	0.38–1.2	5	0.096	52.08
		1.2–1.26	4		41.66
		1.26–1.31	3		31.25
		1.31–1.36	2		20.83
		1.36–1.71	1		10.41
5	Soil organic carbon (SOC) (%)	0.46–0.28	5	0.077	64.93
		0.28–0.25	4		51.94
		0.25–0.23	3		38.96
		0.23–0.19	2		25.97
		0.19 to −0.01	1		12.98
6	Soil ferrous content (lbm/ac)	1.63–1.14	5	0.093	53.76
		1.14–1.08	4		43.01
		1.08–1.02	3		32.25
		1.02–0.95	2		21.50
		0.96–0.77	1		10.75
7	Soil carbonate (lbm/ac)	0.36–0.61	5	0.067	74.62
		0.61–0.67	4		59.70
		0.67–0.77	3		44.77
		0.77–0.96	2		29.85
		0.96–1.42	1		14.92

(continued)

Table 5 (continued)

S. no.	Parameter	Feature class	Assigned weight (AW)	Geometric mean (G)	Normalised weight (*N* = AW/G)
8	Normalized difference moisture index (NDMI)	0.48–0.11	5	0.064	78.12
		0.11-0.04	4		62.50
		0.04 to −0.01	3		46.87
		−0.01 to −0.06	2		31.25
		−0.06 to −0.45	1		15.62
9	Soil salinity dS/m	−0.50 to 0.08	5	0.056	89.28
		0.08–0.11	4		71.42
		0.11–0.13	3		53.57
		0.13–0.16	2		35.71
		0.16–0.46	1		17.85
10	K-factor (mg h MJ^{-1} mm^{-1})	0.10–0.20	5	0.057	87.71
		0.20–0.28	4		70.17
		0.28–0.36	3		52.63
		0.36–0.44	2		35.08
		0.44–0.54	1		17.54
11	Land surface temperature (LST) (°C)	23.04–30.12	5	0.049	102.04
		30.12–32.99	4		81.63
		32.99–35.14	3		61.22
		35.14–37.21	2		40.81
		37.21–43.41	1		20.40
12	LULC	Agricultural land plantation Land with scrub barren Other features (built-ups, waterbodies, wetlands, rocky outcrops, mining and quarries)	5	0.037	135.13
			4		108.10
			3		81.08
			2		54.05
			1		27.02
13	Slope	0–3	5	0.025	200
		3–5	4		160
		5–10	3		120
		10–15	2		80
		15–42.18	1		40

OLI images. These indices, along with coefficient values obtained from on-site measurements, are used to estimate the SOC. This method helps in measuring the spatial distribution of SOC. The equation used for this estimation is as follows,

$$\mathrm{SOC} = 4.24 + 11.90^{*}\,\mathrm{NDVI} - 0.021^{*}\,\mathrm{BSI} \quad (4)$$

The Pearson correlation coefficient method uses numerical values to represent the relationship between two variables, such as in situ SOC and NDVI and BSI data. The indices-based covariates indicate the distribution of SOC (Bhunia et al., 2017).

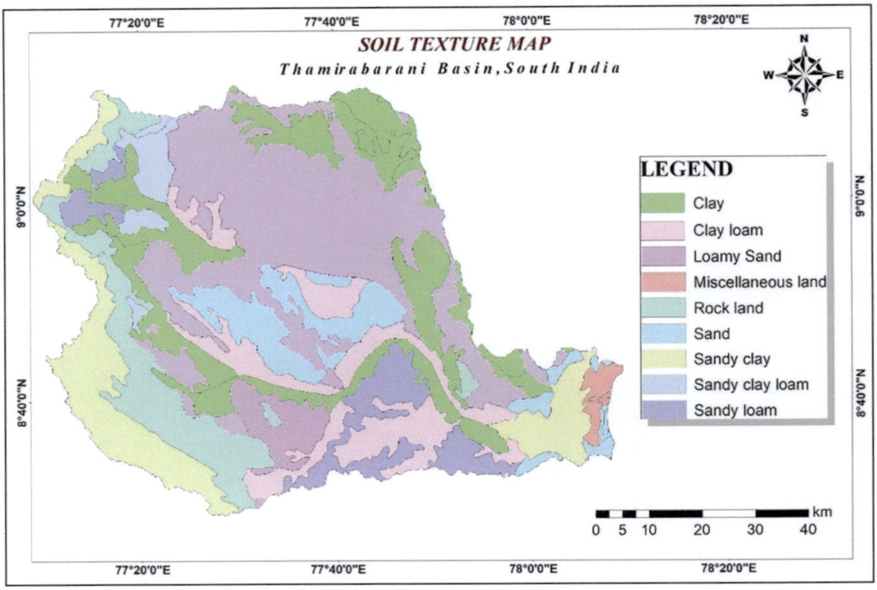

Fig. 4 The soil type of the study area

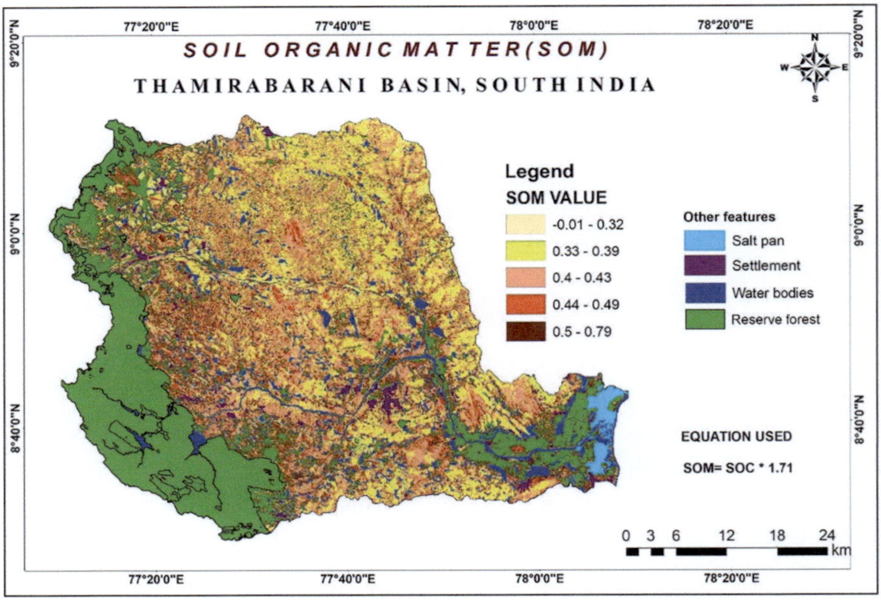

Fig. 5 Spatial distribution of soil organic matter (SOM)

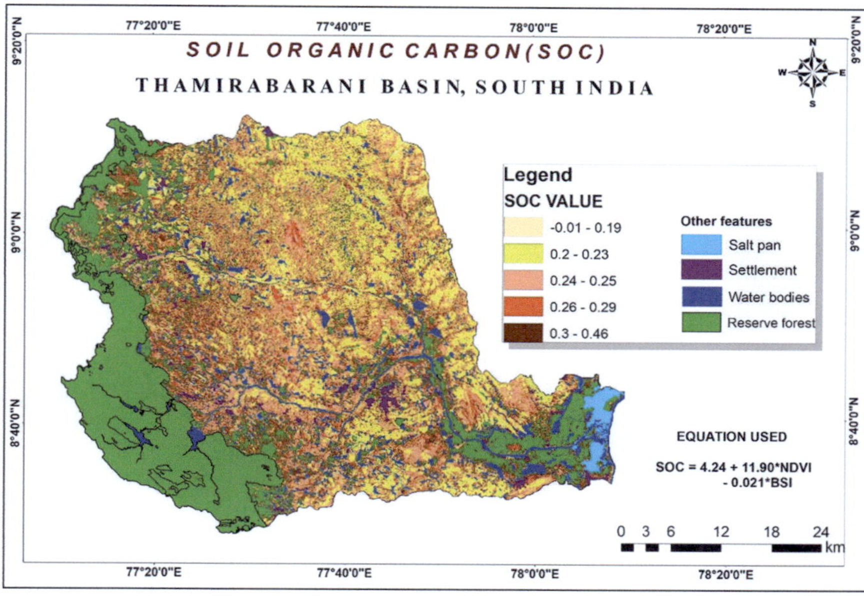

Fig. 6 Soil organic carbon (SOC) characteristics in the basin

Based on the estimation in Fig. 6, the optimum SOC value falls within the range of −0.01% to −0.46% for different soil types. In bare soil areas, SOC is found at a high rate (0.26–0.46%) in irrigated croplands and land with shrubs. In a significant part of the unirrigated lands of the Pedi plains, SOC is estimated to range from 0.20 to 0.25%. On the other hand, SOC is lower (−0.01 to 0.19%) in eroded lands and lands near water bodies due to the removal of topsoil.

5.4 *Soil Clay Mineral*

The clay content in the soil is estimated using the clay mineral index, which is based on the ratio of the SWIR1 and SWIR2 bands of the Landsat 9 OLI image. The equation considers the influence of hydrous minerals, such as clay, on the spectral signature within the bandwidth. For the Landsat 9 OLI image, the clay mineral content is estimated using eq. 6), expressed as,

$$\text{Clay mineral index} = \text{SWIR1} / \text{SWIR2} \tag{5}$$

These two bands' ratios are utilized to determine the clay mineral contents in an area based on their high exchangeable ions and adsorption capacity within the SWIR bands. Figure 7 illustrates the index value ranges, and it is estimated as −0.38 to 1.2, representing low clay mineral regions primarily found in fallows of red loamy soils. The SOC range of −1.21 to 1.26, 1.27–1.31, and 1.32–1.36 mainly

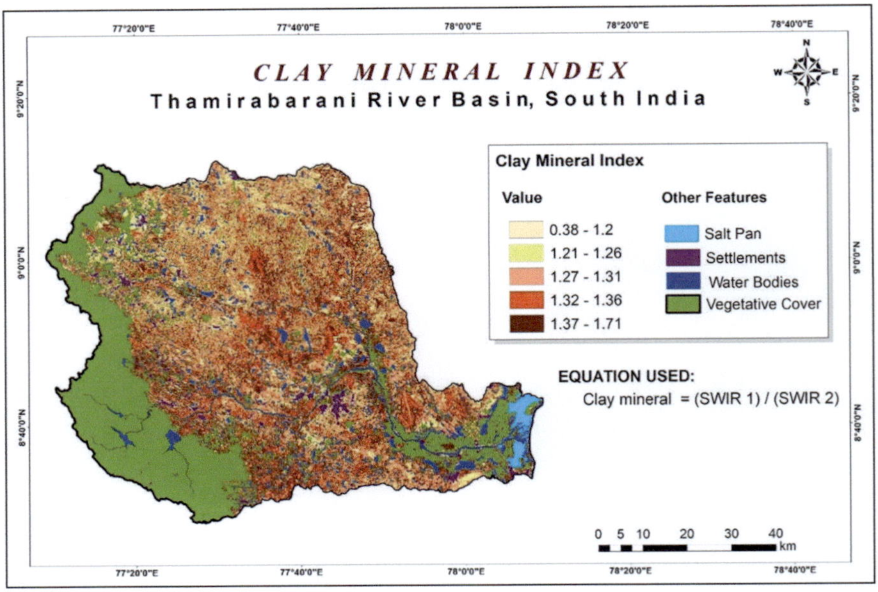

Fig. 7 Soil clay mineral distribution in the basin

covers the regions of barren land and scrub lands. The SOC is noted at a higher value of 1.36–1.71 and is mainly found on exposed clay soils in the northern and north-eastern parts and the area associated with water bodies.

5.5 *Ferrous Mineral Index*

The amount of soil-bearing ferrous material is closely related to the spectral reflectance of SWIR 1 and NIR bands of the Landsat 9 OLI image. This relationship can be expressed using the ferrous ratio index as follows:

$$\text{Ferrous Mineral Ratio} = \text{SWIR1} / \text{NIR} \tag{6}$$

According to Fig. 8, the distribution of ferrous minerals in the basin shows that soils with lower ferrous content (0.77–0.95) are in rocky outcrops and barren lands. Areas with a ferrous content ranging from 0.96 to 1.02 are found in more prominent parts of the Pedi plains, which are covered with fallows and croplands. Moderately higher levels of soil-bearing ferrous content (1.03–1.63) are estimated in significant parts of black clay soils found in fallows and unirrigated croplands, especially in the northern pediplains of the basin.

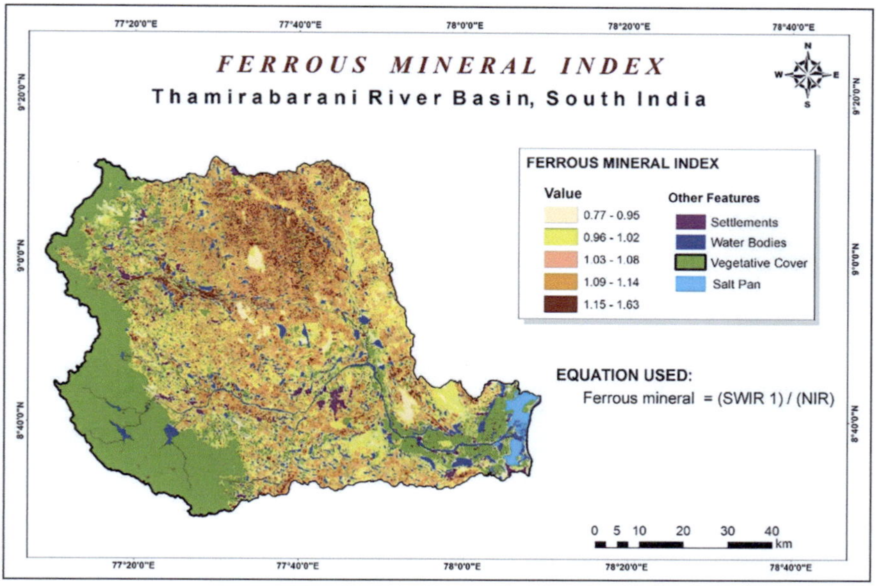

Fig. 8 Ferrous mineral content distribution in the basin

5.6 *Soil Carbonate(SC)*

The soil layer containing carbonate materials exhibits significant surface reflectance characteristics in the blue and NIR bands of the Landsat 9 OLI image (Notesco et al., 2020). The carbonate content can be estimated using the following equation:

$$\text{Soil Carbonate} = \text{Blue} / \text{NIR} \tag{7}$$

Figure 9 illustrates the spatial distribution of soil carbonate matter in the basin. The lower carbonate index value is between 0.36 and 0.61%, while soil along the foothills and pediplains contains 0.62–0.67%. Moderately higher values (0.68–0.77%) are observed in rocky outcrops, bare soil areas, barrens, and land with shrubs, particularly in the north-eastern parts. The highest carbonate content (0.78–1.42%) is found in certain barren lands associated with limestone quarries and discrete patches of salt-affected soils in various parts of the basin.

5.7 *Soil Moisture*

The soil moisture of the basin is estimated using the soil moisture grid data obtained from the Copernicus portal. The inverse distance weighting (IDW) technique is applied to estimate the spatial characteristics of soil moisture using the grid data.

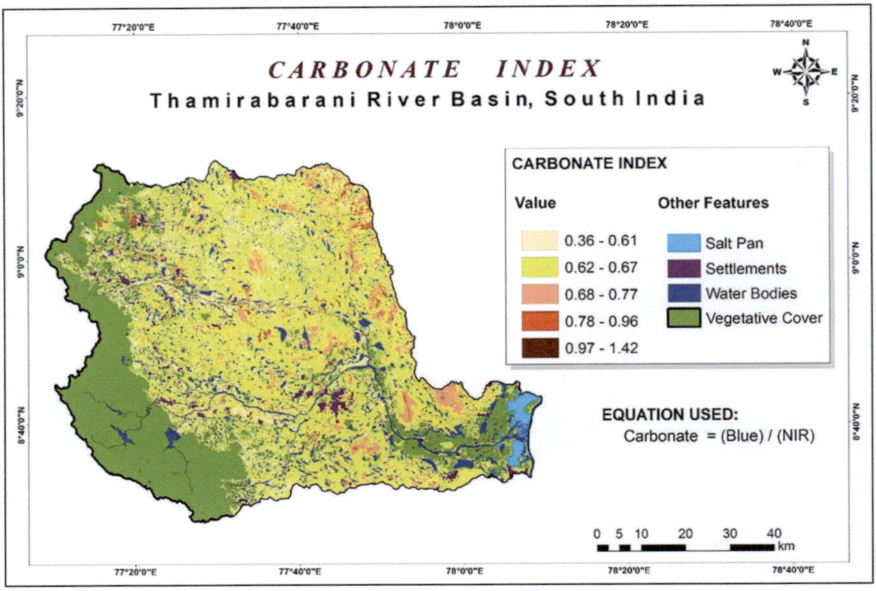

Fig. 9 Soil carbonate distribution in the basin

The Copernicus soil moisture data indicates the soil-water content in a surface soil layer of 2–5 cm depth, which is expressed as a percentage of total saturation. Figure 10 displays the soil moisture index map, with values classified into five classes ranging from very high to very low, depicting its spatial distribution at a site-specific scale. The resulting map showed soil moisture percentages ranging from 47.64 to 67.08%. In comparison, the uplands along the foothills had a soil moisture index value of 47.64–53.06% due to shallower soil layers, while the middle fluvial landforms with thicker soil layers exhibited higher soil moisture values of 59.93–67.08%. The LULC features, i.e., vegetation (forests and plantations), settlements, and water bodies, were masked from soil moisture estimation.

5.8 Normalized Difference Moisture Index

The NDMI indicates the availability of soil-water content in bare soil layers, and it is estimated from the ratio between NIR and SWIR bands of the Landsat 9 OLI image, using the equation (eq. 8),

$$\mathrm{NDMI} = \left(\mathrm{NIR} - \mathrm{SWIR}\right) / \left(\mathrm{NIR} + \mathrm{SWIR}\right) \tag{8}$$

In Fig. 11, the spatial characteristics of the NDMI value are illustrated, ranging from −0.45 to −0.05. This range indicates that the majority of the basin landscapes

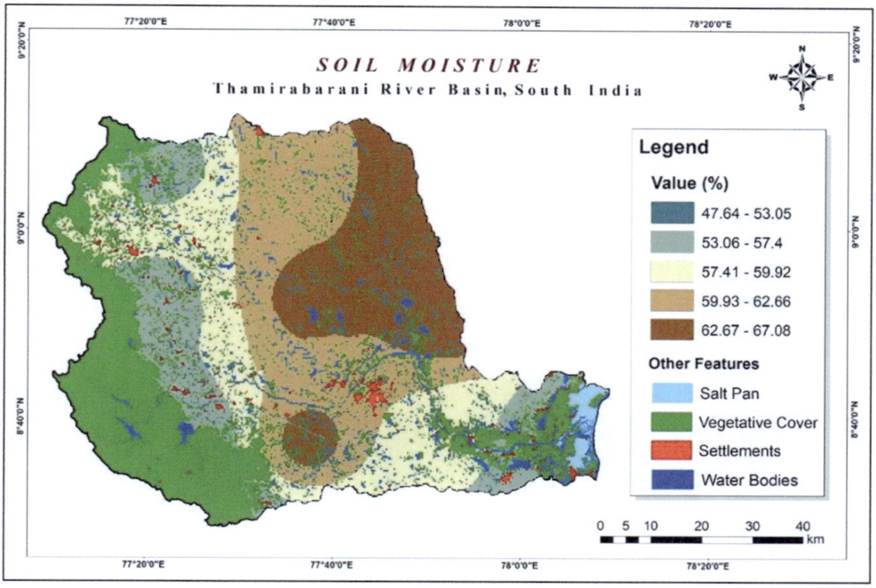

Fig. 10 Soil moisture characteristics in the study area

are experiencing soil moisture stress conditions, signifying the least soil moisture. These conditions are primarily observed in barren rocks, eroded lands, rocky outcrops, and bare soils. The value range of −0.04 to −0.01 indicates a lower moisture rate and is typically found in bare soils, fallows, and unirrigated croplands. For NDMI values ranging from 0 to 0.04, there is moderate vegetation-water content in the land, including shrubs and unirrigated croplands. Finally, the higher NDMI value range of 0.05–0.48 indicates good moisture conditions in dense vegetative covers, plantations, and croplands.

5.9 *Soil Salinity Index (SSI)*

The level of salt in the soil, known as soil salinity, is determined by the amount of salt present in the soil layer. The ratio of near-infrared (NIR) and shortwave-infrared (SWIR) bands of the Landsat 9 Operational Land Imager (OLI) is used to calculate the soil salinity in areas without vegetation. The soil salinity index (SSI) is used to calculate soil salinity using the following equation:

$$\text{Soil Salinity Index}(\text{SSI}) = (\text{SWIR1} - \text{SWIR2}) / (\text{SWIR1} + \text{SWIR2}) \quad (9)$$

In areas with high soil salinity, the NIR spectral reflectance is high, whereas in non-saline areas, it is lower. Saline soils have much lower reflectance in SWIR

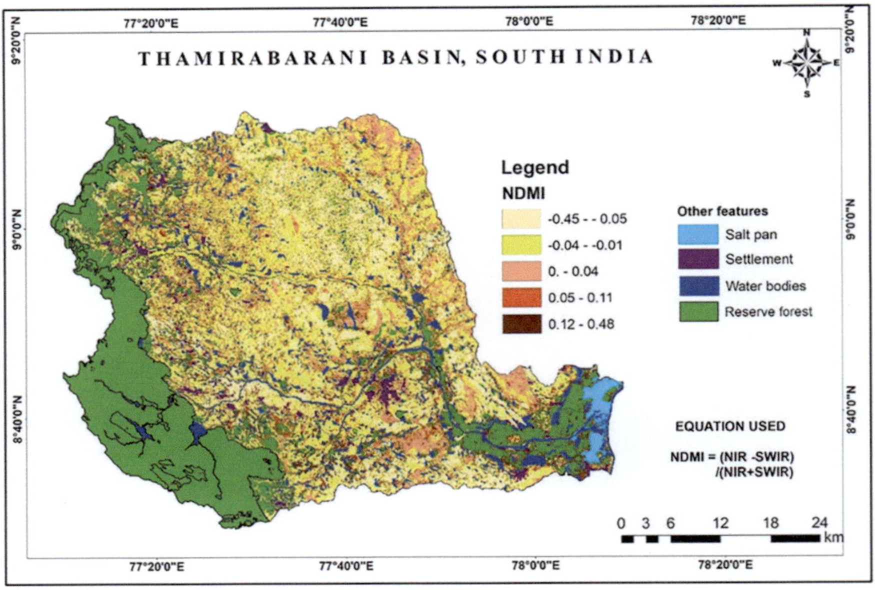

Fig. 11 Spatial characteristics of normalized difference moisture index

compared to non-saline soils due to the strong absorption of carbonate matter found in saline soils. A salinity map, shown in Fig. 12, displays values ranging from −0.5 to 0.46, divided into five classes from very high to very low. It is important to note that high values on this map do not necessarily mean that salinity levels are high in the region; they represent a range of values. When compared with the land use and land cover (LULC) map, low values (−0.5 to 0.11) are found in agricultural lands and land with shrubs, moderate values (0.12–0.13) are observed in fallow lands, and high values (0.14–0.46) are seen in areas with shrubs, rock exposures, and so on. Further analysis of the Soil map reveals that low to moderate salinity values represent loamy soil, while high values are identified in clay, sandy clay loam, and sandy loam regions.

5.10 Soil Erodibility (K-Factor)

Soil erodibility, also known as the K-factor, indicates the potential for soil to erode due to factors such as rainfall, runoff, and overland flow. It is directly linked to soil cohesion, organic content, structural integrity, drainage potential, grain size, and cohesiveness. Soil texture significantly influences the K-factor. Additionally, soil permeability, organic matter, and structure also contribute to soil erodibility. The K-factor of different soils is estimated using rainfall data and represents the average soil loss in tons or hectares. The K-factor value generally remains constant in most

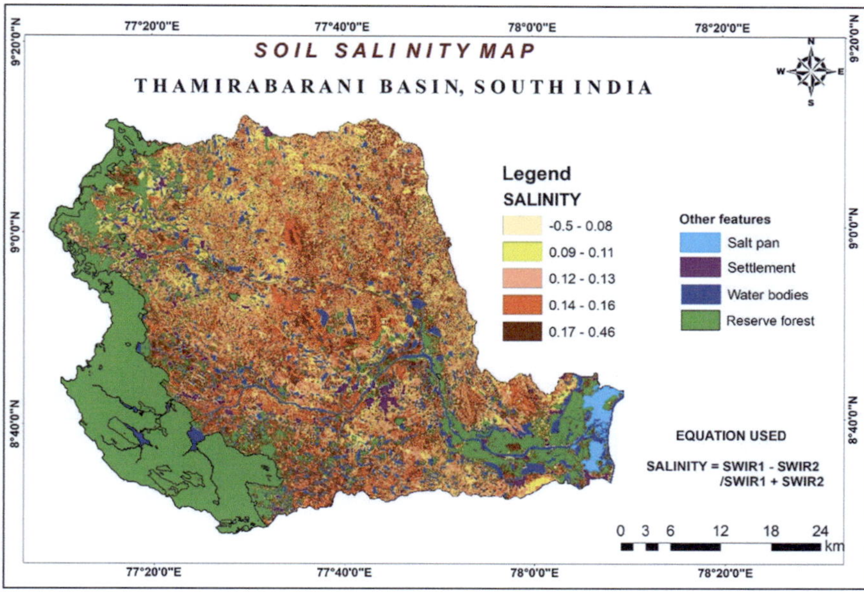

Fig. 12 Soil salinity index characteristics in the basin

places. In this basin, the K-factor values are estimated for various soil types such as clay, clay loam, loamy sand, sand, sandy clay, sandy clay loam, and sandy loam. Coarse-grained soils, like sandy soils, have low K values, ranging from 0.05 to 0.2, due to their higher permeability rate, resulting in a smaller amount of runoff. However, medium-textured soil types like loam, silt loam, and clay soils have higher K-factor values, ranging from 0.25 to 0.4, indicating a higher soil erodibility rate. The spatial characteristics of the K-factor for each soil type are shown in Fig. 13.

5.11 Land Surface Temperature (LST)

The land surface temperature (LST) represents the spatial and temporal variations of the surface temperature equilibrium of the landforms, and it affects the rate of soil-water content (soil moisture) and potential evapotranspiration from the soil layers. It may lead to soil degradation in an area by affecting soil quality characteristics. LST is denoted as the skin temperature of the land and is estimated using the Landsat 9 Thermal Infrared Sensor (TIRS) bands based on its emissivity and brightness values (Pradsad & Mani 2015). It is calculated using the equation:

$$\text{LST} = \text{BT} / \left(1 + \left(W^* \left(\text{BT} / p\right) * \ln\left(e\right)\right)\right) \quad (10)$$

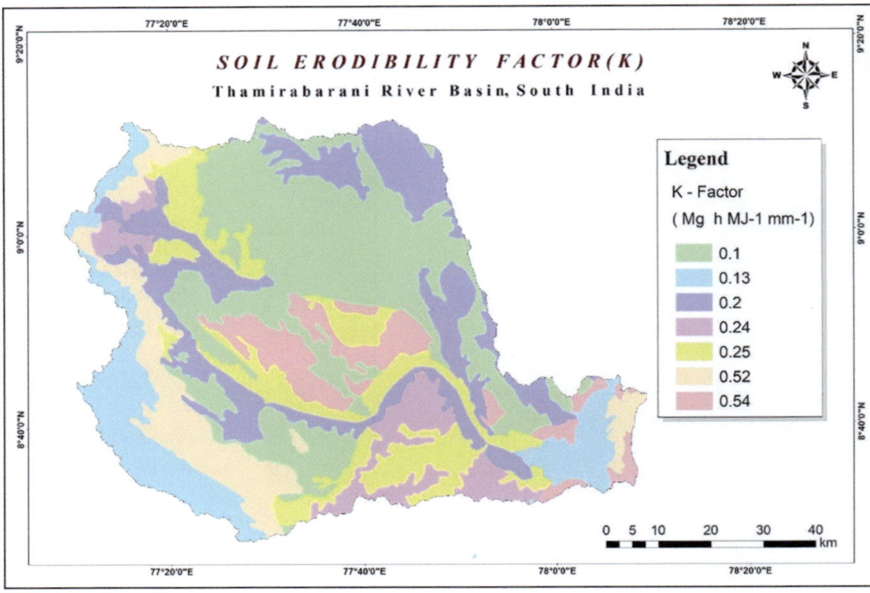

Fig. 13 Soil erodibility (K-factor) characteristics in the basin

where LST refers to the land surface temperature, e is the emissivity, BT is brightness temperature, p is a constant, and W is the wavelength of emitted radiance (11.5 μm). Figure 14 shows the LST for the Thamirabarani river basin of South India. The values range from 23.03 to 43.42 °C. The near-coast areas show a high range of LST, some also with moderate values. The flood plains, wetlands, and foothills showed a lower range of LST. A high range of LST might cause low moisture and high salinity levels, which in turn might affect the vegetal growth of that region, decreasing the soil quality.

5.12 LandUse Land Cover

The land use and land cover (LULC) features were derived from Landsat plate number 1 data using supervised classification techniques in ENVI 5.3 software. In Fig. 15, the LULC features, such as agricultural land, forests, shrubs, scrubs, rivers, water bodies, and settlements (including villages and urbanized areas), are depicted. The basin encompasses approximately 16% of the forested area in the Western Ghats and other upland hills. Within the basin, the middle parts contain 57% of the agricultural lands, while isolated patches along the basin account for 15% of shrubs and scrubs. Water bodies, including rivers, reservoirs, channels, tanks, and ponds, cover 6.9% of the area, while built-up settlements and other infrastructures occupy

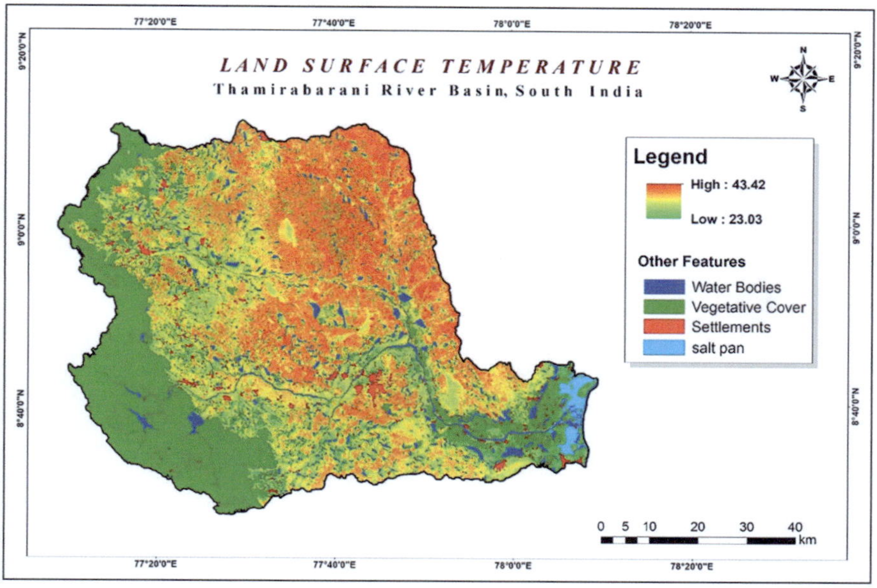

Fig. 14 Spatial distribution of land surface temperature in the basin

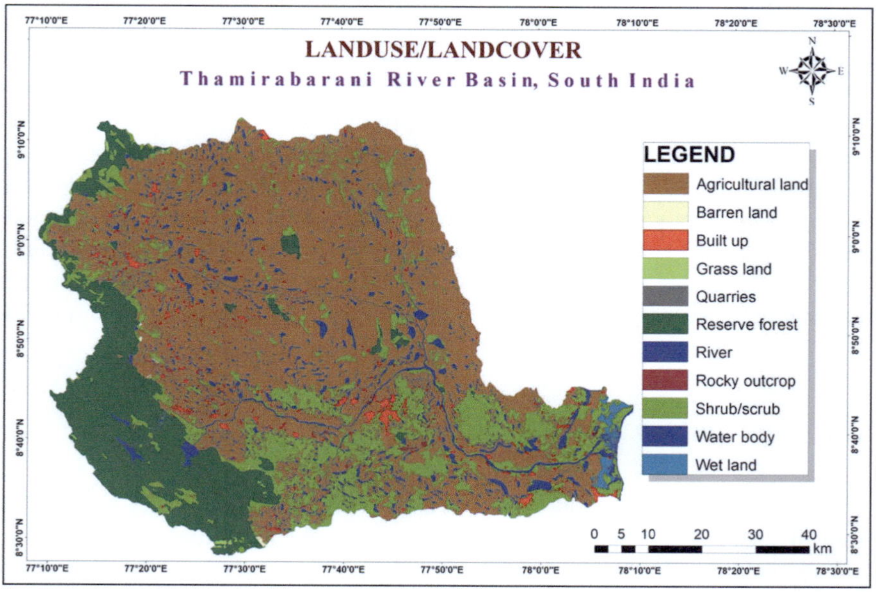

Fig. 15 Landuse/land cover (LULC) features of the basin

3.2%. It is worth noting that many agricultural lands in the basin have been transformed into urbanized habitats.

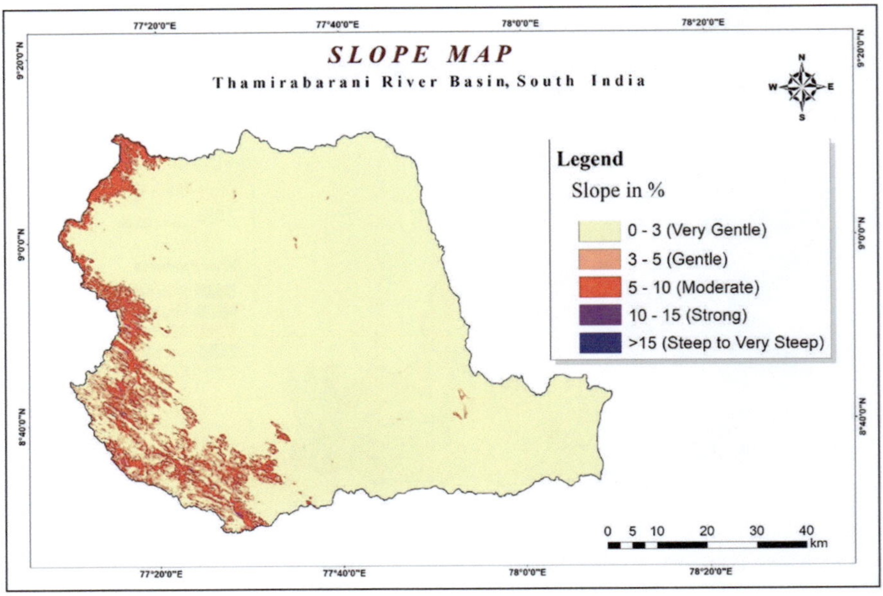

Fig. 16 The slope (topography) characteristics of the basin

5.13 Slope

The slope of the basin is determined using SRTM DEM (30 m) data, which shows that the basin's surface has a gradient sloping from west to east. Figure 16 illustrates the slope characteristics of the basin. The slope varies in different areas depending on the landforms. Steep slopes (>15%) are present in the hilly terrains of the Western Ghats, inland structural hills, and foothills, while the middle pediplains have a moderate slope (5–10%). It is important to note that the slope of the local surface affects soil erosion and surface runoff based on its steepness. Areas with steep slopes are more susceptible to soil degradation due to the loss of nutrients and other properties.

5.14 Soil Quality Zone

The GIS-based soil quality index (SQI) is used to divide the soil quality zones in the basin by analyzing 13 parameters. Soil properties obtained from remote sensing data provide the spatial characteristics of soil physical properties (Chaudhry et al., 2024). The resulting map shows the soil quality based on selected soil properties and soil health and is classified into five classes to assess soil quality at a specific site. A higher SQI value indicates good soil quality zones, which are highly suitable for agricultural activities. On the other hand, a lower SQI suggests poor soil quality zones due to increased salinity, erosion, and degradation, leading to severe impacts on agricultural productivity. Figure 17 illustrates the spatial distribution of soil quality zones in the basin. The resultant map ranged from 36 to 74 as soil quality index values in percentage (%). These values were further broken down into five soil quality zones (very good, good, moderate, poor, and very poor) to facilitate interpretation. Low values, such as 36–44 and 45–51, were classified as poor and poor soil quality zones, covering 1% and 10% of the study area, respectively. The moderate soil quality zone has values of 52–59, covering 46% of the area.

Additionally, values of 60–66 and 67–74 were classified as good and very good soil quality zones, covering 41% and 2% of the area, respectively. Upon comparison with the LULC map, it was observed that agricultural lands and plantations in the study area are associated with good-quality zones, sometimes with moderate quality. Land with shrubs, fallows, and plantations was found to have moderate soil quality. LULC features such as rocky exposures, wetlands, and barren lands were observed in impoverished zones. Further, the soil quality rate is in good to excellent conditions, found with loamy soil, good with loamy, and sandy clay loam soils, respectively. The moderate soil quality zones consisted of sandy loam, clay loam, clay, sandy clay loam, etc. Poor soil quality was found in barren rocky lands and salt-affected areas. Overall, the study shows that the basin mainly comprises moderate to good soil quality, covering 46% and 41% of the area, respectively. These areas are suitable for sustainable land use practices.

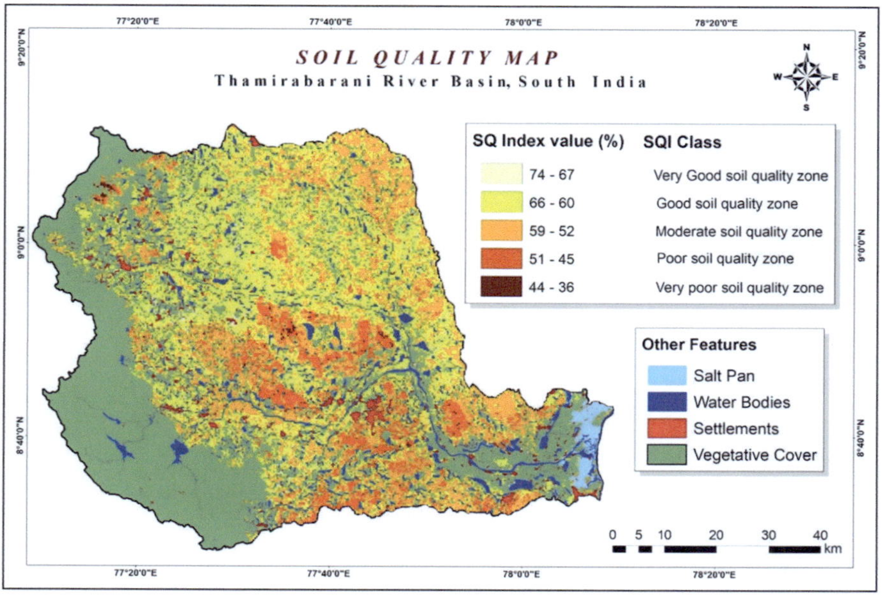

Fig. 17 Spatial distribution of soil quality zone in the basin

6 Field Validation and Accuracy Assessment

The SQI map's results are validated by analyzing the spatial correlation of SOI-derived values at specific sites and their corresponding soil properties, such as ferrous matter, carbonate matter, SOC, and clay content. These soil properties are typically reported as a percentage of topsoil (0–10 cm) per unit area and can be converted into volume for a paddock (parcel of land). Regression analysis for these parameters demonstrates that the field-measured and index-derived values are correlated. Figure 18 depicts the spatial correlation of remote sensing indices derived from soil properties and their corresponding measured data. It is important to note that there are stronger correlations between indices-derived products and measured data for soil carbonate, clay matter, and ferrous matter.

Moreover, the soil quality (SQ) of specific sites was confirmed by comparing them with Google Earth Pro images of the study area. (1) Figure 19a depicts the Arseloor and Sangadredi areas just above the river Thamirabharani. The floodplains along the river exhibit very good to moderate soil quality, as evidenced by the agricultural lands on both sides of the river. The region just above the upper floodplain appears barren, and the land with shrubs indicates poor soil quality. Upon analysis of the soil map, it was found that the floodplains had clay loam, while the poor soil quality zone shows rock land, which might also be a contributing factor to its poor

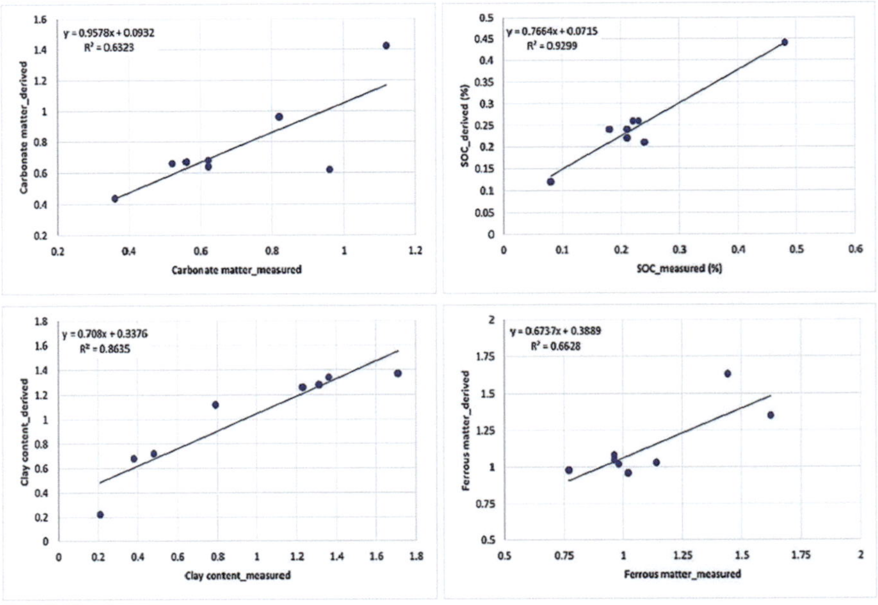

Fig. 18 Validation of soil quality map using soil property data

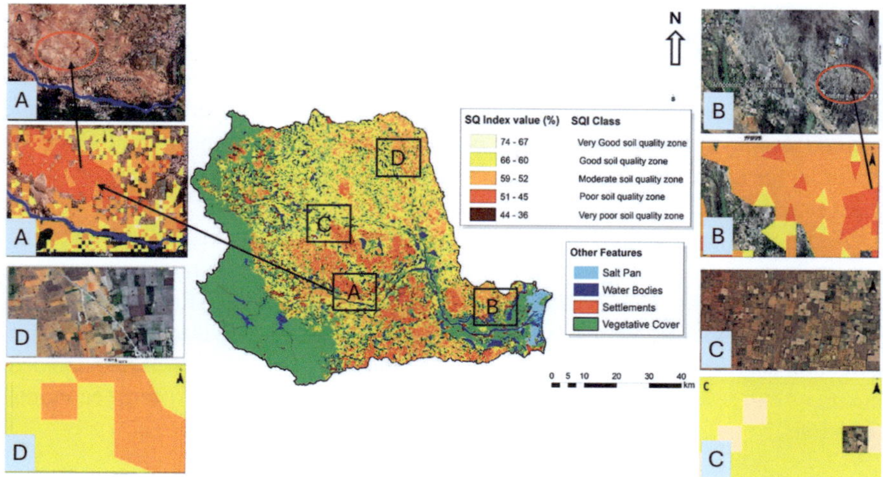

Fig. 19 Soil quality zones comparatively validated with Google Earth Images

quality. (2) Figure 19b shows the regions of Annioperiom and Tannuthu, where the soil quality is at moderate levels. Clay and loamy soils are found in these areas. On analysis with the Google Earth images, it was observed that land with shrubs covers half of the region. Certain areas with good soil quality have associated agricultural land. (3) Figure 19c displays the Melapoine and Nadichakerghudi area above the rivers joining Thamirabarani. The floodplains present along the river show good to moderate soil quality, which can be understood by the presence of agricultural lands on the sides of the river. The region on the eastern floodplain has loamy sand and clay, which might contribute to its good quality. (4) Figure 19d shows that the Karaiyalanur and Solaiseri regions have the most agricultural land. Upon comparison with the results obtained, it is evident that the area shows good to excellent soil quality. The soil found in these regions is loamy soil, which

might be a reason for the soil quality to support agricultural activities.

7 Conclusions

It is crucial to map and assess soil quality for effective land use planning, soil management, crop production, and food security. However, conducting regular soil quality assessments across different scales can be time-consuming and challenging. A study in the Thamirabharani basin in South India demonstrated the effectiveness of remote sensing (RS) and Geographic Information Systems (GIS) in zoning soil quality. The study used 13 parameters, including soil organic matter, organic carbon, clay mineral content, soil iron oxide, soil carbonate, soil salinity, soil moisture, soil erodibility, land use and land cover (LULC), soil texture, land surface temperature (LST), Normalized Difference Moisture Index (NDMI), and slope. These parameters were analyzed using a GIS-based Analytical Hierarchy Process (AHP) to create a comprehensive map for identifying soil quality zones. The study recommended moderate soil quality zones for agricultural and plantation practices, subject to further on-site nutrient analysis. Soil type, other soil properties, land use, and land cover may influence the quality of these areas. The study also highlighted the potential for improving moderate zones through effective agricultural land management practices.

Additionally, the study identified regions with poor soil quality, attributed to factors such as erosion, salinity, loss of organic matter, and surface mining operations. These areas indicate ongoing land degradation processes in the study area. As soils are non-renewable resources, it is crucial to prioritize their preservation and mitigate land degradation. The study emphasized that only qualitative soil management practices can restore degraded soil properties that result from short-term economic decisions.

8 Limitations and Recommendations

- The complexity of soil resources could make the soil assessment process meaningless. According to specific soil communities, education or research must focus on soil quality management practices rather than assessing their quality or health.
- In-situ sampling for various parameters associated with remote sensing index products could provide more accuracy. Field sampling at selected areas could be beneficial in determining the soil quality of the region at various scales and could be meticulously an advanced level of the current project.
- The scientific approach of the study provides soil quality zones at a site-specific scale, which would be significant for the sustainable management of soil resources. Village or rural-level planning for such sustainable management could also be feasible.
- As only a few regions of the study area consist of poor soil quality, these areas could be improved with proper management practices.
- The soil quality could be improved for agricultural purposes by adding organic or bio-fertilizers after extensive field sampling.
- Zones of inferior soil quality could be utilized for other purposes, such as mining operations, proving it is a potential site for such activities. Coastal areas with poor soil quality can be turned to salt cultivation or aquaculture.
- Mitigation strategies for the land degradational processes should be formulated. For example, erosion-prone areas should be designed with architectural measures to prevent soil loss.
- Periodic soil quality monitoring is required to know whether the quality is sustained, aggraded, or degraded. The advent of remote sensing could be significant for soil quality monitoring over a large basin area.

- Land use practices contributing to soil degradation should have strict rules imposed as short-term activities could affect the long-term properties of the soil.

Acknowledgments The corresponding author is thankful to the Director, National Centre for Earth Science Studies (NCESS), Ministry of Earth Sciences (MoES), Government of India, Thiruvananthapuram, for providing continuous support and encouragement to complete this successful research work. The corresponding author is thankful to Dr. K. Maya, Scientist–G, Group Head, BGG, NCESS, for continuous support. The authors thank the anonymous reviewers for their insightful comments and suggestions. The authors acknowledge the IMD, GSI, SOI, USGS-Earth Explorer, and EU-Copernicus Data Space Ecosystem for providing the necessary data sources.

References

Abdellatif, M. A., et al. (2021). A GIS-based approach for the quantitative assessment of soil quality and sustainable agriculture. Sustainability, 13, 13438.

Aksoy, E., et al. (2009). Soil mapping approach in GIS using Landsat satellite imagery and DEM data. *African Journal of Agricultural Research, 4*, 1295–1302.

Anderson, K., et al. (2009). Remote sensing of soil surface properties. *Progress in Physical Geography, 33*, 457–473. https://doi.org/10.1177/0309133309346644

Asfaw et al. (2018). Soil salinity modeling and mapping using remote sensing and GIS: The case of Wonji sugar cane irrigation farm, Ethiopia, Journal of the Saudi Society of Agricultural Sciences, 17(3), 250–258. https://doi.org/10.1016/j.jssas.2016.05.003

Babaeian, E., Sadeghi, M., Jones, S.B., Montzka, C., Vereecken, H., Tuller, M. (2019). Ground, proximal, and satellite remote sensing of soil moisture. *Reviews of Geophysics*, 57, 530–616. https://doi.org/10.1029/2018RG000618

Bhunia, G.S., Kumar Shit, P., Pourghasemi, H.R. (2017). Soil organic carbon mapping using remote sensing techniques and multivariate regression model. *Geocarto international*, 34(2), 215–226. https://doi.org/10.1080/10106049.2017.1381179

Chaudhry, H., Vasava, H. B., Chen, S., Saurette, D., Beri, A., Gillespie, A., Biswas, A. (2024). Evaluating the soil quality index using three methods to assess soil fertility. Sensors, 24(3), 864. https://doi.org/10.3390/s24030864

Chandramohan, K., Vijayalakshmi, G., & Kaliraj, S. (2023). Evaluating the relation of NDVI, NDWI, SMI, and LAI with land and soil degradation zones—A case study of Virudhunagar district, Tamil Nadu, India. In S. Dharumarajan et al. (Eds.), *Remote sensing of soils: Mapping, monitoring, and measurement*. Elsevier. isbn:978-0-44-318773-5. https://doi.org/10.1016/B978-0-443-18773-5.00040-5

de Paul Obade, V., & Lal, R. (2013). Assessing land cover and soil quality by remote sensing and geographical information systems (GIS). *Catena, 104*, 77–92, issn:0341–8162. https://doi.org/10.1016/j.catena.2012.10.014

Dharumarajan, S., Kaliraj, S., Kabindra, A., Lalitha, M., & Nirmal, K. (2023). *Remote sensing of soils mapping, monitoring, and measurement*. Elsevier Publication, isbn:9780443187735.

Else, K., et al. (2018). Soil quality—A critical review. *Soil Biology and Biochemistry, 120*, 105–125, issn:0038–0717.

Fellipe, A. O., et al. (2021). Soil parent material prediction through satellite multispectral analysis on a regional scale at the Western Paulista Plateau, Brazil. *Geoderma Regional, 26*, e00412, issn:2352-0094.

Gholizadeh, A., et al. (2018). Soil organic carbon and texture retrieving and mapping using proximal, airborne, and Sentinel-2 spectral imaging. *Remote Sensing of Environment, 218*, 89–103. https://doi.org/10.1016/j.rse.2018.09.015

Howard, P.J.A., Howard, D.M. (1990). Use of organic carbon and loss-on-ignition to estimate soil organic matter in different soil types and horizons. *Biol Fertil Soils* 9, 306–310. https://doi.org/10.1007/BF00634106

Huete, A. R. (1988). A soil-adjusted vegetation index (SAVI). *Remote Sensing of Environment, 25*(3), 295–309. https://doi.org/10.1016/0034-4257(88)90106-X

Kalinowski, A., et al. (2004). Aster mineral index processing manual. In: *Remote sensing application*.

Kaliraj, S., Reji Srinivas, Palanivel, K., Lakshumanan, C., Chandrasekar, N. (2024a) Remote sensing indices based soil properties measurement—A case study of the Thamirabarani River Basin, South India. In S. Dharumarajan et al., *Remote sensing of soils: Mapping, monitoring, and measurement*. Elsevier. isbn:978-0-44-318773-5. https://doi.org/10.1016/B978-0-443-18773-5.00014-4.

Kaliraj, S., Chandrasekar, N., Parmar, M., Rejith, R. G., Dharumarajan, S., Lalitha, M., & Chandramohan, K. (2024b). Assessment of land degradation vulnerability in the semi-arid region of southern India using GIS-based MEDALUS approach. In S. Dharumarajan et al. (Eds.), Remote sensing of soils: Mapping, monitoring, and measurement. Elsevier. isbn:978-0-44-318773-5. https://doi.org/10.1016/B978-0-443-18773-5.00038-7

Kaliraj, S., Chandrasekar, N., Ramachandran, K. K., & Lalitha, M. (2023). GIS-based NRCS-CN modeling of rainfall-runoff in river Thamirabarani sub-basin, southern India. *Journal of Hydro-Environment Research, 49*, 10–27. https://doi.org/10.1016/j.jher.2023.07.001

Kaliraj, S., Chandrasekar, N., Ramachandran, K. K., Srinivas, Y., & Saravanan, S. (2017). Coastal landuse and land cover change and transformations of Kanyakumari coast, India using remote sensing and GIS. *The Egyptian Journal of Remote Sensing and Space Sciences (Elsevier publication & SCI Journal), 20*, 169. https://doi.org/10.1016/j.ejrs.2017.04.003

Kaliraj, S., Parmar, M., Bahuguna, I. M., & Rajawat, A. S. (2022). Assessment of desertification and land degradation vulnerability in humid tropics and sub-tropical regions of India using remote sensing and GIS techniques. In H. Sajjad et al. (Eds.), *Challenges of disasters in Asia*. Springer. https://doi.org/10.1007/978-981-19-3567-1_2

Kaliraj, S., et al. (2013). Identify potential groundwater recharge zones in Vaigai upper basin, Tamil Nadu, using GIS-based analytical hierarchical process (AHP) technique. *Arabian Journal of Geosciences, 7*, 1385–1401. https://doi.org/10.1007/s12517-013-0849-x

Karale, R., et al. (1983). Soil mapping using remote sensing techniques. *Proceedings of the Indian Academy of Sciences Section C: Engineering Sciences., 6*, 197–208. https://doi.org/10.1007/BF02842884

Karlen, D. L. et al. (2008). *Soil quality assessment: Past, present and future* (pp. 1203). USDA-ARS/UNL Faculty. Retrieved from https://digitalcommons.unl.edu/usdaarsfacpub/1203.

Katebikord, A., et al. (2022). Spatial modelling of soil organic carbon using remotely sensed indices and environmental field inventory variables. *Environmental Monitoring and Assessment, 194*, 152. https://doi.org/10.1007/s10661-022-09842-8

Koushika, S.P., Krishnaveni, A., Pazhanivelan, S., Bharani, A., Arunkumar, V., Periyasamy, D., Ramaswamy, J. (2024). Soil fertility dynamics in the Northeastern agroclimatic zone of Tamil Nadu: a targeted analysis of Tiruvannamalai district. *Plant science today*, 11(sp4), 01–10. https://doi.org/10.14719/pst.5694

Lakhankar, T., et al. (2009). Applications of microwave remote sensing of soil moisture for agricultural applications. *International Journal of Terraspace Science and Engineering, 2*(1), 81–91.

Liu, Y., et al. (2022). A normalized difference bare soil index for remote sensing to improve bare soil mapping accuracy in urban and rural areas. *Catena, 214*, 11. https://doi.org/10.1016/j.catena.2022.106265

Mahmoudabadi, E., et al. (2017). Digital soil mapping using remote sensing indices, terrain attributes, and vegetation features in the rangelands of North-Eastern Iran. *Environmental Monitoring and Assessment, 189*, 1–20. https://doi.org/10.1007/s10661-017-6197-7

Manchanda, M. L., et al. (2002). Soil survey and mapping using remote sensing. *Tropical Ecology, 43*, 61–74.

Mandal, U. K. (2016). Spectral color indices based geospatial modeling of soil organic matter in Chitwan district, Nepal. *International Archives of the Photogrammetry Remote Sensing and Spatial Information Sciences, XLI-B2*, 43–48. https://doi.org/10.5194/isprs-archives-XLI-B2-43-2016

Mehrez, Z., et al. (2011). Editorial remote sensing of soil. *Applied and Environmental Soil Science, 2011*, 904561, 2 pages. https://doi.org/10.1155/2011/904561

Mokarram, M., et al. (2015). Modeling the behavior of vegetation indices in the salt dome of Korsia in north-east of Darab, Fars, Iran. *Modeling Earth Systems and Environment, 1*, 27. https://doi.org/10.1007/s40808-015-0029-y

Muralitharan, J., & Palanivel, K. (2015). Groundwater targeting using remote sensing, geographical information system and analytical hierarchy process method in hard rock aquifer system, Karur district, Tamil Nadu, India. *Earth Science Informatics, 8*, 827–842. https://doi.org/10.1007/s12145-015-0213-7

Nejad, M. F., & Zoratipour, A. (2019). Assessment of LST and NDMI indices using MODIS and Landsat images in Karun riparian forest. *Journal of Forest Science, 65*, 27–32. https://doi.org/10.17221/121/2018-jfs

Nguyen, C. T., et al. (2021). A modified bare soil index to identify bare land features during agricultural fallow-period in Southeast Asia using Landsat 8. *Land, 10*(3), 231. https://doi.org/10.3390/land10030231

Notesco, G., Weksler, S., Ben-Dor, E. (2020). Application of hyperspectral remote sensing in the longwave infrared region to assess the influence of dust from the desert on soil surface mineralogy. Remote sensing, 12(9), 1388. https://doi.org/10.3390/rs12091388

Paul, V., et al. (2014). Assessing the accuracy of soil and water quality characterization using remote sensing. *Water Resources Management, 28*, 5091–5109. https://doi.org/10.1007/s11269-014-0796-7

Pradsad, R., Mani. K (2015). Estimation of spatial variability of land surface temperature using landsat 8 imagery. The international journal of engineering and science (IJES), 4(11), 19–23

Prasad, V. S., et al. (2017). Role of consistency in analytic hierarchy process—Consistency improvement methods. *Indian Journal of Science and Technology, 10*, 1–5. https://doi.org/10.17485/ijst/2017/v10i29/100784

Rakshit, A., et al. (2021). *Soil science: Fundamentals to recent advances*. https://doi.org/10.1007/978-981-16-0917-6.

Rasul, A., et al. (2018). Applying built-up and bare-soil indices from Landsat 8 to cities in dry climates. *Land, 7*(3), 81. https://doi.org/10.3390/LAND7030081

Raza, S. M. H., et al. (2018). Delineation of potential sites for Rice cultivation through multi-criteria evaluation (MCE) using remote sensing and GIS. *International Journal of Plant Production, 12*, 1–11.

Saaty, T.L. (1977) A Scaling Method for Priorities in Hierarchical Structures. *Journal of mathematical psychology*, 15, 57–68. https://doi.org/10.1016/0022-2496(77)90033-5

Saaty, T.L. (1980). The Analytic Hierarchy Process. McGraw-Hill, New York.

Saaty, T.L. (1990). How to Make a Decision: The Analytic Hierarchy Process. *European journal of operational research*, 48, 9–26. https://doi.org/10.1016/0377-2217(90)90057-I

Saaty, T.L. (2005). Analytic Hierarchy Process. https://doi.org/10.1002/0470011815.b2a4a002

Saaty, T.L. and Vargas, L.G. (1991). Prediction, Projection and Forecasting. *Kluwer academic publisher, dordrecht*, 251 p. https://doi.org/10.1007/978-94-015-7952-0

Sahu, N., et al. (2015). High-resolution remote sensing, GPS and GIS in soil resource mapping and characterization—A review. *Agricultural Reviews, 36*, 14. https://doi.org/10.5958/0976-0741.2015.00002.1

Shabou, M., et al. (2015). Soil clay content mapping using a time series of Landsat TM data in semi-arid lands. *Remote Sensing, 7*, 6059–6078. https://doi.org/10.3390/rs70506059

Shankar, D. R. (2017). *Remote sensing of soils*. https://doi.org/10.1007/978-3-662-53740-4.

Simon, T., et al. (2018). Urban soil quality assessment—A comprehensive case study dataset of urban garden soils. *Frontiers in Environmental Science, 6*, 136. issn:2296-665X. https://doi.org/10.3389/fenvs.2018.00136

Strashok, O., et al. (2022). Evaluation and correlation of normalized vegetation index and moisture index in Kyiv. *Journal of Ecological Engineering, 23*(9), 212–218. https://doi.org/10.12911/22998993/151884

Suresh, A. (2016). Estimation of land surface temperature of high range mountain landscape of Devikulam Taluk using Landsat 8 data. *International Journal of Research in Engineering and Technology, 5*, 92–96.

Taloor, A. K., et al. (2021). Retrieval of land surface temperature, normalized difference moisture index, normalized difference water index of the Ravi basin using Landsat data. *Applied Computing and Geosciences, 9*(11), 100051. https://doi.org/10.1016/j.acags.2020.100051

Thaler, A. E., et al. (2019). A new index for remote sensing of soil organic carbon based solely on visible wavelengths. *Soil Science Society of America Journal, 83*(5), 1443. https://doi.org/10.2136/sssaj2018.09.0318

USDA (2006) Keys to Soil Taxonomy. 10th Edition, United States Department of Agriculture, NRCS Soil Conservation Service, Washington DC.

Vani, V. (2017). Comparative study of ndvi and savi vegetation indices in Anantapur district semi-arid areas. *International Journal of Civil Engineering and Technology (IJCIET), 8*(4), 559–566. http://iaeme.com/Home/issue/IJCIET?Volume=8&Issue=4

Livestock Nutrition and Production

Livestock Nutrition: An Issue for Farmers' Friendly Technology

ArkaprabhaShee, Shyamal Kumar Mondal, and Kalyan Sundar Das

Abstract A crop-livestock-based integrated farming method is used to raise a variety of livestock species in tropical regions where livestock is a significant part of the rural economy. Small-scale dispersed product (meat, milk, etc.) production is the distinguishing feature of Indian animal husbandry, which is evolving into a self-sufficient agricultural business. For many families, keeping cattle provides a significant amount of work and a constant source of income, allowing them to raise their level of living. The Indian livestock sector is known for its low productivity. A number of difficulties, including limited grazing supplies as a result of urbanization, industry, and cultivation in fallow areas, prevent realizing livestock production capacity. Traditional feeding management is another important factor. Numerous breakthroughs in animal nutrition research have increased feed's nutritional value while also reducing feed resource waste. Research on the feeding methods that farmers have traditionally used and the feed resources that they have in their hands will be useful to fulfil the nutritional state of livestock as well as identify likely solutions for utilizing the animal's full genetic potential to make livestock farming a profitable endeavour. It is very essential to better understand how farmers are using technology and managing the feeding of their livestock considering the socioeconomic variables as well as to propose recommendations for changing the existing feeding system.

Keywords Animal feed · Livestock feeding · Livestock production · Roughage feed · Feeding technology

---

ArkaprabhaShee
Krishi Vigyan Kendra Murshidabad, (Additional), Ramakrishna Mission Ashrama, Sargachi, Murshidabad, West Bengal, India

Ramakrishna Mission Vivekananda Educational and Research Institute, Belur, Howrah, West Bengal, India

S. K. Mondal · K. S. Das (✉)
ICAR—Agricultural Technology Application Research Institute, Kolkata, Kolkata, West Bengal, India

A. K. Singh et al. (eds.), *Sustainable Agriculture Management in Semi-Arid Climates*, https://doi.org/10.1007/978-3-031-94062-0_12

1 Introduction

One of the primary goals of livestock and poultry farming is to provide consumers with unadulterated and wholesome food in the form of meat, eggs, and milk and to make the enterprise profitable. To achieve this, they must be fed nutritious food. Nutrition entails a variety of chemical reactions and physiological processes that convert meals into energy for tissues and functions (Sagar et al., (2013). It entails ingesting, digesting, and absorbing diverse nutrients, transporting them to all body cells, and eliminating useless materials and metabolic waste products. The animal is provided with feed that precisely meets its nutritional requirements in order to generate higher quality animal products (milk, meat, and eggs), contribute to a cleaner environment, and thus maintain profitability. Nutrient partitioning is a crucial element of productive efficiency that varies among animals.

2 Livestock: A Key Factor for Food Security and Sustainable Development

Livestock contributes significantly to fulfilling human food security in the following ways.

- Supply of important food
- Reduces seasonal food variability
- Provides a source of income
- Generates employment
- Provides energy (draught animal power, source of biogas, dung for fuel, etc.)
- Provides fertilizer and soil conditioner
- Controls weeds
- Controls usury and deposit

3 Scenario of Livestock Feeding

- Farmers face limited resources for feeding their ruminant livestock resulting in insufficient and low-nutritive fodder.
- Residues of crops like wheat and paddy straw are the primary sources of feed.
- Poor nutrition leads to poor productivity, infertility, abortion, and high morbidity.

4 Sources of Roughages

A. *Locally available and low-cost dry roughages*	B. *Seasonally available and costly green fodders*
1. Residues of dry cereal crop-wheat bhusa, paddy straw, maize stover, etc. 2. Pulses straw (Moong straw, pea straw, etc.) 3. Natural grass mixtures in dry form	1. Leucerne, berseem, cowpea, etc. 2. Cereal fodder—oats, maize, sorghum, etc. 3. Green grasses—Durba, sudan, hybrid-napier, etc. 4. Leaves of tree—Pakar, jamun, pipal, etc.

5 Feed Characteristics of Crop Residues

The qualities of crop residues being used as feed are briefly pointed out as follows.

- Deficient in almost maximum nutrients
- High lignocellulose content (resistant to microorganisms which help in digestion)
- Poor palatability
- Low consumption of food
- Low nutrients digestibility
- Maximum chances of fungal toxicity in bajra straw
- Presence of anti-nutritional factors (high levels of silica (8–13%) in sugar cane tops and paddy straw and high calcium-binding oxalate content in paddy straw)

Despite their low nutritional value and high anti-nutritive components, crop leftovers have become the only alternative for ruminant feeding in India due to a scarcity of high-quality roughage supplies. Dry and green fodder are in low supply by 27% and 48%, respectively.

6 Efficiency of Nutrition

Improving nutrient efficiency is a critical goal of the Farm to Fork method, which symbolizes the aspirational goal of creating a circular economy. Nobody is better prepared than the feed and livestock industry to close the nutrient cycle with the 'One Nutrition' concept, which connects human, animal, plant, and microbial populations to reduce nutrient losses while increasing nutrient efficiency.

7 Optimization of Animal Productivity with Poor Quality Roughage

The following means may be used to optimize animal productivity when animals are fed with poor-quality roughages.

7.1 *Supplementation*

The poor-quality roughages must be supplemented to address nutritional imbalances and to provide optimal rumen conditions for effective microbial fermentation.

- Provide high-quality fodder in small amounts for optimal digestion in the rumen
- Offer non-degradable protein sources for lower intestine amino acids and peptides
- Vitamin and mineral supplements

7.2 *Processing of Crop Residues*

The poor-quality crop residues should be processed for the following reasons.

- Increased voluntary feed intake
- Improved palatability
- Improved microbial action by dissociating cellulose and hemicellulose from lignin and silica
- Increased energy availability by reducing digestive losses
- Increased surface area for enzyme and microbe action
- Reduced bulkiness

There are various methods to process dry roughages for its quality improvement which are given in the table below.

Physical methods	Chemical methods	Biological methods
Chaffing	Alkali treatment	White-rot fungi treatment
Bhoosa making	Ammoniation	Enzymes
Water treatment		Mushroom
Steam treatment		
Densification		
Irradiation		

7.2.1 Physical Methods

7.2.1.1 Chaffing

Cut into pieces that are 1–4 cm in length for increasing feed intake and decreasing feed wastage.

7.2.1.2 Bhoosa Making

Threshing converts long straws into smaller particles which are generally 5 cm in length and become comparatively softer. It also increases nutrient absorption from feed.

7.2.1.3 Grinding

To ensure consistent mixing of fodder with other feed ingredients for complete feed, roughage is processed in a hammer mill to make pellets.

7.2.1.4 Irradiation

Radiation sources such as X-rays and gamma rays disrupt the lignocellulose bond and improve digestibility (it increases cost and health risks).

7.2.1.5 Washing by Water

Chaffed straw is soaked in water for about 2 h before being blended. A few livestock owners in the hilly region keep paddy straw on tree branches for washing with rainwater and drying it by aeration. This method removes the water-soluble toxic chemical 'Oxalate', which improves calcium utilization from feed (Ca + Oxalate = Ca − Oxalate).

7.2.1.6 Water Soaking

Soaking dry roughage for more than 3–4 h in water causes dry fodder saturated with absorbed water. It turned dry roughages softer and ultimately increased voluntary feed intake.

7.2.1.7 Densification

Dry roughage may be pressed into compact feeds so that less storage space is required. The following process may be used for densification of dry roughages.

- Bailing: The compression of loose straw using a bailing machine decreases its volume by one-third.
- Briquetting: Compression using pressure and heat. This process is energy intensive.
- Pelleting: The process of chopping, grinding, and boiling before passing through a die. However, this is energy-intensive.
- Block construction: Forage and finish feed blocks

Mechanically solidifying feed is a 'value-adding process'. It decreases transportation costs as well as safety during transit. Fodder in loose form has a low bulk density, making it difficult to supply to remote places quickly and affordably. Fodder densification technology enhances the density of fodder, making it cheaper and lowering handling costs.

Why to Use Complete Feed?

- To ensure balanced nutrition
- To make feed materials storage and transportation easier
- To make space available for expanding stock, pregnant, and breastfeeding animals when space is limited
- To stock feed components in fodder banks for future use (especially during disasters).

Thus, integrating densification with the notion of full feed is a reasonable proposal for fodder banking to achieve sustainable livestock production.

Advantages of Complete Feed Blocks

- Use of unconventional and less appealing feeds to include in the ration of animals for its cost-effective production
- Preparing feed blocks saves one-third more storage space than loose form
- Convenient and trouble-free transportation for such blocks
- Reduces transportation costs
- Reduces storage costs
- Improves nutrient utilization, voluntary intake, and palatability for low-grade roughages
- Little losses during fermentation in a stable rumen environment
- Minimizes nutritional loss from sneezing
- Prevents feed waste by animals at the time of feeding
- Reduces labour costs associated with feeding management

7.2.2 Chemical Treatment of Dry Roughages

7.2.2.1 Alkali Treatment

Alkali disrupts the lignocellulose binding of straw and thus improves its digestibility.

7.2.2.1.1 Sodium Hydroxide Treatment

Four kilograms of sodium hydroxide dissolved in 200 L of water are sprinkled over 100 kg of straw. Then, the treated straw is kept for 3–4 h to respond before feeding. The problem with this technology is (1) sodium hydroxide is expensive and (2) difficult to handle due to its corrosiveness.

7.2.2.1.2 Calcium Hydroxide Treatment

Four kilograms of calcium hydroxide dissolved in 100 L of water are sprinkled over 100 kg of straw. The problem with this technology is also its high cost and poor solubility in water.

7.2.2.2 Ammonia Treatment

7.2.2.2.1 Ammoniation Through Anhydrous Ammonia

In this process, 100 kg of straw is covered in polythene and injected with 3 L of anhydrous ammonia.

Advantages	Disadvantages
• Alkali disrupts the lignocellulose connection (ester linkage) in straw • Increases the digestibility of straw • Improves the nitrogen concentration of straw	• Comparatively too expensive • Unavailable/less available • Transportation of gaseous ammonia is problematic • Needs a lot of attention and a competent individual

7.2.2.2.2 Ammoniation Through Urea Hydrolysis (Urea to Ammonia + Carbon Dioxide)

Initially, 4-kg urea is dissolved in 40 kg of water. Then, the spray is done evenly across 100 kg straw or bhusa. The treated materials are preserved under airtight conditions with plastic sheets for 3 weeks in the hot climate and 4–5 weeks in the cold climate. The following sources of ingredients are used in this process.

- Energy sources: Sugarcane, molasses, and beetroot, whichever is readily available, can be used to produce feed blocks although cane molasses is superior to beetroot molasses. In place of molasses, feed-grade jaggery may be used.

- Nitrogen sources: (1) Non-protein nitrogen like granular urea of fertilizer grade and (2) true protein or vegetable protein like any de-oiled seed cake or meal may be used for this purpose. The best option is de-oiled soybean seed meal, but til cake can also be used in place of mustard cake.
- Mineral sources: (1) Common salt—Feed block can be prepared using any kind of powdered common salt. (2) Mineral mixture—The ideal ratio of vitamins like A, D, E, etc. and minerals like calcium, phosphorus, lysine, methionine, etc. must be used in the feed.
- Structural components: Wheat bran may be used in the making of feed blocks because high-quality soluble carbohydrates present in it can provide structural integrity to the block. Where wheat bran is not available, de-oiled rice bran, a combination of de-oiled rice bran and rice polish, malt sprouts, by-products from bruised barley and malt sprouts, a combination of husk and barley bran and rice polish may also be used.
- Binder: A few inorganic binders like bentonite, cement, gypsum, magnesium, and calcium oxides are used as compact feed block binders. However, in this formulation, an organic binder like guar gum or gum dust can be suggested.

The benefits of urea feeding are (a) increased protein content (9% rice straw and 10% wheat husk), (b) increased straw's palatability, (c) increased digestibility of straw, (d) improved degradability of rumen, (e) the preservation of high moisture materials prevents mould attack, and (f) relatively inexpensive. However, the following precautions must be undertaken while urea feeding is practised.

- Never feed young animals which are less than 6 months old.
- Never exceed the urea limit of 116 g for cow and buffalo and 10 g for goat and sheep.
- Supply enough drinking water.
- Urea must be mixed uniformly to avoid toxicity.
- Avoid introducing urea into your diet quickly.
- Provide adequate soluble carbohydrates (e.g. molasses) for optimal utilization of urea.

The urea toxicity occurs due to (a) urea degrades into ammonia, (b) rumen bacteria use ammonia to produce microbial proteins, (c) microbial proteins are digested in the small intestine to supply protein for animals, (d) concentration of optimal ammonia is 5–8 mg per 100 cc rumen fluid, over which blood ammonia level increases and 1 mg ammonia/100 mL blood is harmful, whereas 3 mg ammonia/100 mL blood is fatal. The reasons for urea toxicity at farmers' doorstep are (1) improper dose of urea feeding, (2) improper urea mixing in feed, and (3) incorrect urea dosage rate, incorrect urea-feed mixture, and unintentional animal usage at the farmer's door, etc. The common signs and symptoms of urea toxicity are bloat, excessive salivation, respiratory difficulty, bellowing, and convulsion. It can be treated with drenching 20–40 L cold water to inhibit ureolytic activity or drenching 4–5 L of 10% acetic acid to bind ammonia.

8 Feed Composition and Value

Farmers must be fully transparent regarding feed composition and nutritional value in order to make informed decisions and create trust with their suppliers. Aside from that, precise and confirmed information acts as the foundation for maximizing the potential of compound feed.

9 Factors for Adopting Technologies

9.1 Livestock Production Profile

Small farmers and landless labourers with limited resources may keep fewer animals due to poor economic conditions. Income, land holding, herd size, and natural resources have a positive relationship in mixed agricultural farming systems, which is consistent with the findings of Kumar et al. (2008).

9.2 Farmers' Socioeconomic Profile

It is clear that larger families have more people willing to share the workload associated with animal production. Over time, the old Indian joint family structure crumbled and the persistence of nuclear families today.

9.3 Literacy

Education and experience help people enhance their knowledge, skills, and attitudes (KSA), which has a big impact on their growth. Income and family education status are positively correlated with mixed farming systems (Kumar et al., 2008). This is because people who are literate (defined as being able to read and write) and educated (defined as having the knowledge to improve one's capacity for thinking and sound judgment) make better use of available income-producing resources. It was learnt that a sizable fraction of knowledgeable persons in India are working in livestock farming as a source of employment and income generation.

9.4 *Occupation*

It was found that most of the farmers were employed primarily in agriculture. Livestock rearing is considered a secondary occupation for them. Crop and livestock production are both examples of small-scale agricultural techniques. It is encouraging to learn that government employees and other service providers are interested in dairying (Devasena et al., 2015). This revealed unequivocally that small and marginal farmers prefer to raise animals for their own consumption and rely on locally sourced agricultural by-products available to them.

10 Feeding Management

Animals graze on natural pastures, which is the cornerstone of livestock feeding systems in maximum areas. The semi-intensive feeding method was definitely the most popular which is followed by a rigorous approach that combined both communal as well as individual stall-feeding. Very few farmers used extensive feeding systems. The food supplies in the farmer's area have a big influence on their preferred feeding strategy. Animals were typically stall-fed, while group feeding was less acceptable in most cases (Devasena et al., 2015). Crop waste was only provided to unproductive animals confined in semi-intensive care once each day. Non-productive animals did not intake concentrated supplements in maximum cases. It was observed that, while farmers made rapid money, only nursing animals were given concentration feeds (oil cakes, brans, pulse chunies, etc.) and cereal remnants. Either all or most of the feed needed was obtained by grazing on native grasslands. Grass or crop waste (straws) served as partial supplementation for animals. It is a matter of surprise that pregnant, dry, and growing animals did not receive any concentrate according to their proper health requirement. Previous observations revealed similar results (Mureda & Mekuriaw, 2008).

The majority of farmers followed semi-intensive feeding techniques, feeding their cattle twice a day before and after allowing them to graze (Devasena et al., 2015). From a managerial perspective, feeding twice or thrice a day has been found to be optimal and cost-effective (Thomas & Sastry, 1991); however, frequent feeding is found to be more beneficial because of its better utilization of nutrients for the improvement of health conditions as well as productivity. Maximum livestock farmers did not maintain the scientific ratio between concentrate and roughage (Devasena et al., 2015). Nutrition has an essential role in determining animal productivity as well as profitability. Equally crucial is identifying the appropriate amount of each ingredient to incorporate into the animal diet. It has been observed that the vast majority of livestock owners are not aware that pregnant animals should be fed concentrate combinations in a very scientific way. The livestock farmers are unaware of scientific feeding practices that may be adopted both for animal welfare and the sustainability of the enterprise. An animal's feed requirements change based on its physiological state; both lactating and non-lactating animals should be fed accordingly. These findings are the same as Mureda and Mekuriaw's (2008) observations that farmers were failing to provide adequate feed to pregnant animals.

The majority of dairy owners are unaware of the correct amount of concentrate combination. When it came time to milk the nursing animals, the amount supplied was usually insufficient. It is observed a limited number of farmers fed rice bran to their calves. Concentrate combinations were rarely offered to calves. Extension organizations should collaborate to influence farmers' perspectives and motivate them to apply the most effective feeding management practices. Meena et al. (2008) reported similar results. A thorough understanding of dairy animal feeding techniques would be beneficial for planning and implementing remedial actions as well as presenting a complete picture of animal farming advances in the study area.

11 Technology Adoption in Feeding Management

Because of their semi-intensive feeding strategy, the farmers fed their livestock on nearby pastures. Despite being aware of the existence of better fodder varieties, farmers were unable to plant enough of them to suit their animals' demands due to a lack of cultivatable land. Green feed, such as local grass, or grown feed made from their own farm products, was fed to the animals based on what was available. One of the most constraint feeding challenges for livestock was a scarcity of high-quality green fodder. This observation is consistent with the findings of Sharma et al. (2007). Sugarcane tops provided an additional source of green feed throughout the harvesting season. In most cases, legume forage supplementation was not used. Few farmers provide premium feeds for their livestock, such as Mulberry (*Morus alba*), Avisa (*Sesbania sesban*), and Subabul (*Luecaenalucocephala*) or locally available good quality forages. Farmers are sometimes unaware of the benefits of feeding legume crops due to a lack of land and difficulty obtaining seeds, limiting their use of these forages. It is also worth mentioning that just a few farmers use machines for chopping, with the vast majority doing so by hand. Chopping of green fodder can reduce waste by 25–30% while preventing selective eating of fodder by the animal. Furthermore, it is feasible to reduce the biological energy waste that happens when an animal chews unchopped material.

Paddy straw was the most frequent agricultural leftover, second only to groundnut straw in terms of quantity as livestock feed is used commonly. Sugarcane tops and sorghum or maize stovers were utilized to a limited extent, based on availability. The majority of farmers collected roughage for animals from their own fields. In reality, some landless farmers bought the dry roughage. Farmers fed their animal concentrates (either cereal grains or oilcake) based on what they had available, commonly using mill wastes (wheat bran, pulse chuni, or rice bran) and handmade concoctions of concentrate. The majority of the farmers used to feed one to two ingredients. Very few animal owners make a homemade concentrate mixture for their animals using rice bran, pulse chuni, and groundnut cake. The ingredient quantities vary according to the availability of raw ingredients. It is not common practice to use ready commercially available concentrate feed. It is apparent that only a few progressive farmers bought commercially available concentrate combinations. The farmer's lack of financial resources was recognized as the greatest barrier to the usage of concentrate supplements.

Farmers did not generally use common salt feeding in livestock feed. Sharma et al. (2007) reported similar results. According to Garg et al. (2005), a few forward-thinking farmers frequently utilize mineral combinations unless they get a veterinarian's recommendation. Farmers who rear livestock are unlikely to have fed mineral mixture to treat their animals because they were ignorant of their benefits. Several farmers thought that the mineral mixture was not used due to its expensive cost. The majority of research in the literature (Chatterjee et al., 2007; Gowda et al., 2008) discovered that using standard minerals or salt was uncommon across the country's states. Minerals are essential nutrients for animal development, reproduction, and production.

It was also noticed that when forages were available, farmers produced hay and silage in restricted quantities as a means of adaptation. Green forages can be preserved and kept for use during lean seasons, significantly reducing seasonal feed shortages. In addition to sugarcane tops, which are abundant during harvest season, green fodder can be found throughout the monsoon season in fields, bunds, roadside vegetation, and beneath forest cover. These fodder kinds can be successfully transformed into silage and fed to animals throughout the year. Feeding ensiled fodder instead of concentrate supplements may be more cost-effective. Crop leftovers and agro-industrial wastes are a valuable but underutilized resource for animal feed. Technology, such as enhancing straw with urea, is not new. Roughages, particularly agricultural leftovers, can benefit from a range of processing or treatment methods that improve crop residue absorption or digestibility, resulting in more effective use. In addition to increasing farmers' income, urea treatment may lower the amount of land needed to produce green fodder. The biggest application limitation was farmers' lack of understanding of the benefits and applicability of these tactics. According to Bharathidasan et al. (2007), the research's benefits do not appear to have been realized in practice.

Only a small percentage of farmers use unconventional roughages and concentrates. Leftovers of many fruits and vegetables like guava, mango, tomato, and papaya wastes are accessible from market yards and the fruit canning sector, and those materials can be fed to livestock. Farmers are not familiar with unconventional animal feeding methods. Mulberry waste is a good feed for cattle. Incorporating non-traditional feeds can minimize feeding expenses while also potentially reducing feed scarcity. Feeding of azolla to livestock has become a good practice but in a slow way, whereas the application of hydroponic fodder cultivation is not accepted rapidly. Appropriate technology selection and implementation are crucial for sustainable animal husbandry. Livestock farmers frequently rely on a semi-intensive rearing method, without total command over feed materials. To exploit ani potentiality, appropriate technologies on economic production and nutritional management must be developed to train livestock farmers.

12 Challenges of Modern Animal Nutrition

Food safety has become a serious issue in the wake of worldwide disasters such as feed contamination (drug, mycotoxin, pesticide, and hormone residues in food items); minor mistakes and consumer faith in food safety are at an all-time low. The principle of 'quality in, quality out' must be recalled, as quality and safe food can only be produced from quality and safe feedstock. The challenge for modern animal nutrition is to produce consumer-acceptable animal products, develop rations free of antibiotics, meat and bone meal, other objectionable feeds formulate rations that do not pollute the environment, and maintain animal health and are economically viable.

13 Strategic Programme in Research and Innovation

To accelerate the shift to more sustainable food systems, the cattle industry requires greater public investment in research than ever before. This includes a better understanding of how animal nutrition research can help in improving animal health and welfare, reducing environmental emissions, and maintaining the highest feed safety standards.

14 Extension Activities for Promotion of Sustainable Livestock Production Through Scientific Feeding

There is an urgent need to carry out different extension activities at farmers' doorstep to promote scientific livestock feeding for sustainable livestock production in the coming days which are discussed below.

- Educating farmers about the unique features and benefits of the technologies
- Creating enthusiasm among new farmers through different extensional activities (demonstrations, field day, exposure visits, etc.)
- Identifying and selecting rural women and marginalized farmers interested in livestock rearing as a means of income generation
- Providing training on scientific nutrition management for livestock and fodder cultivation to enhance their skills and knowledge
- Facilitating access to financial resources, such as loans or grants, to enable the establishment and expansion of livestock enterprises
- Offering vocational training to participants for technical empowerment along with business planning and marketing strategies
- Assisting in the development of business plans customized for every participant's specific goals and resources in feed processing units

- Providing mentorship and ongoing support to help entrepreneurs navigate challenges and seize opportunities
- Development of a range of value-added animal nutrition products and further refinement for betterment
- Establishing partnerships with local veterinary clinics and suppliers to ensure access to quality healthcare products and services for the livestock

However, for developing a long-term strategy, individual, social, and political commitments are very essential.

15 Conclusion

It was concluded that the availability of high-quality feeds and fodders is the main factor that limited livestock productivity. As there is a big difference in the supply and demand for high-quality roughages and the main feed sources for animals are crop leftovers like wheat/paddy straw, establishing and strengthening fodder banks, encouraging fodder production, optimal utilization of available feed resources using appropriate technologies, etc., are very important at this juncture for sustainable livestock production.

References

Bharathidasan, S., Walli, T. K., & Balakrishnan, V. (2007). Feed resources and feeding practices in Tamil nadu and the scope for nutritional interventions for increasing the productivity of dairy animals. *Indian Dairyman, 59*(6), 39–45.

Chatterjee, A., Singh, R. B., Saha, R. C., & Roy, P. K. (2007). Feeding practice and nutritional status in dairy cattle in rural households of old alluvial zone in West Bengal. *Indian Journal of Dairy Science, 60*(6), 408–412.

Devasena, B., Ramana, J. V., Eswara, P. P., & Rama, P. J. (2015). Technology adoption regarding feeding management of crossbred cattle reared in crop-livestock system in chittoor district. *The Haryana Veterinarian, 54*(1), 56–60.

Garg, M. R., Bhanderi, B. M., & Sherasia, P. L. (2005). Assessment of adequacy of macro and micro-mineral content of feedstuffs for dairy animals in semi-arid zone of Rajasthan. *Animal Nutrition and Feed Technology, 5*, 9–20.

Gowda, N. K. S., Prasad, C. S., Selvaraju, S., Reddy, I. J., Anantharam, K., & Sampath, K. T. (2008). Feeding practices and nutrient status of dairy cows under field conditions. *The Indian Veterinary Journal, 85*, 745–748.

Kumar, S., Tripathi, H., & Mandape, M. K. (2008). Income and employment status among self-help group members in dairy husbandry. *Indian Dairyman, 60*(6), 40–46.

Meena, B. S., Kundu, S. S., & Chauhan, J. (2008). Existing feeding patterns of dairy animals in Jhansi district of Bundelkhand region. *Indian Journal of Animal Nutrition, 25*, 63–66.

Mureda, E., & Mekuriaw, Z. (2008). Characteristics and constraints of crossbred dairy cattle production in lowland areas of eastern Ethiopia. *Livestock Research for Rural Development, 20*, 57. Retrieved from http://www.lrrd.org/lrrd20/4/mure20057.htm.

Sagar, V., Anand, R. K., & Dwivedi, S. V. (2013). Nutritional status and reproductive performance of dairy cattle and buffaloes in Sonbhadra district of Uttar Pradesh. *International Journal of Science and Nature, 4*(3), 494–498.
Sharma, P., Pathak, P. S., Meena, B. S., & Pandey, S. (2007). Dairy animal feeding strategies for Bundelkhand region. *Indian Dairyman, 59*(5), 59–71.
Thomas, C. K., & Sastry, N. S. R. (1991). *Dairy bovine production*. Kalyani Publishers.

Capacity Development and Sustainable Livelihood Generation by Successful and Profitable Pigfarming in Semi-Arid Region of India

Madhuchhanda Khan, Kalyan Sundar Das, and Shyamal Kumar Mondal

Abstract Smallholder pig rearing plays a crucial role in sustainable livelihood generation for the disadvantaged classes of society, poor tribal, marginal farmers and landless labourers. The semi-arid agroclimatic region is characterized by low erratic rainfall, intense solar heat and low unstable crop yield. Smallholder pig farming provides insurance against drought-mediated low crop yield, immediate income and secondary livelihood generation under the traditional system of rearing. People raised pigs for two purposes: one is fattening for meat production and another is for piglet production. Tribal people used to keep non-descript pigs especially black ones for household meat consumption, income generation and festival or ceremonial occasions. Nowadays, the rearing of ghungroo pig and improved desi (cross with Large White Yorkshire, Middle White Yorkshire and Tamworth) is started on low input systems like locally available feed material, kitchen waste, agri-byproduct etc. Integration of fish farming with pigs can create a better chance of higher income generation by risk minimization and efficient utilization of resources. Besides cultural and religious taboos, increasing urbanization and industrialization have raised the requirement for pork and high-valued pork products in the form of pork, bacon, ham, sausages, salamis, pickles etc. Strengthening group activity of rural people through the formation of self-help groups, farmer-producer companies and group-based entrepreneurship development in the piggery sector may help to achieve better profit and sustainability as well as women empowerment.

Keywords Breed selection · Marketing · Pig farming · Pig production challenges · Pig production management · Semi-arid region

---

M. Khan
Rathindra Krishi Vigyan Kendra, Palli Siksha Bhavana, Sriniketan, West Bengal, India

K. S. Das · S. K. Mondal (✉)
ICAR—Agricultural Technology Application Research Institute, Kolkata, Kolkata, West Bengal, India

A. K. Singh et al. (eds.), *Sustainable Agriculture Management in Semi-Arid Climates*, https://doi.org/10.1007/978-3-031-94062-0_13

1 Introduction

The livestock sector is intertwined with the Indian agricultural system. This sector plays a pivotal role in the Indian economy by sharing almost 4.9% of Indian GDP and 28.4% of agricultural GDP in the year 2017–2018 (Muthukumar et al., 2021). Pig farming is a sustainable and important venture for the tribal ethnicity of India (Banik et al., 2016; Bharati et al., 2022). The population of pigs in India is 9.06 million which contributes about 1.7% of the total livestock population in India, whereas West Bengal contributes about 0.54 million pigs (20th Livestock Census, 2020). Tribal people used to keep non-descript pigs especially black ones under a nomadic system in their backyard for household meat consumption, income generation and festivals or ceremonial occasions. Over the last few years, tribal rural youths and small and marginal farmers have begun to rear ghungroo pigs as well as improved desi (particularly cross with Large WhiteYorkshire, Middle White Yorkshire and Tamworth) in simple pigsties in semi-arid regions. However, people-reared pigs in developing countries are crossbred with improved breeds or local desi breeds raised under a traditional management system in their backyard (Hide, 2003; Wabacha et al., 2004). Although upgraded pig breeds fetch high economic returns, these breeds also put the resource-poor marginalized farmers at economic risk for their higher investment (Lemke & Zarate, 2008). The socioeconomically weaker sections of the society, particularly the tribal masses of West Bengal, have started increasing pig production. Tribal youth and agri-entrepreneurs small and marginal farmers are considering pig rearing highly profitable because of high prolificacy, fecundity, high feed conversion efficiency, faster growth rate, quite small space requirement and short gestation period. Pigs can thrive well on fibrous low-quality agro byproducts, locally available feed resources, kitchen waste, agri by-products etc., thus requiring low input. Pig production not only provides animal-based nutrition but is also important for higher economic return. People of a different tribal community used to keep non-descript indigenous pigs in their backyard under a traditional rearing system with less scientific intervention. The smallholder pig production system is a major avenue for sustainable livelihood generation in the unorganized pig farming sector in different ways: revenue generation, coverage against adverse climate particularly crop failure, household animal nutrition, manure agricultural land and employment generation for practising farmers and farm women (Lemke & Zarate, 2008). An integrated farming system is an important option for semi-arid regions of India augmenting sustainable and climate-resilient models for higher cash revenue, productivity and employment generation (Rathore et al., 2019). Integration of fish farming with livestock like pigs by small and marginal farmers creates a better chance of higher income generation by diversifying crop production system, risk minimization and efficient utilization of resources (Das et al., 2012). Pigs can be reared in different agroclimatic regions of India with minimum input requirements. Despite of problems and constraints associated with pig farming, this unorganized sector has huge potential to be mobilized as the most profitable livestock sector with proper scientific intervention and extension activities for securing secondary income generation and sustainable livelihood development as well as nutritional security in semi-arid regions of India. To date, the pig production system has not got adequate scope in development programmes in rural India for improving problems of poverty and unemployment of rural youth and the productivity of desi pigs under backyard production systems in semi-arid regions. Arid and semi-arid regions of India cover approximately 166.2 million ha (Srinivasarao et al., 2013) which spread across the length and breadth of the country. Diverse agroecological conditions of this region coupled with low erratic rainfall and intense solar radiation result in drought and low unstable crop yield. Therefore, the implementation of improved technologies not only increases income generation but also raises the socioeconomic status of pig farmers in the region.

2 Pork Production in India

The overall meat production in India is 9.77 million tonnes in the year 2022–2023. In comparison to 2020–2021, meat production has increased by 5.62% although pork production shares around 3.93% of the overall meat production of the country (Basic Animal Husbandry Statistics, 2023). In India, in 2023, total pork consumption is about 292 thousand metric tonnes (Statistica.com, 2024). In India, processed and value-added pork is normally imported. In 2018, nearly 500 metric tonnes of pork was imported in India. The pork exports for the same year were around 270 metric tonnes (Statistica.com, 2024). Pork is consumed in India in two ways—in rural areas, locally raised fresh meat available from farmer's farms; and second, processed meat products like sausage, bacon etc., marketed by a limited number of companies. In India among the pig-producer states, Assam produces the highest number of pigs with 2.10 million pigs followed by Jharkhand (1.28 million), Meghalaya (0.71 million) and West Bengal (0.54 million). Value addition of meat is inadequate in India and less than 2% of total meat is processed into value-added products for trade in India, whereas more than 60% of meat is processed in developed countries (Thomas et al., 2021). The Indigenous breed of pig with their small litter size and slow growth rate produces only 35 kg/animal, whereas the world's average production is approximately 78 kg/animal (Intodia, 2016) (Figs. 1, 2, and 3).

3 Popular Pig Breeds Reared in Semi-Arid Regions of India

The majority of the pigs reared in semi-arid regions are non-descript black pigs with small litter sizes and slower growth rates without any distinct phenotypic character. It is very common in tribal households to rear 4–5 desi pigs in their backyard under a scavenging system of feeding. The predominant tribal community belonging to santhal, oraon and sabar in semi-arid regions rears pigs. Nowadays, agri-entrepreneurs, tribal rural youths, small and marginal farmers and disadvantaged

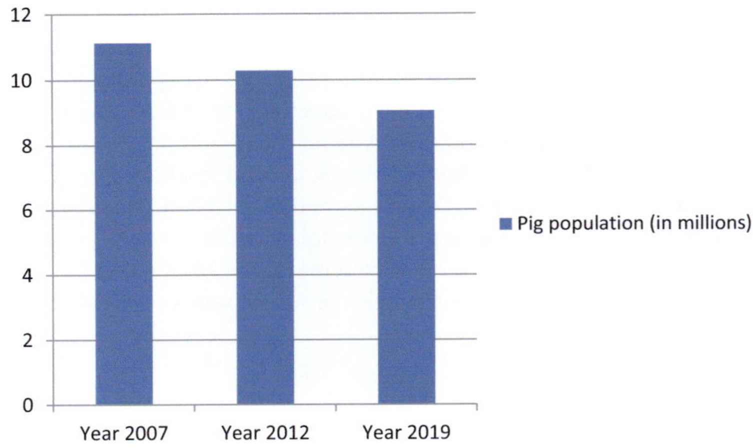

Fig. 1 Pig population (in millions)

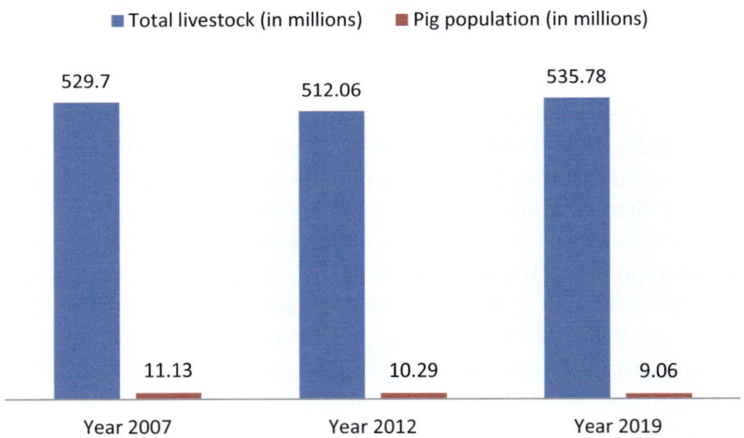

Fig. 2 Comparative pig population (in millions)

classes of society have started to explore the potential of pig farming with their limited capacity. The most common pig breeds reared in a semi-intensive system are crosses with Large White Yorkshire, Middle White Yorkshire and Tamworth mainly. Besides, tribal youth prefer to rear Ghungroo pigs because of their black colour in semi-arid agroclimatic regions. Efficient pig production depends mainly on proper breed selection. Better production performance like higher growth rate, larger litter size, and good meat quality depends on superior parent stock for higher economic return. For improved pig production, genetic improvement through selective breeding, breed up-gradation and cross-breeding to develop the desired level of inheritance is necessary. The development of superior swine germplasm by effective

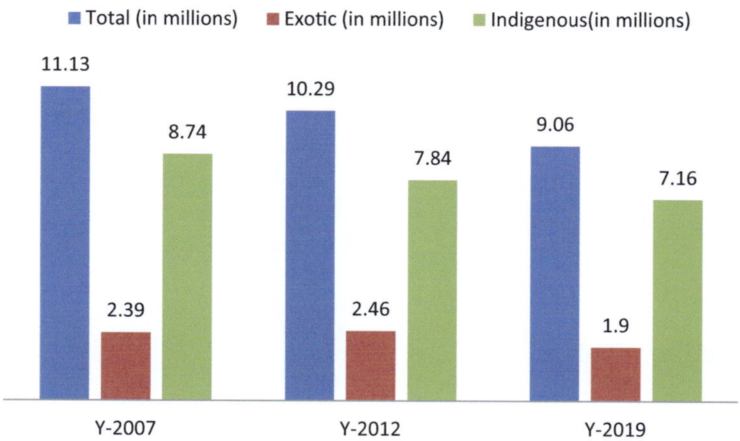

Fig. 3 Change in pig population (in millions)

breeding intervention is essential for the pork industry as well as backyard rearing in India (Banik et al., 2022).

Indian Council of Agricultural Research (ICAR)—National Research Centre (NRC) on Pig, Guwahati, Assam developed the Rani variety of pig by cross-breeding Hampshire and Ghungroo breeds of pig. Jharsuk pig variety was developed by the All India Coordinated Research Project (AICRP) on Pig at Ranchi Veterinary College, Birsa Agricultural University, Ranchi, India by cross-breeding Tamworth and Desi pigs. These pig varieties may be spread for rearing in semi-arid agroclimatic regions because of their faster growth rate, disease resistance, reproductive performance and higher farmer's acceptability (Tables 1, 2, and 3).

Table 1 Production and reproduction performance of Ghungroo pig in semi-arid region

Sl. no.	Particular	Range
1	Litter size	8–12
2	Gestation period (days)	111–115
3	Birth weight (kg)	0.8–1.1
4	Age of puberty (months)	5–7
5	Age of sexual maturity (months)	8–10
6	Furrowing interval (months)	6–7
7	Weaning body weight (kg)	8–9

Table 2 Production and reproduction performance of Large White Yorkshire cross in semi-arid region

Sl. no.	Particular	Range
1	Litter size	7–12
2	Gestation period (days)	111–115
3	Birth weight (kg)	1–1.1
4	Age of puberty (months)	5–6
5	Age of sexual maturity (months)	8–9
6	Furrowing interval (months)	6–7
7	Weaning body weight (kg)	7–10

4 Breeding System

The majority of the resource-poor farmers follow natural service by using their boar in rural areas (Ramesh et al., 2014). Sometimes, very few farmers follow natural service by proven boar. The majority of the stakeholder farmers used to keep male and female pigs together with a sex ratio of 1:3–1:4. Normally, pure breeding, cross-breeding to attain heterosis, breed improvement or breed up-gradation is practised in semi-arid regions of India. Among local breeds, the ghungroo pig is most suitable and well-adapted to the climatic conditions of semi-arid regions. Production potential of this breed is also very promising. Artificial insemination in the piggery sector of semi-arid agroclimatic regions is a very rare practice. Selective breeding, selection based on individual performance, progeny performance, family selection, important economic traits, productive and reproductive performance traits, carcass traits and adaptive traits are necessary for the development of superior swine germplasm (Banik et al., 2022). Constant up-gradation/replacement is required for the smallholder pig production system of the country (Pal et al., 2004). Pure breeding, selection of prized animal with the propagation of pedigreed animal to maintain fecundity and prolificacy, breed improvement by using superior boar to attain higher production potential and cross-breeding to induce hybrid vigour like better growth rate, higher litter size, high birth weight and immunity status must be practised in commercial piggery sector. Artificial insemination may be adapted to induce the desired level of inheritance from a superior pedigree in the commercial piggery sector as well as for smallholder pig production systems in semi-arid regions of India.

Table 3 Production and reproduction performance Tamworth and Desi cross in semi-arid region

Sl. no.	Particular	Range
1	Litter size	7–9
2	Gestation period (days)	111–115
3	Birth weight (kg)	0.9–1.1
4	Age of puberty (months)	6–7
5	Age of sexual maturity (months)	8–9
6	Furrowing interval (months)	6–7
7	Weaning body weight (kg)	7–8

Fig. 4 Common pig breed reared in semi-arid region

Artificial insemination has immense importance in uplifting smallholder pig-rearing systems in rural India (Amin et al., 2010; Visalvethya et al., 2010; Kadirvel et al., 2013). AI in smallholder pig production systems has a significant impact on herd size and profitability in rural areas (Singh et al., 2022). However, non-systematic indiscriminate crossing and selection pose a major threat to achieving the desired blood level inheritance. Small population size is one of the major causes of inbreeding (Yadav et al., 2019), and inbreeding must be kept at an acceptable level per generation to avoid unfavourable consequences and detrimental effects like reduction in mean phenotypic value, premature death etc. (Doekes et al., 2019;Croquet et al., 2006). The rate of inbreeding must be kept below 1% per generation as a general guideline (Meuwissen & Woolliams, 1994; Wakchaure & Ganguly, 2015) (Fig. 4).

5 Housing Practice

Normally, the majority of non-descript pigs are reared under a traditional system of scavenging. Farmers used to provide night shelter made of locally available materials like bamboo, wood and mud. The roof of the sty is made of thatches or tin. In rural India, simple housing systems were designed to deliver feed and water to the animals with minimum fencing and shelter from rain (Pan et al., 2005). However, this backyard system of scavenging is decreasing day by day. Nowadays, agri-entrepreneurs and smallholder pig producers have started to build pig houses with concrete. The roof of the sty is covered by GI sheet, asbestos or tin. The floor is made of cement or sometimes stone. Partly closed or partly opened simple pig sty with a 2–3 m raised platform to avoid dampness or for easy cleaning is observed (Ramesh et al., 2014). Pigs can thrive well under 20–25 °C temperature. Pigs will produce and grow well when protected from extremely low or high temperatures. Shed, air circulation and special arrangements of wallowing, sprinkling and washing should be made to reduce thermal stress in semi-arid regions. In a simple pigsty, an earthen or cement cast manger is used as a feeding trough for pigs. Multiple pens separated by GI pipes can be made to meet the floor space requirement accordingly in a semi-intensive system of pig rearing. Simple low-cost pig houses made of locally available materials like bamboo, wood, mud tin and thatches with low financial input are preferred by the impoverished pig rearers in rural areas. Ghoongroo pigs are reared by practising farmers in both under stall-feeding and stall-feeding-cum-grazing systems although tethered grazing was the common practice in rural India (Pan et al., 2005).

A pig can easily be integrated with fish as IFS is the important strategic solution to adverse climate change particularly crop failure, efficient resource recycling, employment and revenue generation of the farmers as well as good net profit. Wide-scale adoption of IFS will play an important role in the changed agroclimatic situations across arid and semi-arid regions of India (Rathore et al., 2019). Pig manure supplies about 70% of digestible food for fishes with some digestive enzymes for fishes (Bhagawati et al., 2020). Integrated fish-pig farming is a feasible and more profitable scientific intervention with low operational cost as pig dung is used as feed input for fish. Besides, pig dung also fertilizes the pond and enhances the natural productivity of the pond. A pigsty is constructed on the pond dyke with concrete or locally available material with a provision to channel pig manure directly into the pond or a diversion canal to channel the extra manure away to a decomposition pit.

6 Feeding Practices

In smallholder pig farming systems, tribal people usually use broken rice, rice husk, kitchen waste, rice bear waste, colocasia and different seasonal vegetables damaged during transportation or storage as pig feed in semi-arid regions. Pigs are reared on locally available feed resources, agriculture by-products, rice bran, rice bear waste, kitchen waste, bread, chapati, green forages, local grasses, plants like colocasia, damaged vegetables, roots/tuber crops and unconventional feed base to reduce the cost of production. In North-East India, 1.53–3 kg of kitchen waste along with 1 kg of local green grasses were provided per pig as daily average feed (Kumaresan et al., 2009). The most commonly used unconventional feed resources in rural areas are sweet potato (tuber and leaves), pumpkin, banana (Pseudostem, leaves and green banana finger), colocasia (boiled stem and leaves) and tapioca (tubers) crop residue (Kumaresan et al., 2007; Patr et al., 2014). The conventional feed may be replaced by chicken offal, slaughterhouse waste and hotel or kitchen waste (Kannan et al., 2010). For maximum production and better meat quality, concentrate feeding is practised by agripreneurs and rural youth. Chicken offals, fish meal and oil cakes are offered to meet the protein requirement of pigs in some cases. Pig feed should be well-balanced to meet the protein and energy requirements of pigs at different stages of growth, age and physiological conditions. Traditional energy sources of pig feed include maize, wheat, rice, sorghum, oats etc., while different oil cakes like mustard oil cake, ground nut cake, fishmeal, soybean etc. may be provided to meet the protein requirement. To reduce the feed cost, low-cost concentrate feed is prepared by farmers using locally available material. Farmers must be trained to use vitamin, mineral mixture and probiotics in pig ration. Different alternative feedstuffs procured from different industries, including grain milling, baking, brewing, distilling, vegetable and fruit, egg, milk and poultry processing, are used in pig ration to economize feed costs. In diets of grower-finisher pig distillers, dried grains with solubles can be used up to 30% without any harmful effect on pig performance (Cook et al., 2005; De Decker et al., 2005; Xu et al., 2008). Lower inclusion rates may be practised without affecting pig performance (Gowans et al., 2007; Jenkin et al., 2007; Linneen et al., 2008). Precision feeding technology, a crucial part of precision livestock farming is a major breakthrough to make feeding more efficient by giving each individual pig a specific diet and making farming more profitable and environment friendly (Banhazi et al., 2012).

7 Disease Prevalence

Common diseases in the piggery sector observed in semi-arid regions are of bacterial, viral, parasitic and nutritional origin. Due to a lack of awareness, vaccination against swine fever is not generally practised. Diseases like swine fever, FMD, digestive problems, mange and respiratory problems are the common health problems observed in this region. Deworming is usually done by farmers from time to time as and when necessary. Besides, methicillin-resistant *Staphylococcus aureus* infection and *Streptococcus suis* infection, brucellosis etc. are common health hazards in the piggery sector of India (Rajkhowa et al., 2017). Piglet anaemia is a major problem and an important cause of piglet mortality due to deficiency of iron in intensive farming. Supplementation of iron is the only curative measure to prevent piglet mortality by anaemia. Classical swine fever (CSF) or hog cholera is a major threat in the piggery sector. It creates serious economic threats directly due to high mortality and indirectly by creating a ban on exports of pork and pork products (Sarma et al., 2008). Early diagnosis, awareness about the disease epidemiology and vaccination are the only remedies to check the disease. The incidence of African Swine Fever was observed in 2021 in the northeastern states of Nagaland, Mizoram and Manipur, and later on in 2022, cases were reported from Kerala, Assam and

Manipur. No availability of vaccination, high mortality, morbidity and huge economic loss make the disease a deadly and devastating one. Therefore, people need to be vigilant and alert about the occurrence of disease and biosafety, and biosecurity should be maintained at the farmer's level to prevent this contagious disease (Patil et al., 2020).

8 Marketing Linkage

A significant difference between the supply and demand chain of pig and pork products persists in India. In suburbs and metropolises, demand for pig meat and processed pork products is increasing day by day. The increasing growth of population and urbanization in India has created the need for pork and other processed pork products like sausage, bacon etc. Among the tribal community and in certain areas, pork is considered as the most favourite food item. Small pig holder community usually sell their animal on farms to the middle man or traders or individual farmers. They usually face constraints like inadequate access to market details and urban market, skills and interventions in the supply chain, weak institutional provisions and less capacity to process high-quality products needed on the urban markets for pig and pig meat (Lapar et al., 2006; Delgado & Tiongco, 2005; Gulati et al., 2005). Strengthening group activity of rural people through the formation of self-help groups, farmer-producer companies and group-based entrepreneurship development in the piggery sector may help to get credit support and subsidies (Misha Madhavan et al., 2022). Market instability is one of the major problems faced by the small farmers. The effectiveness of livestock marketing depends on the utilization of marketing stores (Adams & Ohene-Yankyera, 2014). Market outlets like retailers, abattoirs, butcher shops, restaurants, hotels, wholesalers, individual consumers and farmer-producer organizations must be explored by the smallholder farmer. The establishment of a centralized pig market may reduce the risk of market volatility. Direct sales through the farmer's association will be more profitable for smallholder farmers. Contract farming with integrating companies is seen to be one of the ways to directly link the smallholder and the market and cut transaction costs (Glover, 1994; Gulati et al., 2005; Costales et al., 2006). Growing demand for pork and pork products in India can be met with value addition and processing of pig meat. Besides cultural and religious taboos, increasing urbanization and industrialization have raised the need for pork and high-valued pork products in the form of pork, bacon, ham, sausages, salamis and pickles at high-end restaurants and hotels. The establishment of a state-of-the-art abattoir and modern pork processing plant is necessary to get maximum profit and mobilization of pig resources in semi-arid regions of the country.

9 Challenges

Major constraints in semi-arid regions of India include lack of superior breeding stock, lack of quality breeding boars, indiscriminate selection, expensive concentrate feed, high rate of production, scarcity of knowledge about bio-security, lack of operational cost, volatile market rate, religious taboo associated with pork consumption etc. Despite its high potential, the smallholder pig farmers faced constraints like minimal technological support including rearing with indigenous pigs, poor nutrition, lack of management and lack of awareness about scientific pig production and marketing systems (Kumaresan et al., 2007, 2009). Stepping up of major challenges, like a poor supply of quality piglets, dearth of improved breed's breeding boar, inbreeding resulting from repetitive use of same boar for a long period, higher feed cost, inadequate training and technical knowledge and non-availability of swine fever vaccine at market, is necessary to make pig production a profitable avenue. Smallholder pig production has the potential for sustainable livelihood development if the farmers could get access to improved pig farming-related technical interventions and new marketing strategies (Munyeche et al., 2011). Throughout developing countries, smallholder pig production system acts as a vital source of animal protein and emergency revenue generation, especially during drought periods (MoAMID (Ministry of Agriculture Mechanization and Irrigation Development), 2010; Halimani et al., 2012). A holistic institutional intervention, like capacity-building training programmes on scientific pig rearing, breed selection, low-cost feed formulation, disease prevention and expert support on vaccination, deworming and disease treatment with respect to pig farming, is essential. Constant surveillance in existing pig farms is necessary to evaluate disease occurrence and spreading (Boyd et al., 2019). Nowadays, precision livestock farming is an essential tool to improve animal well-being, feed conversion, use of antibiotics and result, decrease livestock emissions and increase the net return by minimalizing the yearly costs of the units (Banhazi et al., 2012; Lopes et al., 2016; Pomar & Remus, 2019).

10 Conclusion

In semi-arid regions of India, a demand-supply gap exists in the piggery sector. A huge scope of entrepreneurship development and agribusiness persists in the piggery sector to explore the maximum potential of swine production. Government assistance like the supply of quality piglets, financial assistance including state and centrally sponsored schemes, credit facilities, National Bank for Agricultural and Rural Development (NABARD) sponsored loans and subsidies for entrepreneurship development will promote the piggery sector to develop as an important avenue for sustainable livelihood generation. Besides different capacity-building training programmes and certified courses on pig rearing provided by recognized government organizations will help the farmer to acquire technical know-how on pig farming. Nowadays, the potential of pig production and the processing of pork into high-valued meat products have fascinated agri-entrepreneurs and private organizations to shift traditional farming towards commercial agribusiness. The formation of self-help groups or farmer-producer companies to establish processing units and the development of branded products by landless and small marginal farmers and farm women will help to develop a value chain for the efficient mobilization of pig resources in semi-arid regions of India. This will ensure the sustainable development of the farming community as a whole in semi-arid regions.

References

20th Livestock Census. (2020). Department of Animal Husbandry Dairying and Fisheries, Ministry of Agriculture, Government of India.

Adams, F., & Ohene-Yankyera, K. (2014). Socio-economic characteristics of subsistent small ruminant farmers in three regions of northern Ghana. *Asian Journal of Applied Sciences & Engineering, 3*(3), 351–364.

Amin, N., Tantasupuruk, W., & Teechakumphu, M. (2010). Comparison of artificial insemination with natural mating on small holder farms in Thailand and the effects of boar stimulation and distance of semen delivery on sow reproductive performance. *Tropical Animal Health and Production, 42*, 921–924.

Banhazi, T. M., Babinszky, L., Halas, V., & Tscharke, M. (2012). Precision livestock farming: Precision feeding technologies and sustainable livestock production. *International Journal of Agricultural and Biological Engineering*, 5(4), 54–61.

Banik, S., Naskar, S., Zaman, G., Sarma, D. K., Tamuly, M. K., & Gandhi, R. S. (2016). Doom pig, an indigenous pig germplasm of Assam. *ICAR-National Research Centre on Pig, Rani, Guwahati*, 781131.

Banik, S., Das, P. J., & Kumar, S. (2022). Breeding management in commercial pig breeding farm. In: *Skill for entrepreneurship development in pig husbandry* (pp. 10–16): Edition 2022.

Basic Animal Husbandry Statistics. (2023). Ministry of Fisheries, Animal Husbandry & Dairying. Department of Animal Husbandry and Dairying.

Bhagawati, K., Tamuli, K. K., & Saharia, P. K. (2020). Integrated fish-cum-pig farming system for enhancing farm income in Assam. *KrishiVigyan, 2020*(Special Issue), 209–213.

Bharati, J., De, K., Paul, S., Kumar, S., & Yadav, A. K. (2022). Mobilizing pig resources for capacity development and livelihood security. In A. Kumar, P. Kumar, S. S. Singh, B. H. Trisasongko, & M. Rani (Eds.), *Agriculture, livestock production and aquaculture*. Springer. https://doi.org/10.1007/978-3-030-93262-6_12

Boyd, R. D., Zier-Rush, C. E., Moeser, A. J., Culbertson, M., Rosero, D. S., & Patience, J. F. (2019). Review: Innovation through research in the North American pork industry. *Animal, 13*(12), 2951–2966.

Cook, D., Paton, N., & Gibson, M. (2005). Effect of dietary level of distillers dried grains with solubles (DDGS) on growth performance, mortality and carcass characteristics of grow-finish barrows and gilts. *Journal of Animal Science*, *83*(Suppl. 1):335(Abstr.).

Costales, A., Delgado, C., Catelo, M.A., Lapar, M.L., Tiongco, M., Ehui, S., &Bautista, A. Z. (2006). *Scale and access issues affecting smallholder hog producers in an expanding peri-urban market, southern Luzon, Philippines*. IFPRI research report no. 151. International Food Policy Research Institute.

Croquet, C., Mayeres, P., Gillon, A., Vanderick, S., & Gengler, N. (2006). Inbreeding depression for global and partial economic indexes, production, type, and functional traits. *Journal of Dairy Science, 89*(6), 2257–2267. https://doi.org/10.3168/jds.S0022-0302(06)72297-4

Das, A., Tamuli, M.K., Mohan, N.H & Thomas, R. (2012). Handbook of pig husbandry In Anubrata Das, MK Tamuli,Mohan, NHand R Thomas (Eds.), Today and Tomorrow 's printers and publishers, New Delhi

De Decker, J. M., Ellis, M., Wolter, B. F., Spencer, J., Webel, D. M., Bertelsen, C. M., & Peterson, B. A. (2005). Effects of dietary level of distiller dried grains with solubles and fat on the growth performance of growing pigs. *Journal of Animal Science, 83*(Suppl. 2):79 (Abstr.)

Delgado, C., & Tiongco, M. (2005). Contract farming of milk and poultry in India: Partnerships to promote the environmentally friendly and equitable intensification of smallholder market-oriented livestock production. A project proposal submitted by IFPRI to the Food and Agriculture Organization (FAO) and the International Fund for Agricultural Development (IFAD).

Doekes, H. P., Veerkamp, R. F., Bijma, P., de Jong, G., Hiemstra, S. J., & Windig, J. J. (2019). Inbreeding depression due to recent and ancient inbreeding in Dutch Holstein–Friesian dairy cattle. *Genetics Selection Evolution, 51*(1), 1–16.

Glover, D. (1994). Contract farming and commercialization of agriculture in developing countries. In J. Von Braun & E. Kennedy (Eds.), *Agricultural commercialization, economic development and nutrition*. The John Hopkins University Press.

Gowans, J., Callaahan, M., Yusupov, A., Campbell, N., & Young. M. (2007). Determination of the impact of feeding increasing levels of corn dried distillers grains on performance of growing-finishing pigs reared under commercial conditions. *Advances in Pork Production, 18*, A-22 (Abstr.).

Gulati, A., Minot, N., Delgado, C., & Bora. S. (2005). Growth in high-value agriculture in Asia and the emergence of vertical links with farmers. *Paper presented at symposium on toward high-value agriculture and vertical coordination: Implications for agribusiness and smallholders*, Pusa, New Delhi.

Halimani, T. E., Muchadeyi, F. C., Chimonyo, M., & Dzama, K. (2012). Opportunities for conservation and utilisation of local pig breeds in low-input production systems in Zimbabwe and South Africa. *Tropical Animal Health and Production, 45*(1), 81–90.

Hide, R. (2003). *Pig husbandry in New Guinea: A literature review and bibliography*. Australian Centre for International Agricultural Research.

Intodia, V. (2016). GAIN report. Retrieved from https://apps.fas.usda.govt/newgainapi/api/report/downloadreportbyfilename-pork%20-%202016_New%20 Delhi_India_7-21-2016.pdf.

Jenkin, S., Carter, S., Bundy, J., Lachmann, M., Hancock, J., & Cole. N. (2007). Determination of P-bioavailability in corn and sorghum distillers dried grains with Solubles for growing pigs. *Journal of Animal Science, 85*(Suppl. 2), 113 (Abstr.).

Kadirvel, G., Kumaresan, A., Bujarbaruah, K. M., Venkatasubramanian, V., Das, A., & Ngachan, S. V. (2013). Artificial insemination of pigs reared under smallholder production system in north-eastern India: Success rate, genetic improvement, and monetary benefit. *Tropical Animal Health Production, 45*(2), 679–686.

Kannan, A., Xavier, F., Viswanathan, T. V., & Murugan, M. (2010). Nutrient comparison of different swill materials used for pig rearing in Kerala. *Journal of Veterinary and Animal Sciences, 41*, 73–74.

Kumaresan, A., Bujarbaruah, K. M., Pathak, K., Chetri, B., Das, S. K., Das, A., & Ahmed, S. K. (2007). Performance of pigs reared under traditional tribal low input production system and chemical composition of non-conventional tropical plants used as pigfeed. *Livestock Science, 107*, 294–298.

Kumaresan, A., Bujarbaruah, K. M., Pathak, K. A., Das, A., & Bardoloi, R. K. (2009). Integrated resource-driven pig production systems in a mountainous area of northeast India: Production practices and pig performance. *Tropical Animal Health and Production, 41*(7), 1187–1196.

Lapar, M. L. A., Binh, V. T., Son, N. T., Tiongco, M., Jabbar, M., & Staal, S. (2006). The role of collective action in overcoming barriers to market access by smallholder producers: Some

empirical evidence from northern Vietnam. In: *Research workshop on collective action and market access for smallholders* (vol. 245).
Lemke, U. & Zarate, V.A. (2008). Dynamics and development trends of small holder pig production systems in North Vietnam. *Agricultural Systems*,96,207–223
Linneen, S. K., DeRouchy, J. M., Dritz, S. S., Goodband, R. D., Tokach, M. D., & Nelssen, J. L. (2008). Effects of dried distillers grains with solubles on growing and finishing pig performance in a commercial environment. *Journal of Animal Science, 86*, 1579–1587.
Lopes, S. I., Bexiga, R., & Araújo, J. P. (2016). Precision livestock farming for reproductive performance optimization: A survey. In: *Food futures: Ethics, science and culture. Conference: EurSafe 2016: 13th congress of the European Society for Agricultural and Food Ethics* at Porto, Portugal, pp. 3197–3211.
Meuwissen, T. H. E., & Woolliams, J. A. (1994). Effective sizes of livestock populations to prevent a decline in fitness. *Theoretical and Applied Genetics, 89*(7–8), 1019–1026.
Misha Madhavan, M., Mohan, N. H., Thomas, R., & Jaychitradevi, S. (2022). Entrepreneurship development through self-help groups and farmer producer organizations in piggery sector. In: *Skills for entrepreneurship development in pig husbandry* (pp. 17–22). Edition 2022.
MoAMID (Ministry of Agriculture Mechanization and Irrigation Development). (2010). *Review and stocktaking report on ongoing development efforts in Zimbabwe and their alignment with CAADP targets and principles CAADP implementation in Zimbabwe*.
Munyeche, N., Story, S., Baines, R. N., & Davies. W. P. (2011). Linking small-scale African growers to value markets: Development of a co-owned produce marketing organisation as the standards holder. *ActaHorticulturae, 895*.
Muthukumar, M., Naveena, B. M., Banerjee, R., Singh, V., & Barbuddhe, S. B. (2021). An overview of Indian livestock and meat sector. *Indian Journal of Animal Sciences, 91*(4), 247–254.
Pal, P., Prakash, N., Kumar, R., & Naskar, S. (2004). Performance assessment of improved breed of pig under local condition. *Indian Journal of Animal Sciences, 74*(7), 795–796.
Pan, S., Misra, S. K., & Kundu, M. S. (2005). Ghoongroo pig: A new found animal geneticresource of sub-Himalayan West Bengal, India. *Animal Genetic Resource Information, 37*, 91–96.
Patil, S. S., Suresh, K. P., Vashist, V., Prajapati, A., Pattnaik, B., & Roy, P. (2020). African swine fever: A permanent threat to Indian pigs. *Veterinary World, 13*(10), 2275–2228.
Patr, M. K., Begum, S., & Deka, B. C. (2014). Problems and prospects of traditional pig farming for tribal livelihood in Nagaland. *Indian Research Journal of Extension Education, 14*(4), 6–11.
Pomar, C., & Remus, A. (2019). Precision pig feeding: A breakthrough toward sustainability. *Animal Frontiers, 9*(2019), 52–59.
Rajkhowa, S., Sarma, D. K., & Pegu, S. R. (2017). Virulence markers and antimicrobial resistance of *Streptococcus suis* isolated from diseased pigs. *Indian Journal of Animal Sciences, 87*, 581–583.
Ramesh, J., Gopinathan, A., & Vijayakumar, M. P. (2014). Evaluation of poultry whole carcass meal as an animal protein source in fattening crossbred pig rations. *Advance Applied Research, 6*(2), 190–193.
Rathore V.S., Tanwar S.P.S., Kumar, P. & Yadav, O.P. (2019). Integrated farming system: key to sustainability in arid and semi-arid regions. *Indian Journal of Agricultural Sciences,* 89(2),181–92
Sarma, D. K., Krishna, L., & Mishri, J. (2008). Classical swine fever in pigs and it's status in India. A review. *Indian Journal of Animal Sciences, 78*, 1311–1317.
Singh, M., Mollier, R. T., Pongener, N., Bordoloi, L. J., Kumar, R., & Choudhury, J. K. (2022). Linseed oil in boar's diet during high temperature humidity index(THI) period improves sperm quality characteristics,antioxidant status and fatty acid composition of sperm under hot humid sub-tropical climate. *Theriogenology, 189*, 127–136.
Srinivasarao, C., Venkateswarlu, B., Lal, R., Singh, A. K., & Kundu, S. (2013). Sustainable management of soils of dryland ecosystems of India for enhancing agronomic productivity and sequestering carbon. *Advances in Agronomy, 121*, 253–329.
Statistica.com. (2024). Retrieved from https://www.statista.com/statistics/826720/india-pig-meat-consumption/.

Thomas, R., Singh, V., & Gupta, V. K. (2021). Current status and development prospects of India's pig industry. *Indian Journal of Animal Sciences, 91*(4), 255–268.

Visalvethya, W., Tantasupuruk, W., & Teechakumphu, M. (2010). The development of a model for artificial insemination by backyard pig farmers in Thailand. *Tropical Animal Health and Production, 43*(4), 787–793.

Wabacha, J. K., Maribei, J. M., Mulei, C. M., Kyule, M. N., Zessin, K. H., & Oluoch-Kosura, W. (2004). Health and production measures for smallholder pig production in Kikuyu Division, Central Kenya. *Preventive Veterinary Medicine, 63*, 197–210.

Wakchaure, R., & Ganguly, S. (2015). Inbreeding, its effects and applications in animal genetics and breeding: A review. *International Journal of Emerging Technology and Advanced Engineering, 5*(9), 73–76.

Xu, G. S., Baidoo, K., Johnston, L. J., Cannon, J. E.,Bibus, D., & Shurson, G. C. (2008). Effects of adding increasing levels of corn dried distillers grains with solubles (DDGS) to corn-soybean meal diets on pork fat quality of growing-finishing pigs. *Journal of Animal Science 86*(Suppl. 2):51. (Abstr.)

Yadav, A., Jain, A., Sahu, J., Dubey, A., Gadpayle, R., Barwa, D. K., & Kumar, V. (2019). A reviewon the concept of inbreeding and its impact on livestock. *International Journal of Fauna and Biological Studies, 6*(5), 23–30.

Value Addition in Meat and Fish for Health and Nutrition

Value Addition in Meat and Meat Products for Human Health and Nutrition

Sudip Kumar Das, Olipriya Biswas, E. Angelin Shyona Darwin, Kalyan Sundar Das, and Shyamal Kumar Mondal

Abstract Entailing value addition in meat boosts the economic and functional value of meat products through various processes, resulting in benefits for producers, consumers, and the industry. It includes techniques like marinating, smoking, curing, fermenting, and developing processed foods like sausages, burgers, and ready-to-eat meals. Overall, this improves sensory attributes, extends the shelf life, and increases the convenience of meat products, making them more attractive to consumers. Furthermore, value addition can improve nutritional profiles by including health-promoting ingredients, fortifying with vitamins and minerals, and reducing undesirable ingredients such as fat and sodium. It allows producers and processors to differentiate their products, target niche markets, and increase profitability. Besides, it minimizes waste by utilizing byproducts and less desirable meat cuts. The meat industry sets its standard by providing value addition in specified products and creating new markets, domestically and internationally, promoting innovation and competitiveness. Consumers are reimbursed with a wide array of convenient, tasty, and nutritious meat products that cater to dietary preferences and lifestyles. Overall, the process of value addition in meat is an essential strategy for improving product quality, increasing economic returns, and meeting changing consumer demands.

Keywords Convenience · Functional food · Processing · Value addition

---

S. K. Das
West Bengal University of Animal and Fishery Sciences, Kolkata, West Bengal, India

O. Biswas
Faculty of Fishery Sciences, Department of Fishery Engineering, West Bengal University of Animal and Fishery Sciences, Kolkata, West Bengal, India

E. A. S. Darwin
Department of Livestock Products Technology, West Bengal University of Animal and Fishery Sciences, Kolkata, West Bengal, India

K. S. Das · S. K. Mondal (✉)
ICAR-Agricultural Technology Application Research Institute, Kolkata, West Bengal, India

A. K. Singh et al. (eds.), *Sustainable Agriculture Management in Semi-Arid Climates*, https://doi.org/10.1007/978-3-031-94062-0_14

1 Introduction

Value addition involves various processes and techniques designed to improve the nutritional content, safety, and overall quality of meat products. This approach has garnered significant interest due to a heightened awareness of the link between diet and health, along with a rising demand for foods offering extra health benefits. Meat is naturally rich in essential nutrients, including high-quality proteins, vitamins, and minerals (Sen et al., 2008). However, conventional meat products lack a balanced nutritional profile which has led to the exploration of various methods to fortify and enrich these products (Jadeja et al., 2022). The present focus in processing technology is the upgradation, diversification, and quality assurance. Consumers, nowadays, are more health conscious than ever before, seeking foods that not only taste great but also support overall health and well-being. This shift in consumer preference has driven the meat industry to innovate and offer products that meet these demands. Additionally, the swift urbanization of rural areas, smaller family sizes, an increase in working women, and rising levels of education and awareness about health and hygiene offer valuable insights for expanding domestic markets (Naveena et al., 2018). This review covers various processing strategies and recent advancements in value addition aimed at enhancing the functional quality of meat and meat products.

2 Scope of Value-Added Products

In the global scenario, value-added meat products are expanding rapidly as consumers seek more convenient, high-quality, and diverse meat options. Processed meat products like sausages, bacon, and ready-to-eat are gaining momentum, as they provide improved flavor, an extended shelf life, and additional nutritional benefits (Liu et al., 2023). Increasing living expenses, urbanization, and a preference for protein-rich diets for health are all the key factors in driving the market growth. Innovative processing techniques and the development of new product varieties, such as fortified and functional meat products, are also contributing to this trend (McMillin & Brock, 2005).

In India, the present scenario toward value-added products is also in line with global trends. The rising middle-class population, shifting dietary preferences, and an increasing craving for convenience foods are all maneuvers of this development. Most importantly, India's meat processing industry is evolving, with an emphasis on hygienic production practices and adherence to international quality standards. Also, government initiatives to promote food processing, as well as investments from domestic and international companies, have contributed to the growth in this sector. Despite regulatory hurdles and supply chain inefficiencies, the Indian market has enormous potential, with consumers increasingly seeking ready-to-cook and ready-to-eat meat products that suit their hectic lifestyles and diverse taste preferences (Muñoz-Bas et al., 2023). In addition, the global and Indian markets are also exploring the potential utilization of food waste and byproducts to create value-added products, thereby addressing sustainability concerns and lowering waste. Studies have implored the conversion of food waste into valuable products through bioactive compound extraction, which can be used in various industries, including food and pharmaceuticals (Yadav et al., 2024).

3 Purpose of Value Addition in Meat

Value addition of meat is an approach to processing raw meat with the goal of increasing the convenience, culinary attributes, and commercial value of the food product. This can be accomplished by altering the shape or function of meat or by further processing it. They are beneficial to consumers as they minimize the preparation time and boost the product's value. Value-added products can be broadly categorized based on the method of preparation, variety/convenience, and functionality of the product (Sen et al., 2022). Deboning, size reduction, seasoning, tenderization, tumbling, smoking, retorting, emulsion preparation, and battering are different methods of preparation that are employed in creating various kinds of value-added products. In addition to these, for a successful value-added product to emerge, the essential requirements are the suitable quality of raw materials, accurate formulation, optimal processing, appropriate packaging and storage conditions, nutritional content, labeling requirements, product specifications, and compliance with regulations. The aim of value addition in meat products is to offer valued, diverse, and convenient options that align with consumers' lifestyle needs. There is the utilization of inferior carcass cuts and byproducts effectively with combined technology (Kondaiah, 2004). In this way, it improves the quality, safety concerns, and culinary attributes such as flavor and juiciness of the meat product. Value addition helps in facilitating the export of meat products and competes with imports. Also, it plays a key role in encouraging entrepreneurial ventures and job opportunities. Besides, it helps in developing high-demand value-added products such as low-fat meat products. Furthermore, whenever there is an excess supply of meat, the deficit in marketing channels can be reduced with the help of processed meat products (Bagal, 2024).

4 Value-Added Products

Meat processing involves a variety of technological methods, including particle size reduction techniques (such as chopping, mincing, sectioning, chunking, and flaking), tumbling, and other procedures that modify the biochemical, technological, and microbiological quality of processed meat products. Additionally, new processing methods are continually being developed within industrial settings. In contrast to other types of meat, a significant portion of poultry is transformed into various value-added products, including patties, sausages, breaded nuggets, lollipops, fried drumsticks, and other traditional offerings tailored for different markets. Meat from sheep and goats is typically used to create dishes like kebabs, keema, meatballs, and biryani. Pork is primarily processed into products such as gammon, bacon, sausages, and a selection of traditional meat items. The wide array of meat products available globally can be categorized into various groups based on their characteristics (Table 1). Furthermore, we may categorize meat and meat products according to their functions and intended uses, including functional or designer meat products, geriatric products, and institutional products.

Diverse and value-added meat products can be made through processes such as emulsification, re-structuring, and fortification, among others, thereby improving the nutritional profile, safety, convenience, and sensory qualities of meat, all of which have a significant impact on human health and nutrition.

Table 1 Classification of meat and meat products based on characteristics

Sl. no.	Category	Product example	Characteristics	References
1	Cured meat	Ham, bacon, salami	Treated with salt and nitrites/nitrates for preservation	Warriss (2000)
2	Frozen meat	Frozen chicken breasts, frozen meatballs, burger patties	Preserved by freezing, ready for cooking or reheating	Forrest et al. (1990)
3	Dried meat	Jerky, biltong, pastrami	Low moisture content, long shelf life	McMillin (2018)
4	Canned meat	Corned beef, spam, Vienna sausages, beef stew	Preserved in airtight cans, long shelf life	Fellows (1991)
5	Smoked meat	Smoked salmon, smoked sausage	Treated with smoke for flavor and preservation	Hui et al. (2001)
6	Sausages	Bratwurst, hot dogs	Ground meat mixed with seasonings, stuffed into casings	Sen et al. (2022)
7	Restructured meat and fish	Chicken/fish nuggets, meat patties	Formed from smaller pieces of meat or meat byproducts	Biswas et al. (2016)
8	Dinner meats	Steak with vegetables, seasoned pork roast	Prepared meals, pumped products, battered/breaded meats, precooked or raw, frozen or refrigerated	Hui et al. (2001)
9	Fermented meat	Pepperoni, chorizo	Fermented with beneficial bacteria	McMillin (2018)
10	Low-fat meat products	Chicken patties, nuggets	Fat-substituted meat product	Das et al. (2015)

4.1 Emulsion-Based Products

Emulsion-based meat products are a type of processed meat where a stable mixture is formed by finely dispersing fat and water within a protein matrix. This emulsion provides a uniform texture and consistency, which is characteristic of many popular meat products. These products are produced by mincing boneless meat using a meat mincer, after which an emulsion is created by combining it with salt, condiments, polyphosphates, spices, and binders (Dickinson, 2010). A wide range of meat items, including sausages, patties, nuggets, and kebabs, can be made from the same emulsion. Traditional emulsion meat products are highly esteemed globally and can contain up to 30% fat. This fat is crucial in processing emulsion meat products, as it helps stabilize the meat emulsion, minimizes cooking losses, and enhances viscosity and texture (Kim et al., 2014). Additionally, the development of emulsion-based products facilitates the effective use of byproducts from various processed animals. Moreover, ready-to-eat, heat-and-serve, and traditional convenience products can also be produced.

4.2 Shelf-Stable Products

Shelf-stable meat products are created to maintain their quality at room temperature for long periods without refrigeration. This is done by implementing hurdle technology (HT), which combines various preservation methods, including the management of water activity, humidity reduction, oxidation-reduction potential control, and pH adjustment through the use of acidulants and humectants. These products can be stored safely for more than 3 months. Pickles, meat cookies, and meat spreads are examples of shelf-stable products as they do not require refrigeration. Cooked meat pickle treated with sodium ascorbate was accepted well even after 90 days of storage at ambient temperature (Gadekar et al., 2008). Furthermore, due to the lower initial investment, these products offer a better opportunity for small-scale investors and create jobs for young people. In addition, these products are ideal for long-term storage, emergency preparedness, camping, and transportation.

4.3 Restructured Products

Restructuring involves the partial or complete disassembly of a meat product, which is then reformulated into the same or a different shape. The production of restructured products consists of three main steps: modifying the particle size, blending, and reshaping the meat into the desired form and size. During this process, meat chunks are massaged and tumbled, allowing the pieces to bind together through proteins released as a result of mechanical action. Also, a small quantity of non-meat binders such as alginates may be used in addition to salt, phosphates, and other ingredients (Reddy et al., 2015). The advantages of restructured meat products are the utilization of less desirable cuts and trimmings of meat, offering compatible portion sizes and shapes, tenderness and flavor, low fat and calorific value, and often economical (Boles & Shand, 1998).

4.4 Byproduct-Based Products

Byproduct-based products are made from animal parts that are not typically consumed as whole cuts of meat but still have nutritional and functional benefits. Different byproduct-based products include mechanically separated meat (MSM) wherein residual meat from bones is removed mechanically and infused in products like hot dogs, sausages, and meat nuggets. Organ meat includes the liver, heart, kidneys, and other internal organs, and these are incorporated in products like pâté, liverwurst, haggis, and traditional dishes in various cuisines. Blood can be processed into blood sausage, blood pudding, or used as an ingredient in soups and sauces (Toldrá et al., 2016). Also, bones can be efficiently used to develop value-added products like soup. Meat soup is an excellent source of nourishment for individuals recovering from illness due to its nutritious composition, low-fat content, and easy digestibility. Typically, bone soup is flavored for its potential benefits in the swift recovery from fractures, arthritis, the common cold, and other ailments (Gadekar et al., 2009). Furthermore, they have added benefits as they minimize waste by using more parts of the animal, are often more economical than whole muscle cuts, provide essential nutrients, and promote the texture, flavor, and functionality of processed foods.

4.5 Enrobed Products

Enrobing is the technique of applying edible coatings to processed products to enhance their nutritional value and eating quality. The coating acts as a medium for delivering antimicrobial and antioxidant agents, which helps prolong the product's shelf life. Common enrobing materials include carbohydrates (such as polysaccharides like alginates, carrageenan, and chitosans), proteins (including corn zein, wheat gluten, skim milk powder, soy protein, gelatin, and egg protein), lipids, and resin-based coatings. These materials create a protective seal against moisture, thereby enhancing the product's juiciness. Research has shown that edible coatings can extend the shelf life and improve the quality of fresh, frozen, and processed foods (Biswas et al., 2004). Additionally, this technique enhances the texture of the product while significantly reducing costs and creating opportunities for value addition. Enrobing meat products with edible materials not only increases cooking yields but also provides extra protection against oxidative and microbial

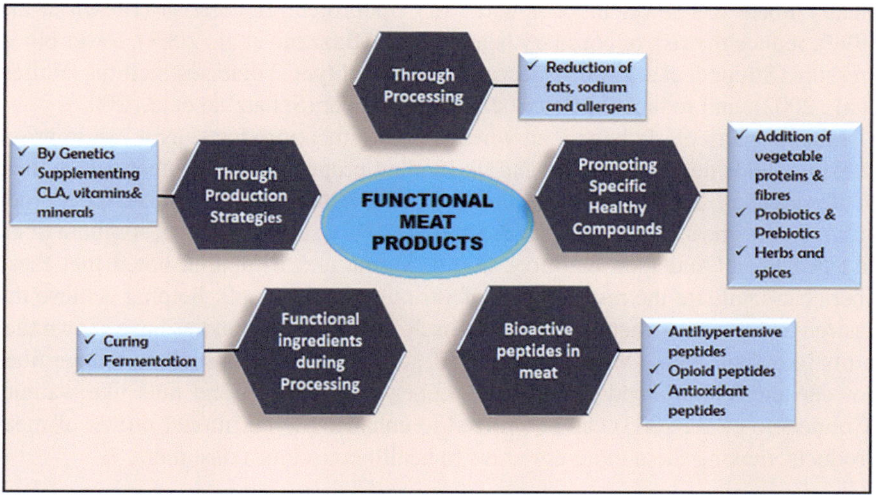

Fig. 1 Different strategies to develop functional meat products

degradation. Furthermore, it enhances the culinary qualities of meat products, making it a highly effective method for value addition that is well-received by consumers (Ahamed et al., 2007).

5 Functional Meat Products for Human Health and Nutrition

Food with contemporary ingredients with a potential advantage for health is termed functional. These are meat and meat-based products that provide health benefits beyond basic nutrition. The positive effects may include improved cardiovascular health, immune function, or digestive health. Value-added products undergo further treatment to increase consumer convenience by reducing processing time and steps and making them convenient, ready-to-eat, and appealing. Functional and healthy meat products are gaining popularity as consumers seek foods that promote their overall health while keeping up with the taste and satisfaction of meat. Meat products can be made functional in the following ways (Fig. 1).

5.1 Fiber-Enriched Meat Products

Dietary fibers are the edible components of plants, including cellulose, hemicellulose, lignin, pectin, and various gums and mucilages, which are indigestible for humans but play a crucial role in digestive health. Research has demonstrated that dietary fibers can lower blood low-density lipoprotein cholesterol (Brown et al., 1999), reduce the risk of coronary heart disease (Bazzano et al., 2003), lower blood pressure (Streppel et al., 2005), decrease the risk of type 2 diabetes mellitus (Willett et al., 2002), and reduce the risk of colorectal cancer (Schatzkin et al., 2007).

Fiber is particularly beneficial when added to meat products, as it can improve cooking yields due to its water- and fat-binding properties while enhancing texture (Cofrades et al., 2000). Common sources of dietary fiber used in cooked meat products include wheat, oats (Talukder & Sharma, 2010), sugar beet, rice, sorghum fiber, and pea fiber (Anderson & Berry, 2000). Verma and Rituparna noted that these fibers can facilitate the restructuring of various meat products, helping achieve the desired texture in restructured ground muscle. Additionally, studies have shown that fruits like apple pulp (Verma et al., 2010), citrus byproducts such as orange fiber powder and lemon albedo (Fernández-López et al., 2004), and nuts like walnuts (Colmenero et al., 2003) can be utilized to enhance the nutritional profile of meat products, making them more attractive to health-conscious consumers.

5.2 Low-Fat Meat Products

Recently, there has been a rising demand for low-fat meat products as it is associated with chronic diseases that disrupt the well-being of human life. Health organizations have proposed that the total fat intake needs to be less than 30% of the total caloric intake (Koutsopoulos et al., 2008). Fats impart flavor attributes and physiological benefits to the product. Incorporation of different plant starches such as sorghum flour, finger millet flour, and carrageenan can significantly substitute the fat during the processing of chicken nuggets without altering their organoleptic quality (Das et al., 2015). However, the reduction of fat causes rubbery and dry textured products affecting the product quality (Colmenero, 2000). The integration of long-chain polyunsaturated n-3 fatty acids (PUFA) is becoming increasingly important in the development of functional meat products (Toldra & Reig, 2004). While the addition of omega-3 fatty acids to food can lead to off-flavors due to the oils' susceptibility to oxidation, various processes have been established to microencapsulate the oil within a gelatin matrix that features double-shell protection, allowing for their incorporation into food products (Kaushik et al., 2015).

5.3 Low Sodium Meat Products

A significant portion of the population is affected by cardiovascular diseases linked to hypertension. The World Health Organization (WHO) recommends that daily dietary salt intake be limited to approximately 5–6 g of NaCl per day (World Health Organization (WHO), 1990). Salt plays a crucial role in meat products, as it enhances water-binding capacity, facilitates the incorporation of fats to create a stable batter, adds desirable flavor and texture, and acts as a preservative. However, due to the growing health concerns associated with excessive salt intake, there is an increasing push to reduce sodium levels in meat products. Research has been conducted to decrease sodium chloride content by substituting it with other chloride salts, such as calcium, lithium, magnesium, and potassium (Hand et al., 1982). Potassium chloride, calcium ascorbate, or their blend could be used as a substitute to a certain level (Kadam et al., 2019).

5.4 *Addition of Natural Antioxidants*

Dietary antioxidants are believed to help prevent oxidative damage in the body and reduce the risk of various diseases, including cardiovascular diseases, certain cancers, Alzheimer's disease, and vision impairment (Holm, 2003). The incorporation of natural antioxidants is emerging as an effective approach to minimizing lipid peroxidation in meat and meat products. Numerous studies have explored the use of herbs, spices, fruits, and vegetable extracts in both raw and cooked meat products, demonstrating a reduction in lipid oxidation, enhanced antioxidant capacity, and improved color stability. Additionally, phenolic compounds with natural antioxidant properties are extracted from plants such as rosemary, tea, soybeans, and citrus peels. These compounds are gaining popularity due to their anti-inflammatory, hepatoprotective, antiallergic, antithrombotic, and anticarcinogenic benefits for human health (Sen & Mandal, 2016).

5.5 *Cardio-Friendly Meat Products*

Functional meat products can be developed by incorporating heart-healthy oils during processing. These include plant oils such as maize, soy, cotton, canola, and linseed, which are rich in α-linolenic acid (ALA) and can replace conventional oils. Another important group consists of geometric and positional isomers of linoleic acid known as conjugated linoleic acids (CLA). CLA is naturally present in meat (Chin et al., 1992) and has been found to exhibit anti-cancer properties in ground beef extract (Ha et al., 1987). In recent decades, there has been increasing interest among researchers in CLA due to its potential benefits for both animal production and human health. However, the results of dietary CLA trials regarding growth, body composition, and meat quality have been inconsistent, likely due to variations in animal species, breeds, age, duration, CLA levels, husbandry conditions, and feed composition. Further research is needed in this area to better understand how to create cardio-friendly meat products.

6 Economics in Value-Added Meat Products

The complexity of the economics of value addition hikes as various factors like consumer demands, production costs, pricing strategies, and economic benefits influence them. Consumers are increasingly inclined to pay a premium for products that offer convenience, high quality, and a diverse selection. This is because of urbanization and lifestyle changes as they opt for more ready-to-eat or easy-to-cook products. Health-conscious consumers stimulate demand for products with added nutritional benefits, such as leaner cuts, low-fat organic options, and products free of artificial additives (Muñoz-Bas et al., 2023). Also, consumers seek diverse tastes and novel flavors boosting demand for innovative functional meat products. This drastic shift supports higher pricing and market expansion.

Investment in advanced technologies and equipment are required to develop meat products with enhanced flavors, improved shelf life, and additional nutritional benefits. Moreover, these also involve more skilled workers to manage these processes. These expenses can be outweighed by the higher prices that value-added products command in the market (Liu et al., 2023). Effective pricing strategies are implemented based on the target market, product type, and competitive position in the place. Also, there is an evolving export market for value-added meat products, particularly from countries with large production capacities to those with high demand but limited production (Coltrain et al., 2000). This sector contributes significantly to economic growth as it creates many job opportunities in rural areas in food processing, packaging, and distribution. Moreover, this sector attracts investment from both domestic and international companies, claiming it as a high-growth area with significant returns (Berwal & Khanna, 2013).

7 Conclusion

The value addition of meat plays a key role in improving human health and nutrition. It requires inputs from commercial, academic, and regulatory interests as it is a multi-step process and also, it has to require acceptance from consumers based on their varied preferences. These aspects must be fully considered in production to enhance the additional health benefits of meat products, which is a great opportunity to heighten the image of the meat industry and fulfill the needs of consumers. These improvements not only respond to growing consumer demand for healthier food choices but also address a variety of public health issues related to malnutrition and chronic disease. There should also be more information and evidence to enable consumers to choose and use these functional foods correctly and achieve the promised health benefits. Marketing surveys, packaging, and advertising play a crucial role in determining the successful movement of a meat product to consumers. In conclusion, the strategic improvement of meat products through value addition not only increases their nutritional value but is also consistent with the broader goals of improving public health and sustainable food production. Continued research and innovation in this area are essential to meet the changing dietary needs and preferences of the global population and to ensure that meat remains an important part of a balanced and healthy diet.

8 Future Perspectives

Value addition in meat is a significant field of research and development with the potential to improve human health and nutrition. Fortification of meat with essential nutrients is gaining high regard in recent times. Incorporating vitamins B12 and D, as well as minerals like iron and zinc, into meat products can help with common nutritional deficiencies. Developing functional meat products by infusing meat with prebiotics and probiotics can promote gut health. Also, adding bioactive compounds such as antioxidants and peptides can enhance well-being. Moving toward healthier meat alternatives by developing low-sodium meat products and promoting lean cuts in processed meat can reduce the risk of obesity and cardiovascular diseases. Another emerging sustainable and nutritious alternative is plant-based meat wherein meat substitutes from plant proteins are developed. They mimic the taste, texture, and nutritional value of meat. These products may be tailored to vegetarians and vegans while providing health benefits. Implementing personalized nutrition by customizing meat products cater to an individual nutritional dietary needs and health conditions. This could be a boon for a specific group of individuals like athletes who require high protein-rich meat, while for those with heart disease, their requirements will be low-fat meats. Enhancing preservation and safety could be done by improvising meat processing techniques, advanced packaging technology, and implementing traceability and transparency. These will provide consumers with detailed information about the origin, production methods, and nutritional content of meat. Also, investing in research and development will aid in collaborative efforts between scientists, industry experts, and policymakers as they can help further bring out innovations in meat value addition. Improving animal welfare and reducing environmental impacts are of great concern globally. Therefore, adopting humane and ethical farming practices and implementing sustainable farming practices to reduce carbon footprint would provide better insights in the near future.

References

Ahamed, M. E., Anjaneyulu, A., Sathu, T., Thomas, R., & Kondaiah, N. (2007). Effect of enrobing on the quality and shelf life of buffalo meat cutlets under frozen storage. *Journal of Muscle Foods, 18*(1), 19–34.

Anderson, E., & Berry, B. (2000). Sensory, shear, and cooking properties of lower-fat beef patties made with inner pea fiber. *Journal of Food Science, 65*(5), 805–810.

Bagal, P. T. A. (2024). Importance of value addition to meat and it's products.pptx [Slideshow]. SlideShare. Retrieved from https://www.slideshare.net/slideshow/importance-of-value-addition-to-meat-and-its-productspptx/267059333#3.

Bazzano, L. A., He, J., Ogden, L. G., Loria, C. M., & Whelton, P. K. (2003). Dietary fiber intake and reduced risk of coronary heart disease in US men and women. *Archives of Internal Medicine, 163*(16), 1897.

Berwal, R. K., & Khanna, N. (2013). Production of value-added chicken meat mince incorporated cookies and their cost economics—benefit ratio. *Pacific Business Review International, 5–12*, 61–70.

Biswas, O., Das, S. K., & Biswas, S. (2016). Restructured fish nuggets from 'wallago attu' extended with soy protein isolate. *Journal of Meat Science, 11*(2), 31–39.

Biswas, A., Keshri, R., & Bisht, G. (2004). Effect of enrobing and antioxidants on quality characteristics of precooked pork patties under chilled and frozen storage conditions. *Meat Science, 66*(3), 733–741.

Boles, J., & Shand, P. (1998). Effect of comminution method and raw binder system in restructured beef. *Meat Science, 49*(3), 297–307.

Brown, L., Rosner, B., Willett, W. W., & Sacks, F. M. (1999). Cholesterol-lowering effects of dietary fiber: a meta-analysis. *American Journal of Clinical Nutrition, 69*(1), 30–42.

Chin, S., Liu, W., Storkson, J., Ha, Y., & Pariza, M. (1992). Dietary sources of conjugated dienoic isomers of linoleic acid, a newly recognized class of anticarcinogens. *Journal of Food Composition and Analysis, 5*(3), 185–197.

Cofrades, S., Guerra, M., Carballo, J., Fernández-Martín, F., & Colmenero, F. J. (2000). Plasma protein and soy fiber content effect on bologna sausage properties as influenced by fat level. *Journal of Food Science, 65*(2), 281–287.

Colmenero, F. J. (2000). Relevant factors in strategies for fat reduction in meat products. *Trends in Food Science & Technology, 11*(2), 56–66.

Colmenero, F. J., Serrano, A., Ayo, J., Solas, M., Cofrades, S., & Carballo, J. (2003). Physicochemical and sensory characteristics of restructured beef steak with added walnuts. *Meat Science, 65*(4), 1391–1397.

Coltrain, D., Barton, D., & Boland, M. (2000). *Value added: Opportunities and strategies*. Arthur Capper Cooperative Center, Department of Agricultural Economics, Kansas State University. Retrieved from https://www.agmanager.info/sites/default/files/VALADD10%25202col.pdf.

Das, S. K., Prabhakaran, P. P., Tanwar, V. K., & Biswas, S. (2015). Effect of some plant starches and carrageenan as fat substitutes in chicken patties. *Journal of Animal Science, 93*(7), 3704–3712.

Dickinson, E. (2010). Food emulsions and foams: Stabilization by particles. *Current Opinion in Colloid & Interface Science, 15*(1–2), 40–49.

Fellows, P. (1991). *Food processing technology: Principles and practice*. Woodhead publishing series in food science, technology and nutrition. Retrieved from http://ci.nii.ac.jp/ncid/BA42985004.

Fernández-López, J., Fernández-Ginés, J., Aleson-Carbonell, L., Sendra, E., Sayas-Barberá, E., & Pérez-Alvarez, J. (2004). Application of functional citrus by-products to meat products. *Trends in Food Science & Technology, 15*(3–4), 176–185.

Forrest, J. L., Aberle, E. D., Hedrick, H. B., Judge, M. D., & Merkel, R. A. (1990). *Principles of meat science*. Retrieved from http://agris.fao.org/agris-search/search.do?request_locale=en&recordID=ES19800511159.

Gadekar, Y. P., Anjaneyulu, A. S. R., Kandeepan, G., Mendiratta, S. K., & Kondaiah, N. (2008). Safe pickle with improved sensory traits: Effect of sodium ascorbate on quality of shelf stable pickled cooked chicken. *Fleischwirtschaft International, 4*, 56–62.

Gadekar, Y. P., Anjaneyulu, A. S. R., Thomas, R., Mendiratta, S. K., & Kondaiah, N. (2009). Quality changes in soup from deboned chicken frames at refrigerated (4 ± 1 °C) and frozen (−18 ± 1 °C) storage. *International Journal of Food Science & Technology, 44*(9), 1763–1769.

Ha, Y., Grimm, N., & Pariza, M. (1987). Anticarcinogens from fried ground beef: Heat-altered derivatives of linoleic acid. *Carcinogenesis, 8*(12), 1881–1887.

Hand, L. W., Terrell, R. N., & Smith, G. C. (1982). Effects of complete or partial replacement of sodium chloride on processing and sensory properties of hams. *Journal of Food Science, 47*(6), 1776–1778.

Holm, F. (2003). *New functional foods ingredients cardiovascular health.* Project no (no. 5). QLK1-CT-2000-00040, SMEs.

Hui, Y. H., Nip, W., Rogers, R. W., & Yoong, O. A. (2001). *Meat science and applications* (pp. 1–710). Marcel Dekker.

Jadeja, R., Teng, X. M., Mohan, A., & Duggirala, K. (2022). Value-added utilization of beef by-products and low-value comminuted beef: Challenges and opportunities. *Current Opinion in Food Science, 48*, 100938. https://doi.org/10.1016/j.cofs.2022.100938

Kadam, V., Gadekar, Y., Shakyawar, D., & Shinde, A. (2019). *Sheep produce: Wool, mutton and milk.* Retrieved from https://krishi.icar.gov.in/jspui/handle/123456789/51581.

Kaushik, P., Dowling, K., Barrow, C. J., & Adhikari, B. (2015). Microencapsulation of omega-3 fatty acids: A review of microencapsulation and characterization methods. *Journal of Functional Foods, 19*, 868–881.

Kim, H., Choi, M., Hwang, K., Song, D., Kim, Y., Ham, Y., Chang, S., Lim, Y., Choi, Y., & Kim, C. (2014). Effect of mixing ratio between Pork Loin and chicken breast on textural and sensory properties of emulsion sausages. *Han-gugChugsanSigpum Hag-hoeji/Korean Journal for Food Science of Animal Resources, 34*(2), 133–140.

Kondaiah, N. (2004). Value added meat products and development of processed meat sector. *Natural Product Radiance, 3*(4), 281–283.

Koutsopoulos, D., Koutsimanis, G., & Bloukas, J. (2008). Effect of carrageenan level and packaging during ripening on processing and quality characteristics of low-fat fermented sausages produced with olive oil. *Meat Science, 79*(1), 188–197.

Liu, Z., De Souza, T. S. P., Holland, B., Dunshea, F., Barrow, C., & Suleria, H. A. R. (2023). Valorization of food waste to produce value-added products based on its bioactive compounds. *Processes, 11*(3), 840.

McMillin, K. (2018). Book review Lawrie's meat science, 8th edition, edited by Fidel Toldrá. *Meat Science, 140*, 157.

McMillin, K. W., & Brock, A. P. (2005). Production practices and processing for value-added goat meat. *Journal of Animal Science, 83*(suppl_13), E57–E68. https://doi.org/10.2527/2005.8313_supple57x

Muñoz-Bas, C., Muñoz-Tebar, N., Candela-Salvador, L., Sayas-Barberá, E., Viuda-Martos, M., Pérez-Alvarez, J. A., & Fernández-López, J. (2023). Development of value-added products suitable for food applications from fresh date fruit (Confitera cv.) and its co-products. *Food and Bioprocess Technology, 17*(5), 1265–1277.

Naveena, B. M., Banerjee, R., Muthukumar, M., & Patil, G. (2018). Opportunities for double farmers income in livestock value chain. In: Summer school on'Innovations in livestock sector for doubling farmers' income: Strategies and oppurtunities in meat value chain'. 25th July to 14th August, ICAR-NRC on Meat, Hyderabad, 1–9.

Reddy, G. B., Sen, A., Reddy, K. S., Reddy, K. K., Rao, T. M., & Kondaiah, N. (2015). Quality characteristics of restructured mutton slices developed by cold-set binding system. *Indian Journal of Small Ruminants, 21*(2), 300.

Schatzkin, A., Mouw, T., Park, Y., Subar, A. F., Kipnis, V., Hollenbeck, A., Leitzmann, M. F., & Thompson, F. E. (2007). Dietary fiber and whole-grain consumption in relation to colorectal cancer in the NIH-AARP Diet and Health Study. *American Journal of Clinical Nutrition, 85*(5), 1353–1360.

Sen, A., & Mandal, P. (2016). Use of natural antioxidants in muscle foods and their benefits in human health: An overview. *International Journal of Meat Science, 7*(1), 1–5. https://doi.org/10.3923/ijmeat.2017.1.5

Sen, A. R., Naveena, B. M., Banerjee, R., & Muthukumar, M. (2022). *Value addition in meat and fish products for human health and nutrition* (pp. 287–303). Springer eBooks. https://doi.org/10.1007/978-3-030-93258-9_15

Sen, A. R., Naveena, B. M., Muthukumar, M., Girish, P. S., & Kondaiah, N. (2008). Functional meat products and their health benefits. *Indian Food Industry, 27*(5), 54–56.

Streppel, M. T., Arends, L. R., Van't Veer, P., Grobbee, D. E., & Geleijnse, J. M. (2005). Dietary fiber and blood pressure. *Archives of Internal Medicine, 165*(2), 150.

Talukder, S., & Sharma, D. P. (2010). Development of dietary fiber rich chicken meat patties using wheat and oat bran. *Journal of Food Science and Technology, 47*(2), 224–229.

Toldrá, F., Mora, L., & Reig, M. (2016). New insights into meat by-product utilization. *Meat Science, 120*, 54–59.

Toldra, F., & Reig, M. (2004). Analysis of meat in food. In L. M. L. Nollet (Ed.), *Handbook of food analysis*. Mercel Dekker.

Verma, A. K., Sharma, B., & Banerjee, R. (2010). Effect of sodium chloride replacement and apple pulp inclusion on the physico-chemical, textural and sensory properties of low-fat chicken nuggets. *Food Science & Technology, 43*(4), 715–719.

Warriss, P. (2000). *Meat science: An introductory text*. CABI Publishing eBooks.

Willett, W., Manson, J., & Liu, S. (2002). Glycemic index, glycemic load, and risk of type 2 diabetes. *American Journal of Clinical Nutrition, 76*(1), 274–280.

World Health Organization (WHO). (1990). *Diet, nutrition, and prevention of chronic diseases*. Technical report series, 797, Geneva.

Yadav, S., Malik, K., Moore, J. M., Kamboj, B. R., Malik, S., Malik, V. K., Arya, S., Singh, K., Mahanta, S., & Bishnoi, D. K. (2024). Valorisation of Agri-food waste for bioactive compounds: Recent trends and future sustainable challenges. *Molecules, 29*(9), 2055.

Value Addition in Fish and Fish Products for Human Health and Nutrition

Supratim Chowdhury, Manasi Chakraborty, Anindya Nayak, and Shyamal Kumar Mondal

Abstract The global demand for fish and seafood products continues to grow due to their nutritional value and health benefits. Fish is a rich source of high-quality proteins, omega-3 fatty acids, essential vitamins, and minerals that play a pivotal role in human health and nutrition. Value addition in fish products extends their shelf life, enhances their palatability, and introduces functional attributes to meet diverse consumer preferences and dietary needs. This chapter explores innovative approaches to value addition in fish, including processing technologies, the development of ready-to-eat and ready-to-cook products, fortified fish-based foods, and nutraceuticals. It also discusses the otential of by-product utilization to promote sustainability and reduce waste in the fisheries sector. Furthermore, the chapter emphasizes the role of value-added fish products in addressing malnutrition and non-communicable diseases. The integration of traditional knowledge with modern processing techniques is highlighted as a pathway to create culturally relevant, health-focused fish products. This chapter aims to provide insights into the strategies and challenges of enhancing the nutritional and commercial value of fish, contributing to the promotion of public health and the sustainable growth of the fisheries industry.

Keywords Fish products · Value addition · Human health · Nutrition · Functional foods

S. Chowdhury
Faculty of Fishery Sciences, Department of Fish Processing Technology, West Bengal University of Animal and Fishery Sciences, Kolkata, West Bengal, India

M. Chakraborty
Ramkrishna Ashram KVK, North 24 Parganas, West Bengal, India

A. Nayak
North 24 Parganas Krishi Vigyan Kendra, West Bengal University of Animal and Fishery Sciences, North 24 Parganas, West Bengal, India

S. K. Mondal (✉)
ICAR—Agricultural Technology Application Research Institute, Kolkata, Kolkata, West Bengal, India

A. K. Singh et al. (eds.), *Sustainable Agriculture Management in Semi-Arid Climates*, https://doi.org/10.1007/978-3-031-94062-0_15

1 Introduction

Value addition in agriculture involves enhancing the worth of products at various stages of the value chain by different actors. This can be related to food quality, cost efficiency, delivery times, flexibility, and innovation. Opportunities for adding value depend on several factors, including market characteristics, social, institutional, and environmental conditions, as well as the technological capabilities of the actors involved (Dos-Santos, 2020). Sathya (2018) referred value addition as the monetary value, health benefits, or any other measure of significance that enhances the utility factors of a food. It can be generated or enhanced through the application of suitable processes. These processes may involve the application of technologies, or tools, applied to living or non-living matter for the purpose of transforming it to value-added products. Such transformation has the possibility to impact the production and processing stages, incorporating innovations and their dissemination. Fish can be processed to develop a variety of value-added products depending on the consumer preference and acceptability. Traditional fish-based products, processed using ethnic methods, are still popular among the ethnic populations and serve to enhance both flavor and nutritional security. Shelf-stable, retortable pouches, which require no refrigeration, are also very popular. Additionally, minced fish can be transformed into various products such as fish sausages, cutlets, patties, balls, pastes, surimi, and texturized items. Advanced nonthermal processing techniques like high-pressure processing, infrared heating, and sous vide cooking are gaining popularity due to their energy-saving and environmentally friendly benefits.

2 Fish in Human Nutrition

Fish and seafood are highly nutritious, providing a rich source of protein, healthy fats, and essential micronutrients. Compared to land animals, aquatic foods are higher in protein, lower in calories, and contain greater amounts of omega-3 long-chain polyunsaturated fatty acids (n-3 LC PUFA). These nutrients are linked to numerous health benefits, including reduced risks of coronary heart disease, cardiovascular disease, inflammatory conditions like arthritis, and even cancer prevention (Tilami & Sampels, 2017). Diets high in polyunsaturated fatty acids, particularly long-chain omega-3s ($\geq$C20), have been shown to support heart health, prevent arteriosclerosis, and help manage autoimmune diseases due to their immunosuppressive properties.

During pregnancy and early childhood, an optimal intake of LC n-3 PUFA is crucial for neural development, as docosahexaenoic acid (DHA), found in the retina and brain, is essential for the nervous system of newborns. It is well-established that omega-3 PUFAs are vital for healthy pre- and postnatal growth (Tilami & Sampels, 2017). Oily fish, rich in n-3 LC PUFA, are an excellent dietary source, and consuming fish 1–2 times per week is linked to a lower risk of coronary heart disease and ischemic stroke (FAO and WHO, 2011).

In addition to omega-3 fatty acids, fish and seafood provide a well-balanced amino acid profile and are rich in essential nutrients like taurine, choline, vitamins D3 and B12, calcium, phosphorus, iodine, and selenium. They also supply important vitamins and minerals like vitamin A, iron, and zinc, which are crucial in regions where these nutrients are deficient. Fish proteins are highly digestible and contain essential amino acids, such as methionine and lysine, which are often limited in terrestrial meats. Research indicates that fish proteins, peptides, and hydrolysates offer health benefits similar to fish lipids. For example, sardine protein has been shown to improve insulin resistance, reduce leptin and TNFα levels, and decrease oxidative stress in rats with metabolic syndrome (Tilami & Sampels, 2017).

Seafood is also an excellent source of selenium, which functions as a cofactor for antioxidant enzymes like glutathione peroxidases and is essential for thyroid function. Low selenium levels have been associated with higher risks of cardiovascular disease, cancer, and kidney disease. Selenium also helps mitigate the toxicity of methylmercury. Fish and fish bones are rich in calcium, with absorption rates comparable to skimmed milk. Aquatic foods provide significant calcium, ranging from 68 to 260 mg/100 g in crustaceans, mollusks, and fish, compared to just 14 mg/100 g in terrestrial meats. Additionally, fish and seafood offer more phosphorus, with an average of 204–230 mg/100 g, compared to 176 mg/100 g in land animals. Phosphorus is essential for bone health and cell membranes, as it is a key component of phospholipids and various intracellular compounds like nucleic acids, creatine phosphate, and ATP (Tilami & Sampels, 2017).

3 Processing Techniques

3.1 Traditional Methods of Fish Processing

Traditionally, various preservation methods, including curing, drying, fermentation, smoking, and conventional storage techniques such as refrigeration and freezing, have been widely used for fish and seafood (Hassoun et al., 2022). Tahiluddin and Kadak (2022) categorized fish processing into traditional and non-traditional methods. Traditional methods include smoking, drying, and salting, while non-traditional methods encompass chilling, freezing, and canning. Traditional processing is typically carried out with minimal facilities, low-cost technology, and is often associated with small-scale fisheries. In contrast, non-traditional methods like canning, freezing, and chilling involve higher investment and are more commonly used in large-scale factories and international trade (FAO, 2021). Drying, smoking, fermentation, and salting are the most widespread traditional techniques for preserving fish (Valu et al., 2023). The traditional and conventional methods for preserving fish are as follows.

3.1.1 Drying

Sun-drying, an ancient and popular method, involves placing fish in sunlight until moisture evaporates, inhibiting microorganism growth and preventing spoilage (Kalita et al., 2020). This method allows fish to be stored for long periods, and fish are required to be rehydrated for later use.

3.1.2 Smoking

It is another common traditional technique, where fish are exposed to smoke from burning wood or other materials, which kills bacteria and helps preserve the fish. The heat from smoking also helps dehydrate the fish, further extending its shelf life (Valu et al., 2023). There are two types of smoking: hot and cold. In hot smoking, fish is smoked at a high temperature of 70–80 °C, resulting in cooked, flavorful products. Cold smoking, performed at temperatures below 30 °C, does not cook the fish and may require additional cooking (Tahiluddin & Kadak, 2022).

3.1.3 Salting

Salting is one of the oldest methods for preserving fish and other seafood (Sampels, 2015). The preservation process through salting works the same for all fishery products, relying on osmosis and diffusion to reduce the moisture content to a level that slows down enzymatic and bacterial activity, thereby extending shelf life. Various salting methods include dry salting (Kench process), wet salting (brining, pickle curing), and popular Filipino techniques such as Kench-cured fish (Binoro), Visayan salted/fermented fish (Tinabal), and shrimp cake (Guinamos) (Tahiluddin & Kadak, 2022).

3.1.4 Fermentation

Fermentation is another traditional method, where fish undergo controlled microbial and enzymatic activity. Naturally occurring bacteria break down proteins and sugars, producing lactic acid and other compounds that preserve and enhance the fish's flavor. Fermented fish products are common in many cultures, known for their unique flavors and textures. In India, examples of traditionally preserved fish include hentak and ngari (fermented fish) from Manipur, tungtap from Meghalaya, and sun-dried and salted products like karati, bordia, and lashim from Assam, as well as shidal from Tripura (Thapa, 2016). These products are also made in eastern Nepal, Darjeeling Hills, Sikkim, and Bhutan.

3.1.5 Pickling or Marinating

This is another preservation technique that combines salt and vinegar/acetic acid. The acidity lowers the pH, preventing the growth of spoilage microorganisms. The effectiveness of these substances increases at higher concentrations. However, pickled or marinated products have a shorter shelf life and are considered semi-preserved unless stored in a chilled environment, where they can last for several months (Sampels, 2015).

3.1.6 Chilling and Freezing

Icing or chilling is a common practice especially for transportation of harvested fish to distant markets or processing factories. The process of chilling fish is often considered the easiest, cheapest, and efficient method for the preservation of fish onboard fishing vessel. Use of ice as a cooling medium reduces the temperature of fish, thus delaying the enzymatic degradation and bacterial spoilage and helps in retention of nutritional quality (Nath et al., 2019). Rapid cooling can be effectively achieved by placing fish in close contact with crushed ice flakes, keeping the fish cool (0–1 °C), moist, and glossy while slowing deterioration. The rate of enzymatic degradation and bacterial spoilage after death is twice as fast at 2.5 °C compared to 1.1 °C. Therefore, chilling the fish immediately after harvesting, followed by icing, helps delay bacterial spoilage. Using ice, fish can be stored and transported in good condition for 3–15 days, until it reaches the consumer or is subjected to further freezing (Nath et al., 2019).

Freezing of fish is done at a temperature of −40 ± 1 °C in an air blast freezer and subsequently stored at −18 ± 1 °C. Smaller fish and shrimps are frozen using contact plate freezers at temperatures between −35 and −40 ± 1 °C in the form of blocks. Karki et al. (2020) indicated that frozen fish can be kept in acceptable condition till 180 days of frozen storage at −18 ± 1 °C. However, freezing and frozen storage can impact the texture of fish as denaturation of protein and weakening of the muscle fibers is observed in the frozen storage.

3.1.7 Canning

Canning refers to the process of packaging food in a container and applying heat to extend its shelf life. This thermal process destroys pathogenic bacteria and controls spoilage organisms. The canning method varies based on the type of food, such as high-acid foods and low-acid foods. Low-acid foods, with a pH above 4.6, include most vegetables, meats, poultry, and seafood. These are prone to spoilage by thermophilic and mesophilic anaerobes, including *Clostridium botulinum*, which causes botulism. The bacteria form heat-resistant spores that require a temperature above boiling for destruction. For low-acid foods, pressure canning at 240 °F at 15 Psi

(116 °C) steam pressure using a steam pressure canner is necessary to prevent botulism by destroying these spores (Rajput et al., 2022).

Traditional methods of preserving food have been used for centuries to extend its shelf life, particularly during times of abundance, ensuring availability during food shortages. Techniques like salting, smoking, sun-drying, and fermentation prevent spoilage by inhibiting harmful microorganisms. In the case of fish preservation, these methods not only preserve the food but also enhance its flavor and texture, varying across different cultures. Each region has developed unique ways of preserving fish, adding distinct tastes, and ensuring longevity. These practices highlight the ingenuity of our ancestors and remain valuable today, promoting sustainability and preserving cultural traditions (Valu et al., 2023).

3.2 *Modern Techniques in Fish Processing*

Seafood products are prone to quality loss due to factors like muscle protein degradation, oxidative processes, and microbial spoilage if not properly preserved (Zhan et al., 2018). Preservation techniques are continuously evolving, leading to new strategies. Freezing is widely used in the industry for seafood preservation, but conventional air blast freezing and cryogenic freezing can compromise quality by causing slow freezing rates that result in large, unevenly distributed ice crystals. This can damage muscle tissues, denature proteins, and alter the texture, water-holding capacity, and color of the product (Zhan et al., 2018). To address these issues, novel freezing methods, including pressure-related freezing techniques (PF), ultrasound-assisted freezing (UAF), electrically assisted freezing, magnetically assisted freezing, and the use of antifreeze proteins, have been developed. However, research on these methods in the seafood sector is limited, suggesting that more scientific studies will evaluate their effectiveness in the coming years (Hassoun et al., 2022).

In recent years, edible films and coatings have gained increasing attention due to their numerous advantages over synthetic materials, traditionally used in food packaging, such as polyethylene, polyamide, and polypropylene. These advantages include the ability to slow down food degradation, extend shelf life, non-toxicity, and environmental sustainability (Abdel-Naeem et al., 2021; Maciel et al., 2020). The development of these edible materials has been driven by technological advancements, particularly in nanotechnology. While films and coatings are sometimes treated interchangeably, they serve distinct purposes and are applied differently. Edible films are typically prepared as solid sheets and then placed over the food, whereas coatings are applied directly onto the food surface (Maciel et al., 2020). Hassoun et al. (2022) envisaged that most edible materials are derived from natural biopolymers, such as polysaccharides (e.g., cellulose, starch, alginates, chitosan) or proteins (e.g., casein, whey, wheat gluten).

The consumer demand for natural alternatives to synthetic additives has led to the use of natural preservatives in recent years to extend the shelf life of seafood. Natural additives derived from plant, animal, or microbial sources, including essential oils, plant extracts, and chitosan, have antimicrobial, antioxidant, and anti-browning properties. They can prevent contamination and preserve the sensory qualities of seafood. These compounds can be applied directly to seafood or incorporated into packaging materials, particularly edible films. Essential oils and plant extracts, such as oregano, thyme, rosemary, grape, clove, and tea, have shown promising antioxidant and antimicrobial effects due to compounds like terpenes, alkaloids, and phenolics (Baptista et al., 2020).

Nanotechnology has also emerged as an innovative field, offering unique properties due to the small size and large surface area of nanostructures. In the food industry, nanotechnology has been applied in food production, processing, preservation, and packaging (Ranadheera et al., 2017; Qiu et al., 2020). In preservation, nanostructures improve the solubility, bioactivity, and stability of compounds, as well as control their release through nanoencapsulation or nanoemulsions (Liu et al., 2021). Additionally, nanocomposites like silver, zinc oxide, titanium oxide, and chitosan nanoparticles have been used to inhibit microbial growth and maintain product freshness (Ranadheera et al., 2017; Qiu et al., 2020).

Thermal processing is widely used in the seafood industry to inhibit the growth of pathogenic and spoilage bacteria and extend the shelf life of products. Traditional methods like boiling, steaming, roasting, and deep-frying have been employed for years but are often associated with challenges such as lengthy processing times and adverse effects on sensory and nutritional qualities. Recently, innovative heating technologies, including microwave and radio frequency heating, ohmic heating, and infrared heating, have gained industrial interest. Unlike conventional methods, which rely on convection and conduction (external heating), these novel technologies use volumetric heating (internal heating) (Hassoun et al., 2021). Ohmic heating has numerous applications such as cooking, thawing, blanching, dehydration, pasteurization, and extraction, offering faster and more uniform heating compared to conventional methods. Additionally, it is environmentally friendly and energy-efficient (Hassoun et al., 2022). Microwave heating, which converts electromagnetic energy into thermal energy, is increasingly used for sterilization, cooking, extraction, and thawing in the food industry. It is valued for its high heating rates, ease of use, and minimal impact on food's sensory and nutritional quality compared to conventional methods. Studies have shown that microwave heating improves properties such as the gelling of surimi gel enriched with fish oil. However, the high cost of equipment limits its widespread adoption in large-scale facilities (Hassoun et al., 2022).

Nonthermal technologies like pulsed electric field (PEF), high-pressure processing (HPP), and ultrasound (US) are among the most widely applied novel techniques in food preservation (Ekonomou & Boziaris, 2021). PEF processing can extend food shelf life while preserving its nutritional value and sensory properties. It involves short pulses (1–100 μs) in strong electric fields (0.3–4 kV/cm). Although PEF has gained popularity for seafood preservation, its industrial use remains limited despite its effectiveness and eco-friendliness (Arshad et al., 2020).

HPP is another emerging nonthermal processing technique with growing applications in the seafood industry. It is primarily used to kill microorganisms and parasites, extend shelf life, and aid in pressure shift freezing or thawing, as well as deboning crustaceans and bivalves. Typically, pressures of 200–600 MPa and holding times of 2–10 min are applied to inactivate microbes while preserving product quality. HPP is used to treat a variety of seafood products, such as fresh or smoked salmon, red mullet, rainbow trout, mackerel, and shellfish like lobster and oyster (Hassoun et al., 2022). Ultrasound (US) is sound waves with frequencies above human hearing (>20 kHz) and used in food processing for preservation, texture modification, extraction, emulsification, freezing, thawing, and microbial inactivation. Ultrasound is categorized into low-energy (5–10 MHz, <1 W/cm^2) and high-energy (20–100 kHz, >1 W/cm^2) groups. Low-energy ultrasound is used for nondestructive food quality analysis, while high-energy ultrasound can induce physical and chemical changes that kill microbes while preserving food quality. US inactivates microorganisms through cavitation, shear forces, microstreaming, and free radical formation and can be used alone or in combination with other methods (heat or pressure) to improve the microbial and sensory safety of seafood (Hassoun et al., 2022).

4 Value-Added Products

Value addition refers to enhancing a product, service, or individual by providing features that exceed standard expectations, often at a higher cost to the customer. In the food industry, value addition involves transforming raw materials into more valuable products through methods like processing, packaging, genetic modification, or introducing new flavors and colors. As consumer demand for health and nutrition grows, technological advancements allow producers to create products that meet these needs. Producers now focus on delivering finished products rather than just raw materials. The USDA defined value addition in agriculture in terms of almost five methodologies that are value added. These include changing the product's physical state, producing it in ways that enhance its value, implementing physical segregation, the source of the base renewable energy of the farm, and whether marketing is achieved as a locally produced agricultural product.

The effective application of advanced technologies is crucial for successfully developing new products that are easy to prepare, package, handle, transport, and consume. Product development refers to the process of generating ideas, designing products, and conducting detailed engineering to bring new items to market, made possible through a solid understanding of food science—specifically food chemistry and nutrition. With knowledge of diverse ingredients and global cuisines, food industries are continuously introducing new products worldwide and are eager to further innovate in line with evolving consumer trends (Sharif et al., 2018).

4.1 Value-Added Fish Products

The rapid growth of the global population and rising living standards have significantly increased the demand for fish. However, overexploitation and depletion of valuable fish stocks, along with declining catches of preferred species, have led to an imbalance between fish supply and demand in many countries. While fish farming is considered a potential solution, aquaculture has not been able to fully meet the growing demand. Consequently, while the supply of commercially important fish species is dwindling, many other species remain underutilized for human consumption (Agh et al., 2014). These underused species hold potential as food sources and are rich in lipids and fatty acids.

Post-harvest losses in fisheries are rising due to the scarcity of high-demand fish, combined with the harvesting of smaller, lower-value species that are less popular due to their unappealing appearance, taste, and texture (Ikbal et al., 2020). These fish are often discarded or used for animal feed and by-products. Despite their lower market value, they are rich in protein and could serve as a critical nutrition source, particularly in developing regions like India, where poverty remains a challenge.

To address these post-harvest losses, transforming underutilized fish into value-added products offers a promising solution (Ikbal et al., 2021). Value addition in seafood includes a wide range of products, from live fish to ready-to-eat meals. Traditional ethnic recipes, such as fish pickles and marinated dishes, present valuable opportunities. Many of these products are preserved using age-old methods like fermentation, drying, and smoking. With the growing demand for ethnic foods driven by globalization, modern processing techniques can further enhance their appeal.

Surimi, a processed fish product, shows great potential for value addition, especially when made from farmed species. This strengthens the link between production growth and resource utilization. Thermal processing, though costly, allows for the creation of high-margin convenience products for export. Traditional methods like curing and drying also hold potential, but challenges such as a lack of standardized procedures and the perception that these methods are only suitable for low-value fish hinder their wider adoption. Improving packaging, introducing energy-efficient drying technologies, and offering better extension services can promote the use of these methods.

Fermentation, which has long been used to preserve fish in tropical climates, now also enhances nutrition, appearance, flavor, and reduces cooking energy. As the global seafood market diversifies, there is growing demand for convenient, nutritious products like nutritional bars and noodles. These low-moisture items have long shelf lives and can be consumed with or without rehydration. Techniques like food extrusion provide flexibility in developing innovative fish-based snacks, especially from farmed species with favorable qualities. Additionally, seafood processing generates significant waste—about 70% of raw material which is expensive to manage. However, this waste, including heads, scales, bones, and viscera, is rich in valuable by-products such as proteins, bioactive peptides, PUFAs, and collagen

(Parvathy et al., 2020). These by-products offer considerable nutritional and medicinal benefits, highlighting the need for sustainable waste management practices in the seafood industry. The different types of fish value-added products are discussed in the following sections.

4.1.1 Cured and Dried Products

Before the advent of ice-making technologies and cold storage, dried fish was the primary method for trading and consuming fish. Historical records reveal that dried fish production, storage, trade, and consumption were widespread for thousands of years. Despite modern advances in refrigeration, dried fish remains a crucial part of diets in many regions, particularly in Sub-Saharan Africa and Southeast Asia, where it is a staple in the market and cuisine.

The classification of dried fish products often lacks consistency due to the use of various overlapping and unstandardized production methods. For example, fish are often salted before drying, and some salted products may be partially fermented. Broadly, fish products can be grouped into dried or fermented categories. Smoked fish, for instance, is dried using smoke instead of sunlight, while fish sauce is a by-product of fish fermentation. Belton et al. (2022) identified six categories of fish products: dried, salted, fermented, smoked, fish sauce, and others. Discussions about dried fish at the global level often overlook significant regional differences. In Asia, dried and fermented fish dominate, whereas in Africa, dried and smoked products are more prevalent. In Europe and the Americas, salted fish, especially salted cod, holds greater importance. Even within Asia, there are further distinctions, with South Asia primarily focused on dried fish and Southeast and East Asia emphasizing fermented fish.

Various dried fish products from South and Southeast Asia illustrate these regional variations. In South Asia, popular items include Sri Lanka's salted and sun-dried fish (*Karawala*), smoked fish (*Dum karawala*), and India's sun-dried Bombay duck, smoked fish (*Archinachepalu*), and sun-dried shrimp (*Jawla*). In Bangladesh, semi-dried and fermented fish (*Shidhol*) and salted, sun-dried fish (*Shutki*) are favored. Southeast Asia features diverse products such as Thailand's sun-dried squid (*Plamueg*) and fermented shrimp paste (*Kapi*), while Cambodia offers fish sauce (*Tuk trey*) and smoked fish (*Trey cha'eur*). Myanmar is known for fermented fish and shrimp paste (*Nga-pi*), and dried fish (*Nga-chauk*) and shrimp (*Pa-zun-chauk*) (Johnson, 2023).

Dried fish processing plays a vital role in providing livelihoods in various environments. A key benefit of this method is its ability to preserve fish, delaying spoilage and allowing it to be stored for months, which is particularly crucial in tropical climates with limited aquatic resources during certain seasons. This preservation technique supports both small-scale production, such as Cambodia's household-level Prahok production, and large-scale trade, such as dried fish exports from India to Bangladesh and from Thailand to Sri Lanka. Dried fish continues to be a critical part of food security, cultural cuisine, and economic activity in many regions, despite the rise of modern cold storage technologies. Its historical and contemporary significance is a testament to its versatility and importance (Johnson, 2023).

4.1.2 Fermented Products

Fermentation is a microbial process that transforms food through the action of bacteria, fungi, or the enzymes they produce. This natural process has long been used to preserve food and enhance its digestibility by reducing harmful substances. Fermented foods are created by controlling microbial growth, allowing the biotransformation of bioactive phenolics and volatile compounds, resulting in new edible forms with improved flavors and aromas (Narzary et al., 2021). These microorganisms, either indigenous or introduced through external 'starter cultures,' cause chemical and textural changes in food (Tamang et al., 2020). While microbial activity can lead to food spoilage, it can also produce safe, edible foods, with the distinction often lying in taste and flavor. What might be considered putrid in one context can become a delicacy in another, such as in fermented foods (Narzary et al., 2021).

The use of fermented fish in traditional cuisines dates back to ancient times across many cultures. The earliest recorded fermented fish sauce, garum, was produced in classical Rome from the viscera and blood of mackerel. Beyond East, Southeast, and South Asia, fermented fish is also consumed in other regions like the Middle East, Africa, and Scandinavia, where dishes like surströmming (fermented herring) and rakørret (fermented trout) are traditional in Scandinavian countries. In France and Spain, fermented anchovy is popular. While Southeast Asia favors fish pastes and sauces, fermented fish in Africa often takes the form of whole fish or large chunks. Asian and African cultures typically rely on traditional, spontaneous fermentation methods, unlike Europe, North America, Australia, and New Zealand, where fermentation usually involves specific starter cultures—microbial inocula used to initiate fermentation. These cultures may consist of a single bacterial strain or a consortium isolated from naturally fermented foods (Narzary et al., 2021).

Microorganisms are essential to all fermented foods. Lactobacillus and other microbes in fermented foods have earned GRAS (Generally Recognized As Safe) status, increasing the commercial value of these products due to the beneficial physicochemical properties they impart. Fermentation, which involves the enzymatic breakdown of organic substances into simpler components, often overlaps with salting techniques. However, some fermented products use only minimal salt. Fermented fish is distinct from salted fish, as the former loses its original shape during partial liquefaction, while the added salt (at a 1:3 ratio) prevents spoilage during fermentation (Tahiluddin & Kadak, 2022; Hassoun et al., 2022).

Fermentation enhances food with a range of tastes, flavors, textures, and sensory qualities, along with therapeutic and nutritional benefits. This ancient process continues to play a crucial role in global food cultures, offering both preservation and enhanced gastronomic experiences (Narzary et al., 2021). Some popular fermented products are given in Table 1.

Table 1 List of fermented fishery products

Country	Fermented product	Fish species
Indonesia	Pedah	Mackerel (*Rastrelliger neglectus*)
	Trasi (Shrimp paste)	Shrimp
	Bekasam	Freshwater fish
Korea	Jeotgal	Various fish species
	Sikhae	Lactic acid fermented fish or clam meat
Thailand	Nam-pla	*Anchoviella* sp., *Stolephorus* sp., *Rastrelliger* sp., *Cirrhinus* sp.
	Plaa-som	Snakehead fish
Malaysia	Belacan	Small shrimps of Acetes and Mysid varieties
	Pekasam	Freshwater fish
Thailand and Malaysia	Budu	Sea water or brackish water fish
Philippines	Bagoong and Patis	*Stolephorus* sp., *Sardinella fimbriata*, and *Decapterus* sp.
	Balao-balao	*Penaeus indicus* or *Macrobrachium* sp.
	Burong isda	Freshwater fish
Myanmar	Ngapi	Fish, shrimp
India	Ngari	*Puntius sophore and Puntius ticto*
	Hentak	*Esomus danricus*
	Tungtap	*Puntius* sp. and *Danio* sp.
India and Bangladesh	Shidal	*Puntius sophore* and *Setipinna phasa*
Japan	Shottsuru (Fish sauce)	Anchovy, opossum shrimp

4.1.3 Marinated Products

Marinating or pickling is a traditional preservation technique that relies on the combined effects of salt and vinegar (acetic acid) to preserve food. The presence of vinegar lowers the product's pH, inhibiting the growth of spoilage-causing microorganisms (Sampels, 2015). The effectiveness of vinegar and salt in preventing spoilage increases with higher concentrations, as they inhibit both bacterial growth and enzyme activity. However, pickled or marinated products generally have a short shelf life and are classified as semi-preserved unless stored in a chilled environment, where they can last for several months (Sampels, 2015). Marinades are categorized into three types: cold, cooked, and deep-fried marinades (Tahiluddin & Kadak, 2022). In cold marinades, fresh fish or fish products are immersed in a solution of salt and acetic acid without applying heat. Cooked marinades involve cooking fish in an acetic acid and salt solution at 85 °C, which helps to kill bacteria and inactivate enzymes. For deep-fried marinades, fish products are fried before being packed in a solution of vinegar and salt (Tahiluddin & Kadak, 2022).

In the Philippines, popular marinated products include pickled anchovies and marinated mussels. These products are prepared by washing the fish, soaking them in a salt solution, removing the bones, and immersing them in a strong acetic acid solution before storing them in a cool room for up to 3 weeks. Marinated fried sardine or mackerel is prepared by soaking the fillets in brine, rolling them in flour, frying, and then packing them in a hot marinade of spices and vinegar. The key ingredients in pickling and marinating are vinegar, salt, spices, and sugar. Recent studies have shown that incorporating natural additives into the marination process can enhance sensory qualities, extend shelf life, and improve the nutritional value of the products (Tahiluddin & Kadak, 2022).

4.1.4 Smoked Products

Smoking is a traditional and effective method of preserving fish, typically performed indoors where fish are exposed to the smoke of burning wood or other organic materials. The smoke contains chemicals that kill bacteria and microbes, while the heat helps to dry the fish, extending its shelf life (Valu et al., 2023). Smoked fish products are known for their pleasant smoky aroma, which is created by the burning of sawdust or wood and the release of volatile compounds, enhancing both the flavor and the durability of the products. There are two main types of smoking: hot smoking and cold smoking. In hot smoking, fish are smoked at high temperatures (70–80 °C), resulting in cooked, juicy, and flavorful products. Cold smoking, on the other hand, keeps temperatures below 30 °C, so the fish remains uncooked and may require additional cooking before consumption (Tahiluddin & Kadak, 2022). Traditionally, various types of smokehouses are used around the world. For example, in the Philippines, people use clay pot smokehouses (called Pelon), drum-type variations, and concrete smokehouses with metal smoking trays. Modern smokehouses, equipped with sophisticated mechanical systems, are also becoming common.

Gnuchi is a traditional smoked fish product made by the Lepcha community of Sikkim, India. Fish species like *Labeo dero*, *Schizothorax richardsonii*, and *Channa* spp. are collected from rivers, cleaned, salted, and rubbed with turmeric powder. The larger fish are placed on racks over an oven, while smaller fish are hung from bamboo strips above it. The fish are smoked for 10–14 days and can be stored at room temperature for 2–3 months. Gnuchi is often consumed in curry form (Valu et al., 2023). Another traditional smoked fish dish from Sikkim is 'suka ko maacha,' made from hill river fish known as Chuchayasala (*Schizothorax progastus*) and Dothayasala (*Schizothorax richardsonii*). The fish are cleaned, gutted, and seasoned with salt and turmeric powder before being smoked above an earthen oven for 7–10 days. This dish is typically cooked into a curry and can be stored for 4–6 months (Thapa, 2016).

4.1.5 Fish Mince and Mince-Based Products

Minced meat refers to the flesh separated from fish in a finely comminuted form, free from bones, skin, and other undesirable parts. While this process can be applied to any fish species, it is particularly beneficial when used on low-cost fish, as it adds significant value. Minced meat can also be produced from filleting waste, making it a versatile and resource-efficient option. The quality and properties of minced meat largely depend on the nature and freshness of the raw materials used. Typically, meat-bone separators, or meat-picking machines, are employed to prepare minced meat. This minced fish meat serves as a base material for creating various high-demand products, such as fish sausages, cakes, cutlets, patties, balls, pastes, surimi, and even texturized products. These value-added items allow for greater utilization of fish resources and can enhance both the economic value and market appeal of lower-cost or underutilized fish species (More & Likhar, 2020).

4.1.6 Surimi and Surimi-Based Products

Fish processing has evolved to meet consumer demands, leading to a variety of products, including surimi—a versatile intermediate product commonly used in imitation seafood products. Surimi is considered a 'high-tech' food product, enabling efficient use of fish as a vital protein source. It has a bland taste and excellent gel-forming properties, making it ideal for creating imitation seafood with similar texture and flavor (Satyaveer et al., 2022). The term 'surimi' refers to minced and water-washed fish muscle tissue, which is a frozen concentrate of fish myofibrillar proteins that provide structure and muscular activity to the flesh. Surimi originated in Japan, where it is used in traditional dishes like kamaboko, and it has gained global popularity, especially in the form of imitation shellfish. White-fleshed non-oily fish, such as Alaska Pollock, is commonly used, but due to stock depletion, alternatives like croakers, lizardfish, and freshwater varieties are increasingly used (Satyaveer et al., 2022). The surimi production process includes several steps:

4.1.6.1 Meat/Bone Separation

Fish are deboned using meat-bone separators to obtain mince.

4.1.6.2 Washing

A crucial step to remove color, odor, and taste from the mince, ensuring maximum gelling potential. The number of washing cycles varies based on species and desired surimi quality.

4.1.6.3 Refining/Straining

The washed mince is refined to remove unwanted materials like connective tissue and skin.

4.1.6.4 Dewatering

Moisture content is reduced to 80–84% using a screw press.

4.1.6.5 Cryoprotectant Addition

Cryoprotectants like sucrose, sorbitol, and polyphosphate are mixed in to preserve surimi quality during freezing.

4.1.6.6 Shaping and Packing

The surimi is shaped into blocks, typically 10 kg, and packed in polythene bags.

4.1.6.7 Freezing

Blocks are rapidly frozen in contact plate freezers until the core temperature reaches −25 °C.

4.1.6.8 Cold Storage

The final product is stored at below −18 °C, with an overall recovery rate of about 24% of the raw materials. These steps ensure that surimi remains a high-quality, versatile ingredient in seafood products worldwide. Several surimi-based products, such as kamaboko, chikuwa, hanpen, fish ham, and sausage, are made by incorporating various ingredients into surimi, with the manufacturing process differentiating them. Among surimi-based imitation shellfish products, fiberized items are in high demand. These typically include surimi, salt, starch, egg white, shellfish flavor, flavor enhancers, and water. The ingredients are mixed into a paste, extruded into sheets, heat-treated, cut into strips, and steam-cooked to create the final product (More & Likhar, 2020).

Here is a breakdown of notable surimi-based products (Satyaveer et al., 2022):

- *Kamaboko*: A traditional Japanese fish paste product, often referred to as 'fish cake.' It forms a gel of myofibrillar proteins, producing a resilient, white product. Steamed versions are called Sumaki or Mushiita, while fried versions are Tenpura

(in West Japan) or Satsuma Age (in East Japan). Ingredients include salt, MSG, sugar, starch, egg white, and more.
- *Chikuwa*: Cylindrical and hollow, this broiled product is made by shaping surimi mixed with seasonings around a tube. It is known for its white interior and brown surface.
- *Hanpen/Naruto*: A boiled surimi product, ground into a fine paste with ingredients like Japanese yam. It has a soft, spongy texture due to air incorporated during kneading. Unlike kamaboko (steamed) and chikuwa (broiled), hanpen is boiled.
- *Fish Ball*: Made from surimi mixed with salt, sugar, MSG, starch, and water. No additional flavors or protein additives are used. The paste is shaped into balls, warmed, then cooked in hot water, and finally chilled and packaged.
- *Fish Ham*: Typically made from Alaskan Pollock or tuna surimi, sometimes with non-fish meat. The surimi is cured with salt, sugar, spices, and smoke flavor, mixed with additional surimi, and formed into a paste. It is then sealed in casing, cooked, and frozen for preservation.
- *Fish Sausage*: An emulsion-based product made by mixing surimi with salt, sugar, starch, spices, vegetable oil, and water. The mixture is stuffed into synthetic casings, heat-processed, cooled, and boiled briefly. Originally from Japan, fish sausage is used as a snack, appetizer, or ingredient in various dishes.

4.1.7 Frozen Products

Individually Quick Frozen (IQF) seafood products in high demand include shrimp/prawn, whole cooked lobster, lobster tails, lobster meat, cuttlefish fillets, squid tubes and rings, boiled clam meat, and skinless, boneless fillets of white lean fish. Shrimp is particularly sought after in various forms, such as whole, peeled, deveined, cooked, headless shell-on, butterfly, fan tail, and round tail-on. Some specialty products that require freezing include the following.

4.1.7.1 Barbecue Shrimp

For barbecue shrimp, the shrimp is washed in chilled water with 5 ppm chlorine, beheaded, deveined, peeled, and washed again. A bamboo stick is inserted from head to tail, and the shrimp is packed in vacuum-sealed thermoformed trays before freezing at −40 °C.

4.1.7.2 Sushi

Sushi (cooked butterfly shrimp) involves a similar process, but after washing, the shrimp is cooked in 1% brine for 2 min at 100 °C. Once cooked, it is cooled in chilled water, peeled (including the tail), and split to form a butterfly shape before vacuum-packing and freezing at −40 °C.

4.1.7.3 Skewered Shrimp

It is prepared similarly to barbecue shrimp, except that 4–5 shrimp are arranged on a skewer in an inverted 'U' shape, vacuum-packed, and frozen.

4.1.7.4 Shrimp Head-on (Central Peeled)

This product is processed after washing and peeling at the center while keeping the head and last two segments intact, the shrimp is deveined, trimmed, washed, packed in vacuum-sealed trays, and frozen at −40 °C.

4.1.7.5 Shrimp Head-on Cooked (Central Peeled)

This also follows a similar process, with the shrimp cooked in 1% brine for 2 min before peeling and freezing.

4.1.7.6 Stretched Shrimp (Nobashi)

It is a newer technique where peeled, deveined shrimp are stretched and cut parallel to minimize curling. Pressure is applied to increase the length using mechanical devices.

4.1.7.7 Fish Fillet

For skinless and skin-on fillets, white meat fish varieties like grouper, red snapper, reef cod, bream, and jewfish are used to create fillets for both domestic and export markets. These fillets are often block frozen or IQF and used for making coated products or ready-to-serve items like fish in sauce or salads.

4.1.7.8 Accelerated Freeze-Dried Products

Accelerated freeze-drying is gaining popularity for preserving high-value seafood like shrimp and squid rings. The frozen product is subjected to high vacuum, causing ice crystals to sublimate, preserving the product's shape and allowing for quick rehydration (up to 95%). In India, this technique is now being applied to various seafood products, offering vast potential for ready-to-eat items.

4.1.7.9 Chilled Items

Chilled fish, especially sashimi-grade tuna, is another key product in international trade. Sashimi, a traditional Japanese delicacy made from raw tuna fillets, mainly uses species like bluefin, bigeye, and yellowfin tuna. The highest quality sashimi is kept chilled from capture to consumption. Other chilled seafood products include pomfret, shrimp, lobster, and crabmeat (More & Likhar, 2020).

4.1.8 Thermally Processed Products

Thermal processing enhances the long-term stability of various seafood products by preserving them in hermetically sealed containers in cooked form for ambient temperature storage, without sacrificing quality. This technology has evolved over time, leading to new products and packaging that consumers accept readily. Tin plate, made from 98% steel and 2% tin, is the most common material for canning, with the steel referred to as can-making quality (CMQ) steel. However, the poor organoleptic qualities of foods in tin containers prompted the introduction of aluminum alloy cans for meat and fish products in 1918. These cans are now widely used in Europe due to the availability of raw materials, electricity, and low production costs. An alternative to tin cans is Tin Free Steel (TFS), which has various brand names like Can Super, Hinac coat, and Hi-Top. TFS was initially developed in Japan by electroplating cold-rolled steel with chromium. Metal cans offer advantages such as availability in different sizes, superior strength, high-speed manufacturing, and ease of filling and sealing. However, they are heavy, difficult to reclose, and challenging to dispose of (Bindu et al., 2014). The concept of pouches for food packaging was developed in the early 1960s by the US Army Natick Laboratories and food packaging companies. Retort-pouched products are shelf-stable and ready to serve, catering to consumer convenience. Low-acid solid foods in retort pouches have gained commercial acceptance for their superior quality and convenience, creating a new segment in canned foods. These three- or four-layer retort pouches consist of an outer polyester layer, a middle aluminum layer, and an inner cast polypropylene layer. Nylon can be added or used instead of aluminum for added strength. The outer polyester layer provides mechanical resistance, while the aluminum foil serves as a barrier layer, extending shelf life. Polypropylene, with a high melting point, ensures seal integrity, flexibility, strength, and compatibility with various products. The layers are bonded with adhesives like modified polyolefins (e.g., ethylene vinyl acetate). Liquid crystal polymers, which offer superior barrier properties, are also being explored. Pouches made with polyvinylidene chloride and nylon instead of aluminum allow for product visibility but have a reduced shelf life due to not being complete barriers to oxygen (Bindu et al., 2014).

Recently, retort pouches coated with nanoparticles of silicon dioxide or aluminum oxide have become commercially available (Bindu et al., 2014). Heat-processed seafood products include ready-to-serve items like fish curry, fish-in-rice, canned sardine curry, and mackerel in natural pack. These can be efficiently processed in retort pouches using an overpressure autoclave. The smaller cross-sectional profile of retort pouches allows for shorter processing times, minimizing temperature-induced changes in product quality (More & Likhar, 2020).

4.1.9 Enrobed/Coated Products

Battered and breaded products, made from various fish and shellfish, are among the most prominent value-added food items, highly regarded by consumers for their convenience. These products undergo a process of applying one or two coats of batter, followed by a layer of bread crumbs, which enhances their bulk while reducing production costs (More & Likhar, 2020). As popular options in the category of fried foods, battered and breaded items are appreciated worldwide for their appealing texture, which combines a dry, crispy crust with a tender, juicy interior. Frying improves the flavor, texture, appearance, and overall acceptance of these products, although it can also result in higher oil absorption. A range of products, including shrimp, squid, clams, fish fillets, and minced meat from low-cost fish, can be battered and breaded, with popular items like fish fingers, fish cutlets, fish nuggets, and fish pakoras (Hauzoukim et al., 2019).

There is a growing demand for healthier, natural food options among health-conscious consumers. Regulatory bodies and the media have emphasized the importance of reducing dietary fat. This concern has driven research to promote healthier products with lower fat content, particularly those affected by industrial pre-frying and frying processes. Hauzoukim et al. (2019) studied three aquatic-derived polymer hydrocolloids—alginate, chitosan, and fish gelatin—in the batter composition of enrobed fish fingers, aiming to reduce oil uptake and improve coating properties. Incorporating these polymers at a 1% (wt/wt) level, they evaluated parameters such as coating pickup, adhesion, frying loss, and cooked yield. The results showed that fish fingers coated with chitosan batter exhibited the lowest oil uptake (3.73%), the lowest hardness, and the highest fat reduction. Chitosan proved to be more effective than alginate and fish gelatin in enhancing oil reduction and physicochemical quality. Developing battered and breaded fish fingers or similar enrobed products with improved functional qualities, such as reduced oil absorption and extended shelf life, is crucial for promoting their consumption while offering greater health benefits (Hauzoukim et al., 2019).

4.1.10 Other Diversified Products

Other diversified fish and fishery are discussed as follows:

4.1.10.1 Fish Surimi Powder

Recent research has shown that surimi can be converted into a dried form known as surimi powder, which can be stored under normal refrigeration without the need for freezing, thereby reducing storage costs. This powdered form offers advantages like ease of handling, lower distribution costs, and utility in dry mixes. Protective sugars or polyols used in surimi also serve as cryoprotectants to prevent protein denaturation during drying (Bakli et al., 2020). Surimi powder can be rehydrated with four times its weight in water, restoring its moisture content to resemble frozen surimi blocks (Santana et al., 2012). Additionally, surimi powder can enhance formulation flexibility in the production of surimi-derived products, promoting uniform blends and protein standardization. Various drying methods, such as freeze drying and spray drying, are employed to create surimi powder while minimizing protein denaturation (Sarkar et al., 2020).

4.1.10.2 Fish Protein Concentrates (FPC)

Fish protein concentrate (FPC) is a product derived from low-cost fish protein for human consumption. There are two types: FPC Type A, which is colorless and odorless with less than 1% fat, produced through solvent extraction but suffers from high preparation costs and poor functionality. FPC Type B is made by drying and grinding fish mince, resulting in a fishy odor and higher lipid content, which has found success as a dietary supplement in several countries (Venugopal & Shahidi, 1995).

4.1.10.3 Fish Protein Hydrolysates (FPH)

FPH are created by digesting fish with proteolytic enzymes at optimal temperatures and pH levels. The digestion process liquefies the protein within hours, and the resulting hydrolysate is centrifuged to remove scales and bones. The concentrated proteins can be spray-dried or precipitated isoelectrically. However, bitterness often arises from peptide formation, which can be mitigated by adding glutamic acid, polyphosphates, gelatin, or glycine. FPH can serve as protein supplements in cereals and soups and as peptone for microbial growth (Venugopal & Shahidi, 1995).

4.1.10.4 Extruded Products

Extrusion processing is increasingly relevant in the food industry for transforming starchy and protein-rich ingredients. It allows for the production of diverse, high-nutrition, and convenient cereal-based products by utilizing fish muscle recovered from bycatch and underutilized fish, thereby adding value to these resources. Incorporating protein-rich seafood into extruded products enhances their nutritional profile and palatability. Innovations in flavors, textures, and packaging techniques, such as modified atmosphere packaging, further extend shelf life (Venugopal & Shahidi, 1995).

4.1.10.5 Katsuobushi

Katsuobushi, or okaka, is a crucial ingredient in Japanese cuisine made from dried, fermented, and wood-smoked skipjack tuna. It is typically sold as delicate shavings that add a rich umami flavor and the unique taste sensation known as kokumi, characterized by heartiness and mouthfulness (Tasteatlas, 2024).

4.1.10.6 Caviar

Beluga caviar, sourced from beluga sturgeon, is one of the world's most prized varieties due to its rarity and high cost, taking up to 25 years for production. Ikura, or salmon roe, consists of nearly transparent salmon eggs, particularly popular in Japan, that undergo curing in salt or brine after extraction. Mentaiko refers to marinated Alaskan pollock roe, often confused with tarako, and features a savory, slightly spicy flavor. Bottarga, an Italian delicacy made from cured roe of tuna, mullet, or swordfish, has a salty taste and silky texture. Masago, or capelin roe, is usually dyed and cheaper than ikura or tobiko. Sevruga sturgeon roe, abundant in the Black and Caspian seas, is renowned for its affordability compared to other caviar types due to faster reproduction rates (Tasteatlas, 2024).

5 Scope

Value addition in fish products refers to the transformation of raw products into more convenient, appealing, and nutritious forms, often through processes such as preservation, packaging, fortification, or the creation of ready-to-eat meals. This concept plays a significant role in addressing both economic and health-related challenges, particularly in terms of ensuring a sustainable food supply, reducing waste, and enhancing the nutritional profile of these products. One of the primary benefits of value addition in fish products is its ability to reduce post-harvest losses. In the fisheries sector, for instance, a significant amount of bycatch and small, low-demand species are often discarded or used for non-human purposes. By utilizing these underutilized fish through value addition, such as producing surimi (processed fish paste) or converting them into processed foods, we can reduce waste while creating economic opportunities. This not only strengthens the seafood industry but also contributes to a more sustainable use of marine resources. Similarly, in the meat industry, value addition helps convert less popular cuts of meat or by-products into processed foods such as sausages, patties, or nutrient-rich products. These methods maximize resource efficiency, create additional income streams, and cater to the rising demand for convenience foods. Furthermore, using advanced processing techniques, such as fermentation, drying, and smoking, helps preserve fish products for longer periods, ensuring a stable food supply and reducing spoilage. Value addition can also have a significant impact on human health. Fish is an essential source of protein, vitamins, and minerals, and the addition of nutrients such as omega-3 fatty acids, fiber, and probiotics through fortification can enhance their nutritional value. For instance, fatty fish are rich in omega-3, which is vital for cardiovascular health, brain function, and inflammation control. By incorporating these healthy fats into processed fish products, consumers can benefit from improved health outcomes. Additionally, the convenience offered by value-added products, such as ready-to-eat or easy-to-prepare meals, encourages healthier eating habits by making nutrient-dense foods more accessible. These products also provide portion control, which can help reduce overconsumption, a factor contributing to obesity and other diet-related health issues. Despite the benefits, value addition must be approached with caution. Some processed meat and fish products may contain high levels of sodium, preservatives, and unhealthy fats, which can contribute to cardiovascular diseases, hypertension, and other health problems. It is important for the food industry to focus on healthier processing techniques that minimize the use of harmful additives while retaining or enhancing the nutritional quality of the products. Efforts should also be made to ensure that these value-added products are affordable and accessible to all socioeconomic groups, particularly in developing countries where food insecurity remains a challenge. Value-added products can bridge the gap between resource availability and nutritional needs, but their impact will be limited if they are priced out of reach for the average consumer.

6 Conclusion

In conclusion, the value addition of fish products offers significant benefits, including reducing food waste, promoting sustainability, and enhancing human health. However, care must be taken to ensure that these products maintain high nutritional standards and avoid harmful additives. With the right approach, value-added fish products can provide an essential source of nutrition for populations worldwide, improve food security, and contribute to a healthier, more sustainable future.

References

Abdel-Naeem, H. H. S., Sallam, K. I., & Malak, N. M. L. (2021). Improvement of the microbial quality, antioxidant activity, phenolic and flavonoid contents, and shelf life of smoked herring (*Clupea harengus*) during frozen storage by using chitosan edible coating. *Food Control, 130*, 108317.

Agh, N., Jasour, M. S., & Noori, F. (2014). Potential development of value-added fishery products in underutilized and commercial fish species: Comparative study of lipid quality indicators. *Journal of the American Oil Chemists' Society, 91*, 1171–1177. https://doi.org/10.1007/s11746-014-2454-x

Arshad, R. N., Abdul-Malek, Z., Munir, A., Buntat, Z., Ahmad, M. H., Jusoh, Y. M. M., Bekhit, A. E. D., Roobab, U., Manzoor, M. F., & Aadil, R. M. (2020). Electrical systems for pulsed electric field applications in the food industry: An engineering perspective. *Trends in Food Science and Technology, 104*, 1–13.

Bakli, S., Nath, S., Chowdhury, S., & Pati, K. (2020). Surimi powder: Processing technology and potential application. *Journal of Entomology and Zoology Studies, 8*(4), 850–859.

Baptista, R. C., Horita, C. N., & Sant'Ana, A. S. (2020). Natural products with preservative properties for enhancing the microbiological safety and extending the shelf-life of seafood: A review. *Food Research International, 127*, 108762.

Belton, B., Johnson, D. S., Thrift, E., Olsen, J., Hossain, M. A. R., & Thilsted, S. H. (2022). Dried fish at the intersection of food science, economy, and culture: A global survey. *Fish and Fisheries, 23*, 941–962. https://doi.org/10.1111/faf.12664

Bindu, J., Mallick, A. K., & Gopal, T. K. S. (2014). Thermal processing of fishery products in flexible and rigid containers. *Fishery Technology, 51*, 137–148.

Dos-Santos, M. J. P. L. (2020). Value addition of agricultural production to meet the sustainable development goals. In W. Leal Filho, A. Azul, L. Brandli, P. Özuyar, & T. Wall (Eds.), *Zero hunger. Encyclopedia of the UN sustainable development goals*. Springer. https://doi.org/10.1007/978-3-319-69626-3_55-1

Ekonomou, S. I., & Boziaris, I. S. (2021). Non-thermal methods for ensuring the microbiological quality and safety of seafood. *Applied Sciences, 11*, 833.

Food and Agriculture Organization (FAO). (2021). *Food loss and waste in fish value chains*. Retrieved from www.fao.org/flw-in-fish-value-chains/en/.

Food and Agriculture Organization (FAO) and World Health Organization (WHO). (2011). *Report of the joint FAO/WHO expert consultation on the risks and benefits of fish consumption*. FAO fisheries and aquaculture report Rome, Italy.

Hassoun, A., Rustad, T., & Bekhit, A. E. D. A. (2021). Bioconversion of marine by-products into edible protein. In A. E. D. A. Bekhit, W. W. Riley, & M. A. Hussain (Eds.), *Alternative proteins* (pp. 297–327). CRC Press.

Hassoun, A., Siddiqui, S. A., Smaoui, S., Ucak, I., Arshad, R. N., Garcia-Oliveira, P., Prieto, M. A., Aït-Kaddour, A., Perestrelo, R., Câmara, J. S., & Bono, G. (2022). Seafood processing, preservation, and analytical techniques in the age of industry 4.0. *Applied Sciences, 12*, 1703. https://doi.org/10.3390/app12031703

Hauzoukim, K. A., Xavier, M., Kannuchamy, N., Balange, A., & Gudipati, V. (2019). Development of enrobed fish products: Improvement of functionality of coated materials by added aquatic polymers. *Journal of Food Process Engineering, 42*, e12999. https://doi.org/10.1111/jfpe.12999

Ikbal, A., Bardhan, A., & Singh, S. M. (2020). Radiation: Food safety and quality control measures in fishery products. In: *Research trends in food technology and nutrition* (vol. 83, pp. 55).

Ikbal, A., Nath, S., & Chowdhury, S. (2021). Role of value-added fish products for boosting entrepreneurship. *Green Report, 2*(5), 28–31.

Johnson, D. S. (2023). Transforming values: Reflections on how dried fish takes value socially. In: E. Thrift, M. Galappaththi, R. Ghosh, D. S. Johnson, W. W. Khaing, M. Rahman, & R. Chuenpagdee (Eds.). *Dried fish matters exploring the social economy of dried fish*. TBTI Global Book Series (pp. 3–20).

Kalita, S., Kalita, K. D., & Das, B. K. (2020). Traditional fish preservation in South Western Assam in India. *IOSR Journal of Humanities and Social Science*, 25(11), 09–21.

Karki, S., Chowdhury, S., Nath, S., & Dora, K. C. (2020). Freezing effects on yield, quality, and microbial load of farmed fish *Neolissochilus hexagonolepis* (Chocolate Mahseer). *Journal of Entomology and Zoology Studies, 8*(6), 2047–2052. https://doi.org/10.22271/j.ento.2020.v8.i6ab.8124

Liu, J., Shao, Y., Yuan, C., Takaki, K., Li, Y., Ying, Y., & Hu, Y. (2021). Eugenol-chitosan nanoemulsion as an edible coating: Its impact on physicochemical, microbiological and sensorial properties of hairtail (Trichiurus haumela) during storage at 4°C. *International Journal of Biological Macromolecules, 183*, 2199–2204.

Maciel, V. B. V., Contini, L. R. F., Yoshida, C. M. P., & Venturini, A. C. (2020). Application of edible biopolymer coatings on meats, poultry, and seafood. In *Biopolymer membranes and films* (pp. 515–533). Elsevier.

More, A., & Likhar, V. (2020). Fish processing: Product and by-product, processing and marketing. *International Journal of Fisheries and Aquatic Studies, 8*(1), 74–77.

Narzary, Y., Das, S., Goyal, A. K., Lam, S. S., Sarma, H., & Sarma, D. (2021). Fermented fish products in south and southeast Asian cuisine: Indigenous technology processes, nutrient composition, and cultural significance. *Journal of Ethnic Foods, 8*, 33. https://doi.org/10.1186/s42779-021-00109-0

Nath, S., Chowdhury, S., & Singh, A. K. (2019). Icing of fish: A way to retain nutritional quality and food safety. *Journal of Entomology and Zoology Studies, 7*(4), 474–478.

Parvathy, U., Binsi, P. K., Visnuvinayagam, S., Zynudheen, A. A., Ninan, G., & Ravishankar, C. N. (2020). Functional peptides from yellowfin tuna (Thunnus albacares): Characterisation and storage stability assessment. *Indian J. Fish,* 67(2), 69–79.

Qiu, L., Zhang, M., Bhandari, B., & Yang, C. (2020). Shelf-life extension of aquatic products by applying nanotechnology: A review. *Critical Reviews in Food Science and Nutrition, 62*(6), 1–15.

Rajput, R., Goswami, D., Arya, M., & Randhawa, A. (2022). Technology for canning. In *Hi-tech horticulture* (Value addition and post-harvest management) (Vol. 6, pp. 135–151). New India Publishing Agency.

Ranadheera, C. S., Prasanna, P. H. P., Vidanarachchi, J. K., McConchie, R., Naumovski, N., & Mellor, D. (2017). Nanotechnology in microbial food safety. In *Nanotechnology applications in food: Flavor, stability, nutrition and safety* (pp. 245–265). Elsevier.

Sampels, S. (2015). The effects of processing technologies and preparation on the final quality of fish products. *Trends in Food Science & Technology, 44*(2), 131–146.

Santana, P., Huda, N., & Yang, T. A. (2012). Gel characteristics of surimi powder added with hydrocolloids. In: *UMT 11th international annual symposium on sustainability science and management*, Terengganu, Malaysia, 9–11 July, 2012.

Sarkar, P., Chowdhury, S., Nath, S., Murmu, P., & Rahman, F. H. (2020). Effect of drying temperature on the quality of dry surimi powder from pangasius. *Current Journal of Applied Science and Technology, 39*(6), 147–155.

Sathya, A. (2018). Role for value addition in processing foods of traditional varieties of grains. In A. M. Grumezescu & A. M. Holban (Eds.), *Handbook of food bioengineering, food processing for increased quality and consumption* (pp. 423–454). Academic Press. https://doi.org/10.1016/B978-0-12-811447-6.00014-X

Satyaveer, S., Bhavanjali, M. R., & Manjunath, N. (2022). *Surimi and minced based fishery products*. AquastarMagazine (pp. 83–89). Retrieved from www.aquastarmagazine.com.

Sharif, M. K., Zahid, A., & Shah, F. H. (2018). Role of food product development in increased food consumption and value addition. In A. M. Grumezescu & A. M. Holban (Eds.), *Handbook of food bioengineering, Food processing for increased quality and consumption* (pp. 455–479). Academic Press. https://doi.org/10.1016/B978-0-12-811447-6.00015-1

Tahiluddin, A. B., & Kadak, A. E. (2022). Traditional fish processing techniques applied in the Philippines and Turkey. *Menba Kastamonu University Faculty of Fisheries Journal, 8*(1), 50–58.

Tamang, J. P., Cotter, P. D., Endo, A., Han, N. S., Kort, R., Liu, S. Q., Mayo, B., Westerik, N., & Hutkins, R. (2020). Fermented foods in a global age: East meets west. *Comprehensive Reviews in Food Science and Food Safety, 19*, 184–217. https://doi.org/10.1186/s42779-021-00109-0

Tasteatlas. (2024). *Top 31 fish products in the world.* Retrieved from https://www.tasteatlas.com/best-rated-fish-products-in-the-world.

Thapa, N. (2016). Ethnic fermented and preserved fish products of India and Nepal. *Journal of Ethnic Foods, 3*(1), 69–77.

Tilami, S. K., & Sampels, S. (2017). Nutritional value of fish: Lipids, proteins, vitamins, and minerals. *Reviews in Fisheries Science and Aquaculture, 26*(2), 1–11. https://doi.org/10.1080/23308249.2017.1399104

Valu, P. A., Bhola, D. V., & Tandel, M. U. (2023). Traditional fish preservation methods and products of India. *Biotica Research Today, 5*(9), 703–705.

Venugopal, V., & Shahidi, F. (1995). Value-added products from underutilized fish species. *Critical Reviews in Food Science and Nutrition, 35*(5), 431–453. https://doi.org/10.1080/10408399509527708

Zhan, X., Sun, D. W., Zhu, Z., & Wang, Q. J. (2018). Improving the quality and safety of frozen muscle foods by emerging freezing technologies: A review. *Critical Reviews in Food Science and Nutrition, 58*, 2925–2938.

Aquatic Macrophytes: Used as Fish Nutrition for Sustainable Aquaculture Production

Neelesh Kumar, Chandra Pal Singh, Abhishek Srivastava, Jai Gopal Sharma, M. J. Dobriyal, and Bijay Kumar Behera

Abstract Feed is critical to the aquaculture industry's development and long-term success. Feeds might be of plant or animal origin. Fish meal has recently dominated fish feed due to a number of variables, including its high protein content, superior essential amino acid profile, improved nutrient digestibility, absence of antinutritional factors, affordability, and accessibility. However, fish meal, the best protein source for fish feed, is presently under threat, jeopardizing feed formulators' capacity to rely on it more extensively. Despite their promise, several plants remain underused in the fish feed sector. These plants' cost effectiveness, availability, and sustainability potential, together with their nutritional content, position them to compete well with currently used plants. Aquatic macrophytes, also known as aquatic plants, are large plants that grow in or near water, including submerged, floating, or emergent species. They are found in both freshwater and marine environments and play vital roles in the aquatic ecosystem. Their use as fish nutrition has gained attention in recent years as a sustainable alternative to traditional fish feed in aquaculture. Utilizing medicinal herbs and plants is a quick, accessible,

N. Kumar (✉) · A. Srivastava
Department of Aquaculture, College of Fisheries, Rani Lakshmi Bai Central Agricultural University, Datia Campus, Datia, Madhya Pradesh, India

C. P. Singh
Department of Fisheries Resource Management, College of Fisheries, ANDUA&T, Kumarganj,
Ayodhya, Uttar Pradesh, India

J. G. Sharma
Department of Biotechnology, Delhi Technological University, Delhi, India

M. J. Dobriyal
College of Horticulture and Forestry, Rani Lakshmi Bai Central Agricultural University, Jhansi, Uttar Pradesh, India

College of Fisheries, Rani Lakshmi Bai Central Agricultural University, Datia Campus, Datia, Madhya Pradesh, India

B. K. Behera
National Fisheries Development Board, Rajendranagar, Telangana, India

A. K. Singh et al. (eds.), *Sustainable Agriculture Management in Semi-Arid Climates*, https://doi.org/10.1007/978-3-031-94062-0_16

economical, effective, and environmentally friendly way to support the socioeconomic and sustainable growth of contemporary aquaculture practices. Due to their affordability and environmental friendliness, aquaculture companies have favored phytotherapy-based methods and substances for increasing culturable fishes' resistance to a range of aquatic diseases. This chapter elaborates on the diverse applications of medicinal herbs/plants and their uses as a nutrition to fulfill the nutrition requirement in aquaculture and a major step toward organic aquaculture.

Keywords Fish nutrition · Sustainable aquaculture · Aqua-industry · Cost-effective · Organic aquaculture

1 Introduction

Aquaculture development is particularly noteworthy when compared to other industries of animal food production. Currently, one out of every three fish ingested by humans originates from aquaculture. It is important to note that aqua-crop's sustainability is determined by a variety of factors, including cost-effective feed. Feed accounts for approximately 60% of overall operational costs in aquaculture; hence, the significant growth of aquaculture will be considerably aided by the development of cheaper aqua feed. The feed formulator's attempts to produce feed at a cheaper cost will have a direct impact on the economy of fish farmers. Fish meal is one of the primary protein constituents in the fish diet; it is rich in protein content, correctly balanced Essential amino acids (EAAs) profile, and great nutritional digestibility, due to the lacking of antinutritional factors (ANFs) (Food and Agriculture Organization, 2016). Until recently, fish feed was made using fish meal as an essential protein element since most people agree that omnivorous fish require 30–40% of their diet, whereas carnivorous fish require more than 40%. However, fish meal inclusion levels for both omnivorous and carnivorous fishes are now declining due to a decrease in fish meal supplies, as well as increased market demand and costs (Hardy, 2010). Many writers have proposed plant-based protein components as a cost-effective alternative to fish meal. However, in order to be considered an acceptable substitute to fish meal, a potential component must possess the above-described criteria. In this regard, one may disagree with the use of plant proteins to substitute fish meal in the fish diet based on the following criteria: Plant components contain ANFs, are lacking in some EAA, have lower nutritional digestibility, bioavailability, and palatability due to high quantities of non-soluble carbohydrates such as fiber and starch (Edwards et al., 2004; De Silva & Hasan, 2007; Hung et al., 2007).

Aquaculture provides high-quality protein at a reasonable cost globally. By 2050, there will likely be 9.7 billion people on the planet, making it extremely difficult to supply wholesome, high-quality food. Fish products are becoming increasingly popular due to increased protein need, rising living standards, and shifting food tastes (Hunter et al., 2017; Hua et al., 2019; United Nations, 2019). By 2025, demand is expected to reach 180 million metric tons (MMT), surpassing current output of 160 million (Food and Agriculture Organization of the United Nations, 2014). Animal-derived foods are essential for global food security since they account for 25% of protein and 18% of calorie intake worldwide (Food and Agriculture Organization of the United Nations, 2014; Béné et al., 2015). Animal-derived foods are essential for global food security since they account for 25% of protein and 18% of calorie intake worldwide (Food and Agriculture Organization of the United Nations, 2014; Food and Agriculture Organization STAT, 2016).

Fish is the only source of high-quality protein, accounting for roughly 16% of the total animal protein ingested by the global population. Fish is a significant source of protein in regions where livestock is scarce. It accounts for up to 10% of animal protein consumed in Europe and North America, 17% in Africa, 22% in China, and 26% in Asia (Food and Agriculture Organization, 2000). Acute or chronic protein energy malnutrition affects 204 million children under five, according to the United Nations, which estimates that 815 million people worldwide suffer from malnutrition. Food insecurity can only be addressed by increasing global protein production. Fish is very important for nutrition and food security (Allison, 2011; Thilsted, 2012; Beveridge et al., 2013; Béné et al., 2015).

2 Aquatic Macrophytes

Aquatic macrophytes are a diverse collection of plants with a wide range of behaviors, morphologies (e.g., immersed, submerged floating, with floating leaves), and development cycles. Aquatic plants, also known as hydrophytes, are vascular and non-vascular plants that have evolved to exist in watery settings. In addition to providing food for some herbivorous animals, substrate for benthic invertebrates, and oxygen generation through photosynthesis, aquatic vegetation in lakes, rivers, and wetlands also serves as a cover for aquatic species including fish, amphibians, and aquatic insects. Waterlilies, lotus, duckweeds, mosquito fern, floating heart, water milfoils, mare's tail, water lettuce, water hyacinth, and algae are all common aquatic plants (Chambers et al., 2007).

3 Uses of Aquatic Macrophytes

3.1 Applications of Aquatic Macrophytes to Reduce Water Pollution

Aquatic macrophytes-dominated ecosystems are among the world's most productive. Aquatic plants efficiently absorb nutrients and promote microbial breakdown of organic waste. This ability may be used to restore natural streams, lakes, and wetlands, as well as for wastewater treatment. This research examines several macrophyte-based wastewater treatment facilities, their treatment capacity, and possible application in a variety of climates. Working experiences in Danish facilities are reviewed. The possibility for resource recovery via harvesting and utilizing the plant material generated can be viewed as a step toward a more holistic approach in which waste products are viewed and used as resources (Brix & Schierup, 1989).

3.2 Use of Aquatic Macrophytes for Managing Water Quality

The main sources of water for human use are relatively shallow groundwater basins, lakes, rivers, and soil moisture. The effects of human activity, eutrophication processes, and natural deterioration all affect the quality of the water in lakes and reservoirs. Though they have received less attention, especially in poorer nations, issues with water quality may sometimes be just as serious as those pertaining to water supply. Worldwide research is currently underway to find more sustainable strategies to slow down the deterioration of water quality. One of the significant attempts to achieve sustainability is phytoremediation. The majority of aquatic macrophytes occur naturally and are well suited to their environments. Aquatic macrophytes may eliminate an excessive amount of nutrients from the water, which would otherwise lead to eutrophication of the body of water. By slowing the water current and encouraging the sedimentation of suspended particles, aquatic macrophytes operate as particulate matter traps and indirectly affect metal retention by absorbing nutritious mineral ions from the water column. By providing wind shelter, aquatic macrophytes help lessen silt resuspension. One of the most studied topics worldwide is the application of aquatic macrophytes to wastewater treatment in order to reduce a range of pollution levels. The way that certain aquatic plant species absorb nutrients is quite unique. Because of this peculiarity, one of the expert jobs when designing a water treatment facility is choosing the aquatic plant species. Since choosing the right aquatic plant species shortens the time and expense of treatment procedures, an attempt has been made to cover the most studied aquatic flora for mitigation purposes as well as their potential application in a mesocosm (Culley Jr & Epps, 1973; Srivastava et al., 2008).

3.3 Use of Aquatic Macrophytes as Contaminants Remover from the Environment

It is commonly known that aquatic plants may be used to phytoremediate a variety of pollutants. With its effective ability to remove a variety of organic and inorganic pollutants through processes like extraction, degradation, or stabilization, phytoremediation has been successfully applied at a number of sites, including military installations, agricultural fields, industrial facilities, mine tailings, and sewage and domestic wastewater treatment plants. The varied group of plants known as aquatic macrophytes has enormous potential for the removal or degradation of a wide range of toxins, such as explosives, heavy metals, inorganic and organic pollutants, and radioactive waste. Regardless of whether they are free-floating, submerged, or emergent, the aquatic macrophytes' function in phytoremediation methods is highlighted in this review with appropriate weight given to each category. Harnessing the extraordinary potential of aquatic macrophytes can significantly expand their use in phytoremediation technologies (Dhir et al., 2009).

Certain plants have the capacity to take in abnormally large concentrations of pollutants from their surroundings, which are then concentrated in their roots, shoots, and/or leaves. Metal hyperaccumulators are plant species that collect extremely high concentrations of heavy metals (greater than 1% of metal in their dry matter) (Appenroth et al., 2000). *Eichhornia crassipes* (Cd, Zn), *Elodea nuttali* (Cu), and *Spirodela polyrhiza* (Cr, Pb) are a few common aquatic plant species that are metal hyperaccumulators. Metal hyperaccumulation is linked to hypertolerance mechanisms, which shed light on the several ways plants have developed to withstand toxicity (Baker & Brooks, 1989).

4 The Fight Against Water Pollution Through Aquatic Macrophytes

Health facilities throughout the world are becoming increasingly concerned about providing their citizens with safe water, particularly in regions that are more vulnerable. While most pollutants may be eliminated by sewage treatment systems, some compounds that remain detectable in large amounts even after normal treatments cannot be effectively removed by these systems. In light of the need to refine methods for eliminating waterborne pollutants, artificial wetland systems have become a viable bioremediation option for contaminant degradation and removal. The selection of plant species that can both tolerate and remove toxins is one of the limiting criteria in the construction of effective artificial wetlands, despite their environmentally attractive appearance and effectiveness in cleaning residual waters (Stefanakis, 2018; Kumar & Dutta, 2019). The selection of plant species that can both tolerate and remove toxins is one of the limiting criteria in the construction of effective artificial wetlands, despite their environmentally attractive appearance and effectiveness in cleaning residual waters. Because the biomass generated must have a designated destination, there have been instances where the plants selected to make up a system have not been demonstrated to improve wetland function (Gikas & Tsihrintzis, 2014). Here, we give a summary of aquatic macrophytes' applications and functions in artificial wetland systems. To determine if plants might improve the removal efficiency of the systems, the removal efficiency of non-planted systems and the ability of plants to remove metals, pharmaceuticals, pesticides, cyanotoxins, and nanoparticles from manmade wetlands were examined (Machado et al., 2017).

5 The Application of Aquatic Macrophytes as a Natural Remedy to Stop Different Impacts on Microalgae

The ability of macrophytes to extract medications from water has shown great promise. Studies on the ecotoxicity of water following phytoremediation are scarce, nevertheless. In this work, the authors assessed the toxicity of ciprofloxacin (Cipro) on *Desmodesmus subspicatus* microalgae cells exposed to Cipro-contaminated water after they had been treated for 96 h by *Salvinia molesta* or *Egeria densa*. Microalgae subjected to Cipro (1, 10, and 100 μg/L) in untreated water exhibited increased oxidative status (hydrogen peroxide concentration) and oxidative damages (lipid peroxidation), along with lower rates of growth, respiration, and photosynthesis (Peltzer et al., 2017). Even when the antibiotic was present at a concentration of 10 μg/L, *S. molesta* had a higher phytoremediation ability than *E. densa*, lowering Cipro concentrations in water to below its hazardous threshold to *D. subspicatus* (2.44 μg/L). Neither *S. molesta* nor *E. densa* emitted any substances that were poisonous to *D. subspicatus* throughout the water treatment process. The new use of *S. molesta* and *E. densa* as a nature-based solution to extract Cipro from tainted water is demonstrated in this paper. We demonstrate for the first time the ecotoxicological safety of this strategy by preventing the harmful effects of Cipro on photosynthetic bacteria and preventing the emergence of antibiotic resistance (Kitamura et al., 2023).

5.1 Use of Floating Aquatic Macrophytes in Wastewater Treatment

The most important natural resource in the world is water. In the World Economic Forum's global risk report, the water problem has been identified as a very significant current and future issue. Two-thirds of the world's population is also expected to experience a water deficit in the next 20 years, according to some predictions. The rapid rise in human activity to meet the expanding population's water needs has made freshwater extraction more severe (Shah et al., 2014; Machineni, 2019). Floating aquatic macrophytes are highly skilled in eliminating a variety of pollutants from wastewater, especially fertilizers. Because they develop so quickly, it is crucial to make sure that the final biomass is removed from the system safely. Lignocellulose makes up the final macrophyte biomass, which has improved nutritional and energetic qualities. As a result, it may be a good source of material for things like making fertilizer, animal feed, and bioenergy. However, there are few in-depth research and their use is still restricted. Here, researcher present a thorough examination of floating aquatic macrophytes and their effectiveness in removing organic contaminants, nutrients, and heavy metals from different kinds of wastewater (Sayanthan et al., 2024).

5.2 *Use of Aquatic Macrophytes as Production of Bioenergy*

An essential part of water bodies are aquatic macrophytes. However, in the event of overgrowth, the aquatic biomass may provide challenges to water body upkeep and recreational usage. The aquatic biomass must therefore be routinely collected and disposed of. The ecological and financial issues of managing aquatic macrophytes are the focus of the Thematic Series "Sustainable use of aquatic macrophytes for the production of bioenergy". The collection focuses on the storage and efficient use of mowed aquatic biomass in anaerobic digesters, in addition to the biomass potential and the supply process (Moeller & Zehnsdorf, 2009).

Due to the "food vs. fuel" controversy and detrimental environmental repercussions, the use of energy crops for biogas production has been a contentious topic in Germany. As a result, materials from landscape management initiatives were the main focus of the search for substitute substrates. Although aquatic biomass is collected during water body management, its potential for energy production has not been explored (Seidel et al., 2015). Through a countrywide assessment, the location and quantity of biomass recovered by aquatic de-weeding activities in rivers and lakes were determined. Furthermore, the amount of aquatic plant biomass that could be present in water bodies was calculated exemplarily for Baden-Württemberg's flowing waterways using information from surveys conducted under the European Water Framework Directive. According to the survey, there were 93 de-weeding sites in standing waterways and 172 in flowing rivers. These sites are mostly found along the lowland rivers of the Upper Rhine Plain and the North German Plain (Zehnsdorf et al., 2017). Most reports of standing water de-weeding operations came from the Alps' foothills. The total amount of fresh biomass gathered annually is 36,244 t. Germany can produce up to 100,000 t of fresh biomass annually. The case study of plant biomass de-weeded from German rivers demonstrates that only a small proportion of total aquatic plant biomass is gathered. In Germany, the amount of biomass acquired and removed from water bodies is significantly smaller than that gained from other substrates harvested by landscape management practices such as mowing meadows or cutting trees and hedges. On the other hand, greater quantities are locally building up, concentrated in particular areas or at particular bodies of water, such reservoirs, for which regional value chains might be created. Changes in the economic and technological framework are necessary to enable the utilization of these local potentials commercially feasible (Szarka et al., 2017; Röhl et al., 2019).

5.3 *Use of Aquatic Macrophytes in Aquaculture*

Macrophytes are known to provide crucial homes for fish and zooplankton, and their existence is frequently linked to clear water conditions, where their effects on zooplankton populations, chemical inhibition, and shading can help to restrict the growth of phytoplankton populations (Mitchell, 1989). They also enhance sedimentation and decrease water velocity, which lowers turbidity (Ginn, 2011). Additionally, macrophytes are thought to be potentially helpful markers of the quality of sediment and/or water. Since light penetration is decreased as a result of excessive phytoplankton development, eutrophication—the presence of excess nutrients in an aquatic system—can cause macrophytes, particularly those submerged, to vanish (Mitchell, 1989).

Approximately 50 aquatic macrophyte species have been documented to either directly or indirectly feed herbivorous fish including Amaranthaceae, Araceae, Azollaceae, Cabombaceae, Ceratophyllaceae, Characeae, Najadaceae, Nymphaeaceae, Polygonaceae, Pontederiaceae, Potamogetonaceae, Salvinaceae, Trapaceae, Convolvulaceae, Haloragaceae, Hydrocharitaceae, Lemnaceae, Lentibuliaraceae, Menyanthaceae, and Typhaceae families (Mandal et al., 2011). Approximately 40 species of fish from the two main families, Cyprinidae and Chichlidae, grow directly on macrophytes in tropical and subtropical regions. Importantly, because of their potential nutritional profile, which includes moisture ranging from 84.1 to 95.9%, dry matter 4.1 to 15.9%, crude protein 8.7 to 26.8%, crude fat 2.2 to 5.1%, carbohydrate 9.3 to 35.6%, ash 8.0 to 25.3%, and crude fiber 15.0–28.1%, these macrophytes may be used as fish food components and substitute expensive commercial feed. They also have a caloric content of 2.47–4.2 kcal/g. They are mixed with feedstuffs and given to fish in order to provide them with a significant nutritional status. Herbivores have thus demonstrated adequate growth in terms of relative growth rate, feed conversion ratio, and food consumption (Mandal et al., 2011). Approximately half of plant protein is transformed into fish protein, and adding as much as 30% of macrophytes to prepared diet has improved the amino acid balance. Since aquaculture is one of the fastest-growing agricultural sectors, using macrophytes to prepare fish feed offers rural residents a means of subsistence and has enormous potential as fish food components (Mandal et al., 2011).

The ability of the floating aquatic macrophytes to remove contaminants through the phytoremediation process can be regarded as having been demonstrated during the first 24 h of the experiment. The effluent treated with *Salvinia molesta* had the largest reductions in COD, with 74.1% elimination (de Vasconcelos et al., 2021). All treatments showed adequate turbidity when compared to the effectiveness of CONAMA resolutions n° 357/2005 and 430/2011; only the Pistia stratiotes plus *Salvinia molesta* treatment failed to address BOD; only the control and Pistia stratiotes treatment were appropriate for disposal in a lentic recipient body with regard to total nitrogen, and no treatment satisfied this resolution with regard to the total phosphorus parameter (de Vasconcelos et al., 2021).

6 Freshwater Aquatic Macrophytes as Fish Nutrition in Aquaculture

Tropical and subtropical nations are home to large populations of freshwater macrophytes, the aquatic plants with the quickest rate of growth. They thrive in water that is rich in nutrients. Based on where they are found in the water body, these macrophytes may be loosely divided into four groups: emergent (*Potamogeton* spp.), buried (*Hydrilla* spp.), marginal (*Ipomoea* spp.), and surface floating (*Azolla* spp.). Global recognition has been given to freshwater macrophytes' nutritional importance. In many bodies of water, the unrestricted spread of freshwater macrophytes causes issues. From the perspective of human and animal nutrition, the legal utilization of these nutrient-rich plants may pave the way for new developments. Humans and non-ruminant animals can both benefit from the leaf protein that is derived from freshwater macrophytes (Hasan & Chakrabarti, 2009). Various freshwater aquatic macrophytes are currently using in the aquaculture sector (Table 1).

Macrophytes are rich in minerals, lipids, protein, fatty acids, and amino acids. The duckweeds *Lemna minor* and *Spirodela polyrhiza*, which have distinct mineral compositions than terrestrial plants, were recently investigated for their amino acid and fatty acid profiles. Aquatic plants contain more calcium, iron, and manganese than terrestrial plants do. Minerals have crucial roles as catalysts in a variety of biological events. These are necessary components for metabolism, growth, and development, as well as helping animals deal with changing environmental conditions (Chakrabarti et al., 2018; Sharma et al., 2019).

Each mineral has its own optimal dosage. Low/high concentrations may have an effect on organismal physiology. Toxic minerals such as arsenic, mercury, antimony, cadmium, and others are necessary by the body in trace amounts, but high levels of helpful minerals such as sodium, potassium, magnesium, calcium, and iron can be detrimental. Polyunsaturated fatty acids (PUFAs) in the diet provide a variety of health advantages for human and animals. The analysis of fatty acid profiles in feed components confirms the quality of the diets. Fish are incapable of producing two important fatty acids: n-6 and n-3. Thus, these fatty acids should be included in fish diets (Horrobin & Manku, 1990; Skalnaya & Skalny, 2008). Regular use of antibiotics during fish culture should not be done to lower illness risk since they can damage the native microbiota of juveniles or adults and raise the likelihood of antibiotic-resistant bacteria (Semwal et al., 2023). The prickly chaff flower, *Achyranthes aspera*, is a prevalent weed in India and other tropical nations. Human ailments such as inflammation, diabetes, hypertension, pneumonia, diarrhea, dysentery, asthma, and more are treated using different portions of the plant (Kumar et al., 2022; Kumar et al., 2023). When snow trout were given various diets that included 0.5% *Achyranthes aspera* seeds, it was shown that their development was noticeably greater than that of the control fish. Additionally, enhanced diets altered the composition of fish meat, and the expression of fads2d6 was much greater than that of the control fishes (Kumar et al., 2021).

Table 1 Nutrition composition (in %) of various aquatic macrophytes (dry matter basis) useful for sustainable aquaculture

Macrophyte	Crude protein	Crude fat/ Lipid	Carbohydrate	Crude fiber	Ash	Moisture	Caloric content (kcal/g)	References
Allmania nodiflora	22.6	4.9	–	22.3	14.7	–	–	Banerjee and Matai (1990)
Alternanthera philoxeroides	25.9	4.7	–	22.3	10.5	–	–	Banerjee and Matai (1990)
Azolla africana	28.90	4.60	46.00	–	15.00	–	–	Fasakin et al. (2001)
Azolla caroliniana	18.80	3.90	–	–	16.70	–	–	Datta (2011)
Azolla filiculoides	19.70	4.20	–	–	18.50	–	–	Datta (2011)
Azolla mexicana	18.60	3.80	–	–	17.20	–	–	Datta (2011)
Azolla microphylla	24.33	6.07	47.31	–	14.73	7.56	–	Kumar et al. (2022)
Azolla pinnata	29.40	5.43	40.42	–	17.43	7.32	–	Kumar et al. (2022)
Azolla rubra	19.00	4.10	–	–	15.50	–	–	Datta (2011)
Carex hudsonii	–	–	–	–	3.0	–	4.05	Fischer (1973)
Ceratophyllum demersum	15.8	5.97	27.9	20.7	20.6	–	3.71	Boyd (1968); Banerjee and Matai (1990)
Chara sp.	17.5	1.63	23.8	–	35.8		2.58	Boyd (1968)
Commelina benghalensis	16.9	4.9	–	23.4	20.2	–	–	Banerjee and Matai (1990)
Eichornia crassipes	17.93	16.62	49.05	–	08.85	–	–	Adelakun et al. (2016)
Elodea densa	20.5	3.27	29.2	–	21.1	–	3.35	Boyd (1968)
Enhydra fluctuans	33.04	4.56	32.28	–	18.26	11.86		Kumar et al. (2022)
Gentiana lutea	17.2	2.2	–	22.7	12.9	–	–	Banerjee and Matai (1990)
Glyceria fluitans	–	–	–	–	10.0	–	4.11	Fischer (1973)
Hottonia palustris	–	–	–	–	10.0	–	4.06	Fischer (1973)
Hydrilla spinosa	20.6	3.4	–	21.3	18.3	–	–	Banerjee and Matai (1990)
Hydrilla verticillata	24.11	4.94	40.14	–	21.00	9.81	–	Kumar et al. (2022)
Hydrocotyle asiatica	12.3	4.1	–	27.6	17.9	–	–	Banerjee and Matai (1990)

Ipomoea aquatica	28.96	7.16	40.37	–	12.68	10.83	–	Kumar et al. (2022)
Ipomoea reptans	32.20	6.00	31.80	–	30.00	–	–	Kalita et al. (2007)
Juncus effusus	–	–	–	–	13.0	–	3.93	Fischer (1973)
Jussiaea repens	18.1	3.5	–	25.5	12.4	–	–	Banerjee and Matai (1990)
Lactuca sativa	–	–	–	–	15.0	–	3.86	Fischer (1973)
Lemna minor	36.94	8.13	26.77	–	19.98	8.19	–	Kumar et al. (2022)
Lemna polyrhiza	18.60	1.50	94.90	–	2.50	–	–	Bairagi et al. (2002)
Limnanthemum cristatum	24.4	2.2	–	22.4	14.5	–	–	Banerjee and Matai (1990)
Limnophila heterophyla	15.4	4.3	–	23.3	12.6	–	–	Banerjee and Matai (1990)
Lippia geminate	19.1	3.8	–	24.6	18.3	–	–	Banerjee and Matai (1990)
Marsilea quadrifolia	19.62	5.97	52.03	–	11.33	11.06	–	Kumar et al. (2022)
Mikania scandens	22.5	4.5	–	22.5	17.7	–	–	Banerjee and Matai (1990)
Myriophyllum spicatum	9.8	1.81	18.8	–	40.6	–	2.47	Boyd (1968)
Najas guadelupensis	22.8	3.75	30.56	–	18.7	–	3.55	Boyd (1968)
Nelumbo nucifera	21.9	2.8	–	–	9.9	–	–	Banerjee and Matai (1990)
Nymphaea lotus	17.23	16.47	43.92	–	13.60	–	–	Adelakun et al. (2016)
Nymphaea nouchali	16.8	2.8	–	26.3	18.7	–	–	Banerjee and Matai (1990)
Nymphaea stellata	16.7	2.6	–	24.0	14.1	–	–	Banerjee and Matai (1990)
Ottelia alismoides	14.3	2.9	–	–	16.5		–	Banerjee and Matai (1990)
Phragmites karka	13.2	4.2	–	26.4	9.4	–	–	Banerjee and Matai (1990)
Pistia stratiotes	16.79	5.15	46.63		20.69	–	–	Kumar et al. (2022)
Polygonum barbatum	26.8	4.6	–	22.0	8.0	–	–	Banerjee and Matai (1990)
Potamogeton crispus	10.9	2.85	37.2	–	16.0	–	3.61	Fischer (1973)
Salvinia auriculata	14.6	2.9	–	25.0	12.5	–	–	Banerjee and Matai (1990)

(continued)

Table 1 (continued)

Macrophyte	Crude protein	Crude fat/ Lipid	Carbohydrate	Crude fiber	Ash	Moisture	Caloric content (kcal/g)	References
Salvinia cuculata	11.00	7.00	50.80	–	31.20	–	–	Kalita et al. (2007)
Salvinia molesta	28.52	5.27	43.26	–	15.78	7.18	–	Kumar et al. (2022)
Salvinia natans	17.81	4.79	50.20	–	18.43	8.77	–	Kumar et al. (2022)
Spirodela polyrhiza	36.65	7.62	30.03	–	18.19	7.52	–	Kumar et al. (2022)
Trapa bispinosa	10.8	5.1	–	27.7	13.6	–	–	Banerjee and Matai (1990)
Trapa natans	11.40	8.00	67.30	–	13.30	–	–	Kalita et al. (2007)
Typha latifolia	10.3	3.91	32.2	28.1	6.9	–	3.61	Banerjee and Matai (1990); Boyd (1968)
Vallisneria spiralis	17.3	3.1	–	16.4	22.6	–	–	Banerjee and Matai (1990)
Wolffia arrhiza	20.40	4.63	–	–	17.60	–	–	Chareontesprasit and Jiwyam (2001)

The larger duckweed *Spirodela polyrhiza* (floats freely) was added to the diet of common carp (*Cyprinus carpio*) fry in order to assess its impact (Shrivastav et al., 2022). The main source of protein in the control diet was soybean meal. Amylase, trypsin, chymotrypsin, and lipase activity all exhibited linear correlations when higher duckweed was added to the fish diets (Shrivastav et al., 2022). Five distinct diet-fed fish have comparable compositions of critical amino acids. Fish fatty acid levels were impacted by the higher duckweed supplemented diets. The amount of duckweed included in the meal was inversely correlated with the monounsaturated fatty acids (MUFA). Fish given the SP0 diet had the greatest MUFA concentration. The fish given the 20% plant diet had the highest concentration of linoleic acid. With the rising amount of duckweed, fish's n-3 PUFA contents increased, with 20% plant diet having a noticeably higher level than the others (Shrivastav et al., 2022). The expression of fas and fads2d6 was much greater in SP5 and SP10 than in the other groups. When compared to other diet-fed fish, the expressions of *elovl2* and *elovl5* were considerably greater in fish given the 5%, 10%, and 15% plant diets. The nutritional value and growth performance of common carp were enhanced when more duckweed was added to their diets (Shrivastav et al., 2022).

Duckweed (*Lemna minor*) was tested for its possible nutritional benefit in relation to common carp fry (Goswami et al., 2022). Duckweed was added to fish diets in five different grades: 0%, 5%, 10%, 15%, and 20%. Compared to other diets, carp fed diets that were 15% and 20% plant-based had substantially increased protein and essential amino acid levels. Compared to the control group, fish fed diets based on duckweed had a much greater lipid content. There was a clear correlation between the amount of duckweed included in the diet and the carp's n-3 long-chain polyunsaturated fatty acid concentration. With an increase in dietary duckweed, the contents of elongated and desaturated linolenic acid derivatives, such as 20:4n-3, 20:5n-3, 22:5n-3, and 22:6n-3, rose in a graduated fashion (Goswami et al., 2022). Fish fed duckweed had far lower levels of MUFA and n-6 PUFA. Carp given 10%, 15%, and 20% plant diets showed increased expression of *fads2d6*, *elovl2*, *elovl5*, and *fas* than other carp. By raising the amounts of protein, fat, amino acids, and n-3 PUFA, *L. minor* improved the nutritional value of carp (Goswami et al., 2022).

7 Conclusion

This chapter discusses how aquatic macrophytes, such as submerged, floating, and emergent plants, can serve as a vital source of nutrition for fish in aquaculture systems, offering a potential and sustainable approach for enhancing fish nutrition. These plants provide important elements such as proteins, carbs, lipids, vitamins, and minerals, which help fish grow, stay healthy, and reproduce. The use of aquatic macrophytes in aquaculture can help to enhance sustainability by lowering dependency on conventional feed components like fishmeal and soy, which have substantial environmental consequences. Incorporating macrophytes into aquaculture systems can help to minimize the carbon footprint, reduce eutrophication, and promote ecosystem health by improving water quality. Aquatic macrophytes might assist reduce feed costs and enhance the economic viability of aquaculture operations, especially in underdeveloped countries where high-quality feed is scarce. Their use may decrease environmental consequences, lower feed costs, and increase the long-term viability of aquaculture methods. However, further research and development are required to address current problems and maximize their incorporation into commercial fish farming systems.

References

Adelakun, K. M., Kehinde, A. S., Amali, R. P., Ogundiwin, D. I., & Omotayo, O. L. (2016). Nutritional and phytochemical quality of some tropical aquatic plants. *Poultry, Fisheries and Wildlife Science, 4*(2), 1–4. https://doi.org/10.4172/2375-446X.1000164

Allison, E. H. (2011). *Aquaculture, fisheries, poverty and food security*. Working paper, 60p. Penang: World Fish Center. Retrieved from https://hdl.handle.net/20.500.12348/1163.

Appenroth, K. J., Bischoff, M., Gabrys, H., Stockel, J., Swartz, H. M., Walczek, T., & Winnefeld, K. (2000). Kinetics of chromium (V) formation and reduction in fronds of duckweed Spirodela polyrhiza—a low frequency EPR study. *Journal of Inorganic Biochemistry, 78*(3), 235–242. https://doi.org/10.1016/S0162-0134(00)00018-0

Bairagi, A., Ghosh, K. S., Sen, S. K., & Ray, A. K. (2002). Duckweed (*Lemna polyrhiza*) leaf meal as a source of feedstuff in formulated diets for Rohu (*Labeo rohita* Ham.) fingerlings after fermentation with a fish intestinal bacterium. *Bioresource Technology, 85*(1), 17–24. https://doi.org/10.1016/S0960-8524(02)00067-6

Baker, A. J. M., & Brooks, R. R. (1989). Terrestrial higher plants which hyperaccumulate metallic elements. A review of their distribution, ecology and phytochemistry. *Biorecovery, 1*, 81–126.

Banerjee, A., & Matai, S. (1990). Composition of Indian aquatic plants in relation to utilization as animal forage. *Journal of Aquatic Plant Management, 28*, 69–73.

Béné, C., Barange, M., Subasinghe, R., Pinstrup-Andersen, P., Merino, G., Hemre, G. I., & Williams, M. (2015). Feeding 9 billion by 2050–Putting fish back on the menu. *Food Security, 7*, 261–274. https://doi.org/10.1007/s12571-015-0427-z

Beveridge, M. C., Thilsted, S. H., Phillips, M. J., Metian, M., Troell, M., & Hall, S. J. (2013). Meeting the food and nutrition needs of the poor: The role of fish and the opportunities and challenges emerging from the rise of aquaculture. *Journal of Fish Biology, 83*(4), 1067–1084. https://doi.org/10.1111/jfb.12187

Boyd, C. E. (1968). Fresh water plants: A potential source of protein. *Economic Botany, 22*, 359–368.

Brix, H., & Schierup, H. H. (1989). The use of aquatic macrophytes in water-pollution control. *Ambio, 18*, 100–107. Retrieved from http://www.jstor.org/stable/4313539.

Chakrabarti, R., Clark, W. D., Sharma, J. G., Goswami, R. K., Shrivastav, A. K., & Tocher, D. R. (2018). Mass production of Lemna minor and its amino acid and fatty acid profiles. *Frontiers in Chemistry, 6*, 479. https://doi.org/10.3389/fchem.2018.00479

Chambers, P. A., Lacoul, P., Murphy, K. J., & Thomaz, S. M. (2007). Global diversity of aquatic macrophytes in freshwater. In E. V. Balian, C. Lévêque, H. Segers, & K. Martens (Eds.), *Developments in hydrobiology: Freshwater animal diversity assessment* (Vol. 198, pp. 9–26). Springer. https://doi.org/10.1007/978-1-4020-8259-7_2

Chareontesprasit, N., & Jiwyam, W. (2001). An evaluation of Wolffia meal (*Wolffia arrhiza*) in replacing soybean meal in some formulated rations of Nile tilapia (*Oreochromis niloticus* L.). *Pakistan Journal of Biological Sciences, 4*, 618–620. https://doi.org/10.3923/pjbs.2001.618.620

Culley, D. D., Jr., & Epps, E. A. (1973). Use of duckweed for waste treatment and animal feed. *Journal (Water Pollution Control Federation), 45*(2), 337–347.
Datta, S. N. (2011). Culture of Azolla and its efficacy in the diet of *Labeo rohita*. *Aquaculture, 310*, 376–379. https://doi.org/10.1016/j.aquaculture.2010.11.008
De Silva, S. S., & Hasan, M. R. (2007). Feeds and fertilizers: The key to long-term sustainability of Asian aquaculture. In: M. R. Hasan, T. Hecht, S. S. De Silva, & A. G. J. Tacon (Eds.) *Study and analysis of feeds and fertilizers for sustainable aquaculture development*. FAO fisheries technical paper. No. 497 (pp. 19–47). FAO.
de Vasconcelos, V. M., de Morais, E. R. C., Faustino, S. J. B., Hernandez, M. C. R., Gaudêncio, H. R. D. S. C., de Melo, R. R., & Bessa-Junior, A. P. (2021). Floating aquatic macrophytes for the treatment of aquaculture effluents. *Environmental Science and Pollution Research, 28*(3), 2600–2607. https://doi.org/10.1007/s11356-020-11308-8
Dhir, B., Sharmila, P., & Saradhi, P. P. (2009). Potential of aquatic macrophytes for removing contaminants from the environment. *Critical Reviews in Environmental Science and Technology, 39*(9), 754–781. https://doi.org/10.1080/10643380801977776
Edwards, P., Tuan, L. A., & Allan, G. L. (2004). *A survey of marine trash fish and fish meal as aquaculture feed ingredients in Vietnam* (pp. 1–56). ACIAR© Australian Centre for International Agricultural Research.
Fasakin, E. A., Balogun, A. M., & Fagbenro, O. A. (2001). Evaluation of sun-dried water fern, Azolla africana and duckweed, Spirodela polyrhiza in practical diets for Nile tilapia, *Oreochromis niloticus* fingerlings. *Journal of Applied Aquaculture, 11*, 83–92. https://doi.org/10.1300/J028v11n04_09
Fischer, Z. (1973). Physiology and bioenergetics of grass carp. *Polish Archives of Hydrobiology, 20*, 529–557.
Food and Agriculture Organization. (2000). *The state of world fisheries and aquaculture*. FAO.
Food and Agriculture Organization. (2016). *The state of world fisheries and aquaculture*. Report of the Fisheries and Aquaculture Department.
Food and Agriculture Organization of the United Nations (2014). *Eurospan distributor* (pp. 39–44).
Food and Agriculture Organization STAT. (2016). *Food and agriculture data 2016*. Food and Agriculture Organization of the United Nations (FAO).
Gikas, G. D., & Tsihrintzis, V. A. (2014). Municipal wastewater treatment using constructed wetlands. *Water Utility Journal, 8*, 57–65.
Ginn, B. K. (2011). Distribution and limnological drivers of submerged aquatic plant communities in Lake Simcoe (Ontario, Canada): Utility of macrophytes as bioindicators of lake trophic status. *Journal of Great Lakes Research, 37*, 83–89. https://doi.org/10.1016/j.jglr.2011.03.015
Goswami, R. K., Sharma, J., Shrivastav, A. K., Kumar, G., Glencross, B. D., Tocher, D. R., & Chakrabarti, R. (2022). Effect of *Lemna minor* supplemented diets on growth, digestive physiology and expression of fatty acids biosynthesis genes of *Cyprinus carpio*. *Scientific Reports, 12*(1), 3711. https://doi.org/10.1038/s41598-022-07743-x
Hardy, R. W. (2010). Utilization of plant proteins in fish diets: Effects of global demand and supplies of fishmeal. *Aquaculture Research, 41*(5), 770–776. https://doi.org/10.1111/j.1365-2109.2009.02349.x
Hasan, M. R., & Chakrabarti, R. (2009). *Use of algae and aquatic macrophytes as feed in small-scale aquaculture: A review*. Fisheries Aquaculture Technical Paper 531 (pp. 123). Food and Agricultural Organization, FAO. Retrieved from http://www.fao.org/docrep/012/i1141e/i1141e.pdf.
Horrobin, D. F., & Manku, M. S. (1990). Clinical biochemistry of essential fatty acids. In D. F. Horrobin (Ed.), *Omega-6 essential fatty acids: Pathophysiology and roles in clinical medicine* (pp. 21–54). Wiley Liss.
Hua, K., Cobcroft, J. M., Cole, A., Condon, K., Jerry, D. R., Mangott, A., Praeger, C., Vucko, M. J., Zeng, C., Zenger, K., & Strugnell, J. M. (2019). The future of aquatic protein: Implications for protein sources in aquaculture diets. *One Earth, 1*(3), 316–329. https://doi.org/10.1016/j.oneear.2019.10.018

Hung, L. T., Truc, L. T. T., & Huy, H. P. V. (2007). *Case study on the use of farm-made feeds and commercially formulated pellets for pangasiid catfish culture in the Mekong Delta, Vietnam.* FAO Fisheries Technical Paper, 497:363.

Hunter, M. C., Smith, R. G., Schipanski, M. E., Atwood, L. W., & Mortensen, D. A. (2017). Agriculture in 2050: Recalibrating targets for sustainable intensification. *Bioscience, 67*, 386–391. https://doi.org/10.1093/biosci/bix010

Kalita, P., Mukhopadhyay, P. K., & Mukherjee, A. K. (2007). Evaluation of the nutritional quality of four unexplored aquatic weeds from northeast India for the formulation of cost-effective fish feeds. *Food Chemistry, 103*, 204–209. https://doi.org/10.1016/j.foodchem.2006.08.007

Kitamura, R. S. A., Fusaro, T., Marques, R. Z., Brito, J. C. M., Juneau, P., & Gomes, M. P. (2023). The use of aquatic macrophytes as a nature-based solution to prevent ciprofloxacin deleterious effects on microalgae. *Water, 15*(12), 2143. https://doi.org/10.3390/w15122143

Kumar, S., & Dutta, V. (2019). Constructed wetland microcosms as sustainable technology for domestic wastewater treatment: An overview. *Environmental Science and Pollution Research, 26*, 11662–11673. https://doi.org/10.1007/s11356-019-04816-9

Kumar, G., Sharma, J., Goswami, R. K., Shrivastav, A. K., Kumar, N., Chandra, S., & Chakrabarti, R. (2021). The study of effect of vitamin C and *Achyranthes aspera* seeds enriched diets on the growth, biochemical composition, digestive enzyme activities and expressions of genes involved in the biosynthesis of fatty acids in snow trout *Schizothorax richardsonii* (Gray, 1832). *Journal of Applied Aquaculture, 35*(3), 489–509. https://doi.org/10.1080/10454438.2021.1985679

Kumar, G., Sharma, J., Goswami, R. K., Shrivastav, A. K., Tocher, D. R., Kumar, N., & Chakrabarti, R. (2022). Freshwater macrophytes: A potential source of minerals and fatty acids for fish, poultry, and livestock. *Frontiers in Nutrition, 9*, 869425. https://doi.org/10.3389/fnut.2022.869425

Kumar, N., Sharma, J., Mittal, P., & Chakrabarti, R. (2023). Effect of leaves and seeds of *Achyranthes aspera* as feed supplements on the immunological and stress parameters and related gene expressions of Asian catfish (*Clarias batrachus*). *Veterinary Research Communications, 47*(1), 99–109. https://doi.org/10.1007/s11259-022-09932-5

Machado, A. I., Beretta, M., Fragoso, R., & Duarte, E. (2017). Overview of the state of the art of constructed wetlands for decentralized wastewater management in Brazil. *Journal of Environmental Management, 187*, 560–570. https://doi.org/10.1016/j.jenvman.2016.11.015

Machineni, L. (2019). Review on biological wastewater treatment and resources recovery: Attached and suspended growth systems. *Water Science and Technology, 80*(2013), 2026. https://doi.org/10.2166/wst.2020.034

Mandal, R. N., Datta, A. K., Sarangi, N., & Mukhopadhyay, P. K. (2011). Diversity of aquatic macrophytes as food and feed components to herbivorous fish—a review. *Indian Journal of Fisheries, 57*(3), 65–73.

Mitchell, S. F. (1989). Primary production in a shallow eutrophic lake dominated alternately by phytoplankton and by submerged macrophytes. *Aquatic Botany, 33*(1–2), 101–110. https://doi.org/10.1016/0304-3770(89)90023-5

Moeller, L., & Zehnsdorf, A. (2009). Sustainable use of aquatic macrophytes for the production of bioenergy. *Energy, Sustainability and Society*. Retrieved from https://www.biomedcentral.com/collections/sampb.

Peltzer, P. M., Lajmanovich, R. C., Attademo, A. M., Junges, C. M., Teglia, C. M., Martinuzzi, C., Curi, L., Culzoni, M. J., & Goicoechea, H. C. (2017). Ecotoxicity of veterinary enrofloxacin and ciprofloxacin antibiotics on anuran amphibian larvae. *Environmental Toxicology and Pharmacology, 51*, 114–123. https://doi.org/10.1016/j.etap.2017.01.021

Röhl, M., Roth, S., Schütz, W., Zehnsdorf, A., & Herbes, C. (2019). Biogas production from submerged macrophytes—a case study of regional biomass potentials in Germany. *Energy, Sustainability and Society, 9*(21), 1–12. https://doi.org/10.1186/s13705-019-0204-5

Sayanthan, S., Hasan, H. A., & Abdullah, S. R. S. (2024). Floating aquatic macrophytes in wastewater treatment: Toward a circular economy. *Water, 16*(6), 870. https://doi.org/10.3390/w16060870

Seidel, D., Busch, G., Krause, B., Bade, C., Fessel, C., & Kleinn, C. (2015). Quantification of biomass production potentials from trees outside forests—a case study from Central Germany. *Bioenergy Research, 8*, 1344–1351. https://doi.org/10.1007/s12155-015-9596-z

Semwal, A., Kumar, A., & Kumar, N. (2023). A review on pathogenicity of Aeromonas hydrophila and their mitigation through medicinal herbs in aquaculture. *Heliyon, 9*(3), e14088. https://doi.org/10.1016/j.heliyon.2023.e14088

Shah, M., Hashmi, H. N., Ali, A., & Ghumman, A. R. (2014). Performance assessment of aquatic macrophytes for treatment of municipal wastewater. *Journal of Environmental Health Science and Engineering, 12*, 1–12. https://doi.org/10.1186/2052-336X-12-106

Sharma, J., Clark, W. D., Shrivastav, A. K., Goswami, R. K., Tocher, D. R., & Chakrabarti, R. (2019). Production potential of greater duckweed *Spirodela polyrhiza* (L. Schleiden) and its biochemical composition evaluation. *Aquaculture, 513*, 734419. https://doi.org/10.1016/j.aquaculture.2019.734419

Shrivastav, A. K., Kumar, G., Mittal, P., Tocher, D. R., Glencross, B. D., Chakrabarti, R., & Sharma, J. (2022). Effect of greater duckweed *Spirodela polyrhiza* supplemented feed on growth performance, digestive enzymes, amino and fatty acid profiles, and expression of genes involved in fatty acid biosynthesis of juvenile common carp *Cyprinus carpio*. *Frontiers in Marine Science, 9*, 788455. https://doi.org/10.3389/fmars.2022.788455

Skalnaya, M. G., & Skalny, A. V. (2008). *Essential trace elements in human health: A physician's view* (p. 224). Publishing House of Tomsk State University.

Srivastava, J., Gupta, A., & Chandra, H. (2008). Managing water quality with aquatic macrophytes. *Reviews in Environmental Science and Biotechnology, 7*, 255–266. https://doi.org/10.1007/s11157-008-9135-x

Stefanakis, A. I. (2018). Introduction to constructed wetland technology. In *Constructed wetlands for industrial wastewater treatment* (pp. 1–21). Wiley.

Szarka, N., Eichhorn, M., Kittler, R., Bezama, A., & Thrän, D. (2017). Interpreting long-term energy scenarios and the role of bioenergy in Germany. *Renewable and Sustainable Energy Reviews, 68*, 1222–1233. https://doi.org/10.1016/j.rser.2016.02.016

Thilsted, S. H. (2012). The potential of nutrient-rich small fish species in aquaculture to improve human nutrition and health. In: R. P. Subasinghe, J. R. Arthur, D. M. Bartley, S. S. De Silva, M. Halwart, N. Hishamunda, C. V. Mohan, P. Sorgeloos, (Eds.), *Farming the waters for people and food, proceedings of the global conference on aquaculture 2010, Phuket, Thailand. 22–25 September 2010* (pp. 57–73). FAO, Rome and NACA, Bangkok.

United Nations. (2019). *World population prospects*. Retrieved from https://population.un.org/wpp/.

Zehnsdorf, A., Moeller, L., Stärk, H. J., Auge, H., Röhl, M., & Stinner, W. (2017). The study of the variability of biomass from plants of the elodea genus from a river in Germany over a period of two hydrological years for investigating their suitability for biogas production. *Energy, Sustainability and Society, 7*(15), 1–7. https://doi.org/10.1186/s13705-017-0117-0

Nutrient Resource Management in Fisheries and Aquaculture

Satya Narayan Parida, Neelesh Kumar, Partha Sarathi Tripathy, Ajaya Kumar Rout, Charan Singh Kushwaha, and Bijay Kumar Behera

Abstract As the aquaculture operation expands and increasingly contributes to global food supply, it is essential to guarantee its growth remains socially, economically, and ecologically sustainable, in accordance with the UN Sustainable Development Goals. Although the technological development of these alternatives differs across and within the new component categories, it is essential to evaluate them within the whole feed ingredient array. Historical, current, and future perspectives are used to show significant operational changes that have progressively been adopted and new methods that are emerging as part of the march towards sustainable intensive aquaculture production systems. Aquaculture persistently decreases its carbon footprint by lowering greenhouse gas emissions, enhanced feed management techniques, understanding of nutrient needs and innovative farming methodologies. In this chapter, different aspects of nutrition and feed management in the aquaculture sector have been covered. A framework for the aquaculture sector aimed for effectively managing to satisfy the production to ensure food safety was discussed.

Keywords Aquaculture · Feed · Fisheries · Nutrition

1 Introduction

The successful operation of aquaculture is heavily dependent on nutrient resource management. The fisheries industry is the most rapidly expanding, with considerable potential to satisfy the food and nutritional requirements of an increasing global population. Aquaculture around the world is moving towards more intensive ways.

S. N. Parida (✉) · N. Kumar · P. S. Tripathy · A. K. Rout · C. S. Kushwaha
College of Fisheries, Rani Lakshmi Bai Central Agricultural University, Jhansi, Uttar Pradesh, India

B. K. Behera
National Fisheries Development Board, Hyderabad, India

A. K. Singh et al. (eds.), *Sustainable Agriculture Management in Semi-Arid Climates*, https://doi.org/10.1007/978-3-031-94062-0_17

The nutritional dynamics in aquaculture depend on natural fish food sources, including plankton and fish biomass, inside the cultured environment. Further, the sustainability of this culture with good production will rely on feed-based aquaculture, which constitutes around 50–70% of the operating costs (FAO, 2021; Hammarlund et al., 2024; Pailan & Biswas, 2022). Feeding aquatic species from aquaculture perspective requires nutritionally adequate feed that is rapidly digested to minimize waste. It is essential to create novel feed components that are abundantly accessible to serve as alternatives, hence reducing prices or eliminating the unsustainable sources. Understanding more about the nutritional needs of different fish and prawn species, along with better feed-making technology and feeding methods, has been key to the growth of modern aquaculture (Goddard, 1996). Aquatic organisms, like fish, eat just the necessary quantity of feed, and once their energy requirements are met, they stop feeding. This comes with a strategic management of nutrition for the sustainable management of fish (Manam, 2023). Each feed formulated for aquaculture is meticulously engineered to achieve optimal development in a short timeframe. The feed must be formulated cost-effectively with constrained resources while simultaneously improving the productivity and health of the organisms. The biochemical composition of the animal determines the composition of a balanced diet, which varies from species to species. Therefore, it is imperative to consider nutrient management for both the culture environment and the growing animals.

It is essential to balance the use of fertilizers and the provision of formulated feeds since both have direct and/or indirect impacts on the nutrient status of water. Fertilization sustains an ideal aquatic environment that enables the cultivated species to use the supplied feed effectively. To efficiently manage the aquaculture industry in response to the rising fish demand, the availability of both nutrient supplies and feed must be used simultaneously (Pailan & Biswas, 2022). The food requirement can be determined for a sustainable culture angle by calculating the total culture area, stocking density, expected yield, number of harvests per year, and expected feed conversion ratio. Establishing a feeding schedule and ratio is essential to optimize yield. The live body weight of the fish determines the feeding ratio. Fish with a heavier body mass consume less food than those with a lighter body mass. 5–10% of the total body weight is the recommended ratio. It is recommended that culture organisms or fish be fed twice daily, preferably at the same time and in the same location each day (Manam, 2023).

Fish exhibit significant variation in nutritional composition based on species-specific traits, phylogenetic relationships, and environmental factors (Robinson et al., 2022). Typically, fish are classified based on their eating habits in their natural habitats, which include detritus, phytoplankton, zooplankton, or zoo benthos. The determination of an appropriate stocking size for each species may significantly influence the success of polyculture. The distinction between various dietary patterns is not always absolute, as numerous fish species may alter their food preferences in response to the abundance of various food items (Kumar et al., 2005). The metabolism of proteins, carbohydrates, and lipids is the source of energy for any organism. Energy is required for the execution of daily activities, such as respiration, digestion, maturation, and reproduction.

2 Nutritional Demands of Aquaculture

There exists a correlation between dietary intake and nutritional needs. Individuals who consume plant-based diets and those who eat both plant and animal-based foods have lower energy requirements than carnivores. Dietary protein is a crucial element influencing the development and feed expenses of fish. Among the three biochemical components, protein constitutes 55–70% of body tissues, making it an essential ingredient in fish nutrition. Exceeding a certain threshold is harmful; similarly, protein or amino acids over the ideal limit become expensive and hazardous to the environment due to the excessive excretion of nitrogenous waste into the medium (Glencross et al., 2024). Feeding fish a higher dietary protein increases energy expenditure and elevates nitrogenous waste excretion. Generally, bigger fish need less protein, while smaller fish necessitate a higher protein intake. Commercially formulated feed contains the requisite protein and amino acids (Wilson & Halver, 1986). Consequently, a balance must be maintained in non-protein energy sources such as carbs and fats. Different stages of lifecycle in fish demands different feed type; Fig. 1 indicates the life cycle a fish has. The dietary protein need is contingent upon many parameters, including species, fish size, availability of protein, quality, and non-protein energy in the formulated feed. Protein consists of chains of amino acids, each displaying distinct characteristics. Enzymes, hormones, and

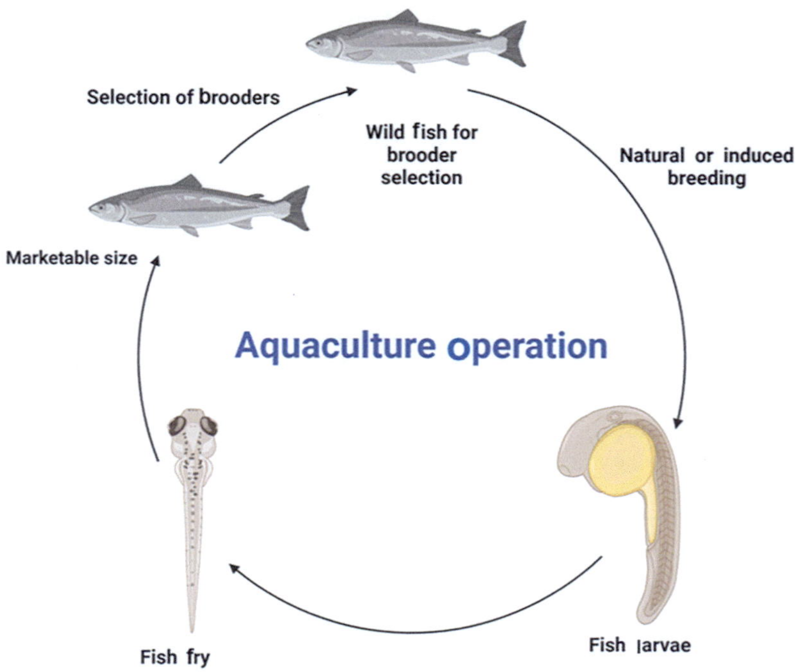

Fig. 1 Different stages for lifecycle in fishes for management of feed source in aquaculture

immunoglobulins are proteins essential for the regular functioning of the animal body. Amino acids obtained from food are classified as necessary amino acids, whereas those synthesized by the body are termed non-essential amino acids. Aquaculture cannot achieve a well-balanced feed without protein or amino acids. Amino acids serve as the fundamental components in the synthesis of proteins. For instance, skeletal deformity among juvenile fish is a common reason for the lack of amino acids, which results in death. Additionally, to promote optimal development in fish, it is necessary to establish a balance with an amino acid, regardless of whether the protein availability is optimal or suboptimal (Xing et al., 2024). Similarly, when the fish eats a lower quantity of protein from its food, it experiences a reduction in development, a fall in body weight, and the removal of protein from the vital organs of the body to preserve the functionality of those organs. It is also important to be careful to add free amino acid since fish do not always respond to the diet that only whole proteins (Prabu et al., 2021).

In contrast to proteins, fish relies less on carbohydrates as an energy source. When carbohydrates are used, the liver absorbs and stores them as glycogen. The capacity of carnivorous fishes that live in marine environments to consume, digest, and absorb carbohydrates is weaker than that of other fish species. Consuming carbohydrates more than the required limit results in a low FCR and a reduction in growth. Formulated feed for carnivores has fewer than 20% carbs, while feed for omnivores comprises about 30–40% carbohydrates (Stone, 2003). Incorporating carbohydrates into prepared feed enhances its stability and structure. Carbohydrates included in the designed feed serve as a binder, decrease feed preparation costs, and provide dietary starch for the fish (Qu et al., 2022). Similarly, lipids are essential for fish to ingest and effectively use a prepared diet. Fish need vital fatty acids, phospholipids, sterols, fat-soluble vitamins, steroid hormones, and prostaglandins to sustain physiological activities and preserve cellular membrane integrity. Aquatic organisms, particularly fish, get energy from triglycerides as well. A food supplement provides an important fatty acid the body cannot synthesize. Lipid is included in the formulated feed to leverage the protein-sparing effect, decreasing feed costs. Carnivorous fish inhabiting temperate regions need a higher lipid intake in their diet compared to their tropical counterparts. Like protein and carbs, increased fat consumption results in less appetite, a reduced growth rate, and enhanced fat deposition in the fish carcass. Further, excess lipid intake by fish results in fat accumulation in their tissues, rendering them fatty and unsuitable for human consumption. When the fish consumes lipids, it is broken down into two types of fatty acids, i.e. essential and non-essential fatty acid (Tocher, 2003). Essential fatty acids are unable to be synthesized by the body. As a result, they have been included in the diet feed (Mohanty et al., 2014). Previous research has also shown that young and juvenile fish need a greater quantity of essential fatty acids than mature adult fishes do (Rahimnejad et al., 2021).

Vitamins are organic substances necessary to the body and are required in very small quantities to carry out fundamental processes such as development, reproduction, and overall health (El-Sayed & Izquierdo, 2022). A deficiency in these vitamins or an inadequate quantity may lead to a reduction in appetite, a retardation in growth, and the development of abnormalities in the tissues and organ systems of the body. Fish readily absorb the water-soluble vitamins via their intestines, where they are then stored in the tissues of their bodies, with any excess vitamins being eliminated by excretion (Lall & Kaushik, 2021). Hypervitaminosis is the result of increased absorption of nutrients by the body. For this reason, the addition of more significant quantities of vitamins than the fish need has an impact on the health of the fish (Gasco et al., 2018). Minerals are inorganic substances or compounds that are necessary for their lives. Although fish can take in some of the minerals available in the water via their gills and digestive system, the remaining minerals are taken in through the food. It is vital for fish to consume minerals, particularly to strengthen bones, maintain the acid–base balance, ensure enzyme activity, and regulate osmotic activities. Minerals preserve the physical and chemical characteristics of the diet, improve the quality of the final product, maintain the quality of the aquatic environment, and improve health (Lall, 2022).

3 Management of Feed During Different Stages of Culture

During the nursery phase, as the newly introduced spawn begins consuming the native live feed organisms in the pond, these planktonic creatures depleted within 2–3 days. Consequently, natural feed becomes supplemented by artificial feed. Finely powdered mustard oil cake or GOC and rice polish (1:1) is administered at four times the weight of spawn during the first 5 days. From the sixth to the 12th day, the amount may be doubled. For optimal development and maximal survival, the food must consist of 40–45% protein and 26–30% carbs and should be supplemented with vitamins, minerals, and micronutrients. The first meal for spawn consists of rotifers, namely Branchionus, measuring 100–150 μm in size. At this time, the water should have a rotifer with water of 2000–5000 per liter (Turano et al., 2000). The nourishment of most aquaculture species on rearing phage continues to depend on live feeds throughout their early developmental phases. Regardless of their nutritional content, live feeds are readily identifiable and may be easily caught owing to their swimming motions in the water. The predominant live feeds used in aquaculture are rotifers (Brachionus sp.) and brine prawns (Artemia sp.), owing to the availability of cost-effective techniques for their mass production (Conceição et al., 2010). Given the abundance of live food, the daily feed ratio in the rearing pond is 10%, 8%, and 6% of body weight during the first, second, and third months, respectively. The feeding frequency of 3–4 times daily is advantageous, and the whole daily feed quantity is placed on a tray suspended from a bamboo pole at various locations 3–4 m from the pond's edge. Throughout a culture period of around one year, many feed forms, including dry mash, wet ball, cooked paste, cooked balls, or dry pellets, may be used for nourishing fingerlings and growers in a composite fish culture system (Manam, 2023).

4 Fish Health Management Through Diet

Functional feeds include nutrients or dietary elements that possess biological properties, enhancing the resilience of fish against environmental and diseases. This is a direct approach to health management of infectious illnesses and disorders, since it entails preventative or therapeutic mass treatment of fish in a natural manner (Waagbø & Remø, 2020). In addition to digesting and absorption, the intestine operates as the most significant immune organ and acts as the primary barrier against infections, making it essential for fish development and physiological processes. Thus, it is essential to investigate various functional additives to enhance the intestinal health of aquatic organisms (Yang et al., 2024). Previous studies highlighted the use of plant source material as a feed additive to tackle recent rise in aquatic pathogen and at the same time induce the immunity (Behera et al., 2023; Kumar et al., 2022; Zhang et al., 2023). The use of probiotics and prebiotics in aquafeeds enhance health status, increase disease resistance, boost growth performance, decrease malformations, and promote gut microbiota balance. Probiotics are live microorganisms that confer protection by creating an unfavorable environment for pathogen growth, competing for vital nutrients, lowering gut pH and adhesion sites, secreting substances with bactericidal properties against other microbial populations, or enhancing the immune response.

A diverse array of microalgae, yeasts, and bacteria has been assessed as probiotics; nevertheless, most probiotics used in aquaculture are bacteria. Similarly, prebiotics are non-digestible dietary components, such as oligosaccharides and dietary fiber, that foster the development of helpful gut microorganisms while inhibiting the proliferation of harmful ones and enhance intestinal immunity (Oliva-Teles, 2012). Further, the recent rise of using immunostimulants in feed to enhance non-specific defenses rather than strengthening immune responses, leading to disease resistance, shows an alternative to directly minimizing the role of antibiotics in aquaculture. Due to fish relying more on non-specific defense systems than mammals, the efficacy of immunostimulants is more significant in fish than in higher vertebrates. Furthermore, immunostimulants may enhance the immune response by functioning as adjuvants when administered with an antigen in fish vaccines (Kumar et al., 2024). For example, compounds have immunostimulant properties, including beta-glucans, chitin, lactoferrin, and levamisole. A detailed list of feed additives along with its role is mentioned in Table 1. Till now, beta-glucansuses as potential activators of macrophages, lysozyme, complement activation, and the oxidative capacity of phagocytic cells, seem to be the most promising immunostimulant compounds for use in aquaculture (Meena et al., 2012).

Table 1 Different feed additive with their potential role in aquaculture

Name of feed additive	Role	References
Phloroglucinol	Defended against AHPND induced by fish pathogen *V. parahaemolyticus* and vibriosis induced by *V. harveyi*	Roy et al. (2019)
Achyranthes aspera	Immunostimulatory and disease resistance properties in *Labeo rohita*	Kumar et al. (2019)
Protexin (A multi-strain probiotics)	Increased growth and nutrition absorption	Riaz et al. (2024)
Vitamin C	Improved performance of growth, enhanced antioxidant status and disease resistance	Ibrahim et al. (2020)
Thymol	Improve digestive system	Alagawany et al. (2021)
Spirulina platensis meal	Better growth and total protease activity in oscar fish	Mohammadiazarm et al. (2021)
Terpenes and organosulfur	Antibacterial, immunostimulatory, antioxidant, anti-inflammatory, and sedative properties	Firmino et al. (2021)
Propolis	Increased immune response and disease resistance	Farag et al. (2021)
Kera-Stim®50 (free amino acid mix)	Digestive enzyme activities of Nile tilapia	Wangkahart et al. (2022)
Melatonin	Decreased concentrations of biological stress indicators such as plasma cortisol and lactate	Acharyya et al. (2021)
Phenol-rich fulvic acid	Immunostimulation via gills and stress resistance	Lieke et al. (2021)
Selenium nanoparticle and selenomethionine	Enhanced blood biochemical profiles	Jahanbakhshi et al. (2021)

5 Understanding on Aquatic Ecosystems

The ecology of a water body is significantly influenced by its landscape context, including geology, climate, hydrology, and physical structure. Further, it also depends on the morphometry, depth, shoreline, and seasonal changes. The hydrological regime is a crucial element, particularly for reservoirs, since it significantly impacts fish production. During management of resource, an essential objective is to regulate water quality and the trophic state of the water body by preventing excessive loads of plant nutrients (particularly phosphorus and nitrogen), organic pollutants, and toxic substances, thereby mitigating the risk of hypertrophic conditions, blooms of toxin-producing cyanobacteria, anoxia, and subsequent fish mortality. Inland watery systems are affected by many factors from the nearby region, and the water body itself. It is becoming clearer that all of these factors must be taken into account in order to figure out what causes and affects the fishery. This is because fishery dynamics are closely linked to other services that the lake or pond provides as well as to fishing communities with their social economies (Schiemer et al., 2024). The trophic status is mostly influenced by phytoplankton production, which is regulated by external nutrient inflow from the catchment and adjacent wetlands, together with internal nutrient recycling and food web dynamics. A variety of trophic status indicators has been established mostly based on data from temperate aquatic systems. Basic limnological measures, including Secchi disc depth and the concentrations of total phosphorus and chlorophyll-a, together with visual assessments of floating aquatic vegetation (e.g., Salvinia, water hyacinth), serve as valuable diagnostic markers to track ecosystems where the aquaculture practice is going on. Human activities put a lot of stress on the ecosystems of reservoirs and lakes, which has a huge effect on their stability. This is especially true for eutrophication. Climate change adds to the stressors that tend to make the environment even worse. Therefore, limnological factors and forces need to be at the core of management (Jeppesen et al., 2012). Many of these metrics have been formulated based on confined waterbodies like ponds thus insufficient for applicability in huge waterbodies like lakes. An effective method is satellite-based monitoring of lake levels and nutrient conditions, a technology that is rapidly advancing (Downing et al., 2014).

6 Feed Management in RAS and Biofloc

Recirculating aquaculture systems (RAS) are the most common super-intensive aquaculture methodology. RAS has evolved over four decades, facilitating high-density fish production via several water treatment processes, including decantation systems, physical and biological filtration, which enable water reuse in these closed systems (Hisano et al., 2019). Although RAS are regarded as an efficient eco-friendly system, they incur high installation and operational costs. Biofloc technology (BFT) has proliferated in recent years due to its ability to provide high productivity rates of fish and prawns with low or no water exchange. In this system, an important factor for maintaining water quality is the management of the microbial community, particularly heterotrophic bacteria, whose proliferation is enhanced by the addition of organic carbon compounds that facilitate the removal of inorganic nitrogen from the water and can be redirected towards growth of microbial biomass (Crab et al., 2012). In comparison of both the techniques based on popularity, BFT is outperforming the RAS (Luo et al., 2014).

To enhance aquaculture beyond the limits of photoautotrophic systems and to improve environmental control, intensive pond-based systems have been established where oxygen is predominantly supplied by mechanical aerators and waste is concurrently eliminated by communities of autotrophic and heterotrophic microorganisms in bioflocs technology. The primary characteristic of these mixotrophic systems is the development of bioflocs, which are suspended aggregates of algae, bacteria, protozoa, fecal material, and uneaten feed. Besides their waste-treatment function, bioflocs serves as a healthy food supply derived totally from recycled waste materials. Intensive mixing and aeration are essential for the optimal functioning of biofloc systems, since flocs must remain suspended in the water column to work correctly and successfully mitigate the elevated respiration rates. These methods seem to be most efficient and relevant to the culture of animals like penaeid shrimp, tilapia, and the nursery stage of certain fish, which are cultivated at higher densities and use bioflocs as a nutritional supplementary to formulated feed (Boyd et al., 2020). Bioflocs consist of aggregated organic debris and various microorganisms, including bacteria, microalgae, ciliates, flagellates, amoebas, rotifers, nematodes, and annelids. This method has a high protein content, ranging from 17.2 to 47.1%, attributable to the prevalence of microorganisms with superior nutritional quality. Moreover, bioflocs have a balanced composition of lipids, carbohydrates, ash, and vitamin C. For instance, in Tilapia culture, Bioflocs serve as a supplementary natural resource of high biological value, allowing for a reduction in dietary protein content in Nile tilapia fingerling diets to 28% CP and 13.39 MJ kg^{-1}, hence decreasing feed expenses and minimizing nitrogen loss in the aquatic environment (Hisano et al., 2021).

7 Role of Commercialization on Feed Production

The rapid expansion and advancement of farmed finfish and crustacean production in the last 20 years may be linked to market accessibility and the external supply of feed inputs at a comparable pace in the top aquaculture-producing nations. In 2016, the finfish and crustacean aquaculture industry accounted for more than 69% of total worldwide fishmeal production and 75% of total global fish oil production accessible in the marketplace (Hua et al., 2019). The statistical data from the FAO indicates that approximately 61.8 million tons of farmed fish and crustaceans, representing 55.2% of total global aquaculture production in 2017, rely on external sources for nutrient supplies, predominantly through commercially manufactured feeds. From 2000 to 2017, the commercial aquaculture feed industry expanded more than thrice, increasing from 13.84 million tons to 51.23 million tons. This enhancement indicates an average annual percentage rate of 8.0% since 2000, to reach 58.85 million tons by 2020 and 73.15 million tons by 2025, respectively (Boyd et al., 2020). The feed industry has achieved significant advances in the use of cost-effective, locally available feed components, often characterized by reduced nutritional content and better absorption. The availability of feed ingredients persists in providing conventional sources, which encompass plant oilseed meals, protein concentrates, and oils. By-products from captured and aquaculture fisheries; terrestrial animal by-product meals and fats; cereal by-product meals, protein concentrates, and oils as well as microbiologically produced single-cell proteins (SCP), including those derived from algae, bacteria, and yeast. The effective decrease in the dietary inclusion of fishmeal and fish oil in aquaculture feeds has resulted from the understanding that fish and crustaceans need a diet comprising 40 or more essential nutrients, including essential amino acids, fatty acids, sterols, vitamins, minerals, and trace elements, rather than specific feed components like fish meal or fish oil. The feed ingredient business has been implementing new manufacturing processes aimed at enhancing the nutritional content of traditional feed components using recent advancement in research. Historically, selection of the ingredient was predicated simply on nutritional levels, nutrient digestibility, and cost. However, several aquaculture-producing nations and customers in significant export markets are increasingly contemplating additional selection criteria. Consequently, the sustainability of feed ingredient production relies on specific considerations regarding feed and food safety such as the utilization of potential feed adulterants, genetically modified feed ingredients concern the import dependency and food security (Tacon et al., 2022).

8 Required Action on Policy Level

The integration of suitable technologies and management practices, including the implementation of appropriate regulatory measures, fish health management, and the management of external environmental costs through guidelines, pricing mechanisms, and funding schemes, will result in a more profitable, responsible, and environmentally sustainable aquaculture while simultaneously safeguarding wild fisheries. In aquaculture, the issue that has to be answered is how sustainable fisheries and aquaculture can be developed in the long term for the purpose of ensuring food security, nutrition, as well as providing livelihoods. Effective and suitable policy support facilitates the objectives and the resolution of appropriate issues concerning the advancement of international fish farming, ensuring that the emphasis is on critical topics and is guided by the developmental requirements of producers and consumers. Implementing the Farmer First Approach (FFA) presents significant prospects for improving agriculture and empowering farmers to lead change (Tacon & De Silva, 1997). The use of information technology has also facilitated information interchange among farmers, and transitioning from top-down directives to farmer-to-farmer collaboration and collective learning for effective management of fish feed is really a game changer. Asia has consistently increased global aquaculture production (91.11%) to satisfy the growing demand. A fall in Asian aquaculture productivity will undoubtedly impact the stability of the world fish supply. Nevertheless, better collaboration in research, development, and extension initiatives, bolstered by robust governmental policy regulations, is essential to address nutrition requirement of fish and the effective formulation of feed. Complete survey is required to serve as a framework for the formulation of country-specific strategic plans and relevant policy development to meet the unique concerns of farmers of each nation for successful nutrient management (Kumar, 2014).

9 Conclusion

Effective sustainable resource utilization requires co-management among diverse stakeholders, grounded on scientifically valid data, which encompasses indigenous local knowledge and addresses imbalanced power dynamics. Feeding management practices must ensure that feed is nutritionally sufficient and well digestible while reducing waste production, including innovative feed components that are abundantly accessible to partly or wholly fulfil nutritional requirements. At the same time substitute the feed components that are not economically and ecologically sustainable. Diverse techniques focused on the in situ use of aquaculture waste, rather than releasing it into natural waterways, such as the integration of adaptable species, using biofloc technology are being extensively adopted soon.

References

Acharyya, A., Das, J., & Hasan, K. N. (2021). Melatonin as a multipotent component of fish feed: Basic information for its potential application in aquaculture. *Frontiers in Marine Science, 8*, 734066. https://doi.org/10.3389/FMARS.2021.734066

Alagawany, M., Farag, M. R., Abdelnour, S. A., & Elnesr, S. S. (2021). A review on the beneficial effect of thymol on health and production of fish. *Reviews in Aquaculture, 13*, 632–641. https://doi.org/10.1111/RAQ.12490

Behera, B. K., Parida, S. N., Kumar, V., Swain, H. S., Parida, P. K., Bisai, K., Dhar, S., & Das, B. K. (2023). Aeromonas veronii is a lethal pathogen isolated from gut of infected Labeo rohita: Molecular insight to understand the bacterial virulence and its induced host immunity. *Pathogens, 12*, 598. https://doi.org/10.3390/PATHOGENS12040598/S1

Boyd, C. E., D'Abramo, L. R., Glencross, B. D., Huyben, D. C., Juarez, L. M., Lockwood, G. S., McNevin, A. A., Tacon, A. G. J., Teletchea, F., Tomasso, J. R., Tucker, C. S., & Valenti, W. C. (2020). Achieving sustainable aquaculture: Historical and current perspectives and future needs and challenges. *Journal of the World Aquaculture Society, 51*, 578–633. https://doi.org/10.1111/JWAS.12714

Conceição, L. E. C., Yúfera, M., Makridis, P., Morais, S., & Dinis, M. T. (2010). Live feeds for early stages of fish rearing. *Aquaculture Research, 41*, 613–640. https://doi.org/10.1111/J.1365-2109.2009.02242.X

Crab, R., Defoirdt, T., Bossier, P., & Verstraete, W. (2012). Biofloc technology in aquaculture: Beneficial effects and future challenges. *Aquaculture, 356–357*, 351–356. https://doi.org/10.1016/J.AQUACULTURE.2012.04.046

Downing, A. S., Van Nes, E., Balirwa, J., Beuving, J., Bwathondi, P., Chapman, L. J., Cornelissen, I. J. M., Cowx, I. G., Goudswaard, K., Hecky, R. E., Janse, J. H., Janssen, A., Kaufman, L., Kishe-Machumu, M. A., Kolding, J., Ligtvoet, W., Mbabazi, D., Medard, M., Mkumbo, O. C., Mlaponi, E., Munyaho, A. T., Nagelkerke, L. A. J., Ogutu-Ohwayo, R., Ojwang, W. O., Peter, H. K., Schindler, D., Seehausen, O., Sharpe, D., Silsbe, G. M., Sitoki, L., Tumwebaze, R., Tweddle, D., Van De Wolfshaar, K. E., Van Dijk, H., Van Donk, E., Van Rijssel, J. C., Van Zwieten, P. A. M., Wanink, J. H., Witte, F., & Mooij, W. M. (2014). Coupled human and natural system dynamics as key to the sustainability of Lake Victoria's ecosystem services. *Ecology and Society, 19*, 31. https://doi.org/10.5751/ES-06965-190431

El-Sayed, A. F. M., & Izquierdo, M. (2022). The importance of vitamin E for farmed fish—A review. *Reviews in Aquaculture, 14*, 688–703. https://doi.org/10.1111/RAQ.12619

FAO. (2021). *Maximizing nutrition in fisheries and aquaculture using a food systems approach.*

Farag, M. R., Abdelnour, S. A., Patra, A. K., Dhama, K., Dawood, M. A. O., Elnesr, S. S., & Alagawany, M. (2021). Propolis: Properties and composition, health benefits and applications in fish nutrition. *Fish & Shellfish Immunology, 115*, 179–188. https://doi.org/10.1016/J.FSI.2021.06.010

Firmino, J. P., Galindo-Villegas, J., Reyes-López, F. E., & Gisbert, E. (2021). Phytogenic bioactive compounds shape fish mucosal immunity. *Frontiers in Immunology, 12*, 695973. https://doi.org/10.3389/FIMMU.2021.695973

Gasco, L., Gai, F., Maricchiolo, G., Genovese, L., Ragonese, S., Bottari, T., & Caruso, G. (2018). Supplementation of vitamins, minerals, enzymes and antioxidants in fish feeds (pp. 63–103). https://doi.org/10.1007/978-3-319-77941-6_4.

Glencross, B., Ling, X., Gatlin, D., Kaushik, S., Øverland, M., Newton, R., & Valente, L. M. P. (2024). A SWOT analysis of the use of marine, grain, terrestrial-animal and novel protein ingredients in aquaculture feeds. *Reviews in Fisheries Science & Aquaculture, 32*, 396–434. https://doi.org/10.1080/23308249.2024.2315049

Goddard, S. (1996). Feeds in intensive aquaculture. In: *Feed management in intensive aquaculture* (pp. 1–22). https://doi.org/10.1007/978-1-4613-1173-7_1.

Hammarlund, C., Andersson, A., & Nordström, J. (2024). Nutrient policies and the performance of aquaculture in developed countries—A literature review. *Aquaculture Economics & Management, 28*, 208–237. https://doi.org/10.1080/13657305.2024.2314511

Hisano, H., Barbosa, P. T. L., de Arruda Hayd, L., & Mattioli, C. C. (2021). Comparative study of growth, feed efficiency, and hematological profile of Nile tilapia fingerlings in biofloc technology and recirculating aquaculture system. *Tropical Animal Health and Production, 53*, 1–7. https://doi.org/10.1007/S11250-020-02523-Z

Hisano, H., Barbosa, P. T. L., Hayd, L. A., & Mattioli, C. C. (2019). Evaluation of Nile tilapia in monoculture and polyculture with giant freshwater prawn in biofloc technology system and in recirculation aquaculture system. *International Aquatic Research, 11*, 335–346. https://doi.org/10.1007/S40071-019-00242-2

Hua, K., Cobcroft, J. M., Cole, A., Condon, K., Jerry, D. R., Mangott, A., Praeger, C., Vucko, M. J., Zeng, C., Zenger, K., & Strugnell, J. M. (2019). The future of aquatic protein: Implications for protein sources in aquaculture diets. *One Earth, 1*, 316–329. https://doi.org/10.1016/J.ONEEAR.2019.10.018

Ibrahim, R. E., Ahmed, S. A. A., Amer, S. A., Al-Gabri, N. A., Ahmed, A. I., Abdel-Warith, A. W. A., Younis, E. S. M. I., & Metwally, A. E. (2020). Influence of vitamin C feed supplementation on the growth, antioxidant activity, immune status, tissue histomorphology, and disease resistance in Nile tilapia, Oreochromis niloticus. *Aquaculture Reports, 18*, 100545. https://doi.org/10.1016/J.AQREP.2020.100545

Jahanbakhshi, A., Pourmozaffar, S., Adeshina, I., Mahmoudi, R., Erfanifar, E., & Ajdari, A. (2021). Selenium nanoparticle and selenomethionine as feed additives: Effects on growth performance, hepatic enzymes' activity, mucosal immune parameters, liver histology, and appetite-related gene transcript in goldfish (Carassius auratus). *Fish Physiology and Biochemistry, 47*, 639–652. https://doi.org/10.1007/S10695-021-00937-6

Jeppesen, E., Søndergaard, M., Lauridsen, T. L., Davidson, T. A., Liu, Z., Mazzeo, N., Trochine, C., Özkan, K., Jensen, H. S., Trolle, D., Starling, F., Lazzaro, X., Johansson, L. S., Bjerring, R., Liboriussen, L., Larsen, S. E., Landkildehus, F., Egemose, S., & Meerhoff, M. (2012). Biomanipulation as a restoration tool to combat eutrophication: Recent advances and future challenges. *Advances in Ecological Research, 47*, 411–488. https://doi.org/10.1016/B978-0-12-398315-2.00006-5

Kumar, M. S. (2014). Sustainable management of fisheries and aquaculture for food security and nutrition: Policies requirements and actions. *Agricultural Research, 3*, 97–103. https://doi.org/10.1007/S40003-014-0111-0

Kumar, M. S., Burgess, S. N., & Luu, L. T. (2005). Review of nutrient management in freshwater polyculture. *Journal of Applied Aquaculture, 16*, 17–44. https://doi.org/10.1300/J028V16N03_02

Kumar, V., Das, B. K., Swain, H. S., Chowdhury, H., Roy, S., Bera, A. K., Das, R., Parida, S. N., Dhar, S., Jana, A. K., & Behera, B. K. (2022). Outbreak of Ichthyophthirius multifiliis associated with Aeromonas hydrophila in Pangasianodon hypophthalmus: The role of turmeric oil in enhancing immunity and inducing resistance against co-infection. *Frontiers in Immunology, 13*, 956478. https://doi.org/10.3389/FIMMU.2022.956478

Kumar, V., Parida, S. N., Roy, S., Dhar, S., Bisai, K., Behera, B. K., & Das, B. K. (2024). An overview of modern biotechnological tools in aquatic food production—A review. *Annals of Animal Science, 24*, 13–25. https://doi.org/10.2478/AOAS-2023-0034

Kumar, N., Sharma, J. G., Singh, S. P., Singh, A., Hari Krishna, V., & Chakrabarti, R. (2019). Validation of growth enhancing, immunostimulatory and disease resistance properties of Achyranthes aspera in Labeo rohita fry in pond conditions. *Heliyon, 5*, 1246. https://doi.org/10.1016/J.HELIYON.2019.E01246

Lall, S. P. (2022). The minerals. In: *Fish nutrition* (pp. 469–554). https://doi.org/10.1016/B978-0-12-819587-1.00005-7.

Lall, S. P., & Kaushik, S. J. (2021). Nutrition and metabolism of minerals in fish. *Animals, 11*, 2711. https://doi.org/10.3390/ANI11092711

Lieke, T., Steinberg, C. E. W., Pan, B., Perminova, I. V., Meinelt, T., Knopf, K., & Kloas, W. (2021). Phenol-rich fulvic acid as a water additive enhances growth, reduces stress, and stimulates the immune system of fish in aquaculture. *Scientific Reports, 11*, 174. https://doi.org/10.1038/s41598-020-80449-0

Luo, G., Gao, Q., Wang, C., Liu, W., Sun, D., Li, L., & Tan, H. (2014). Growth, digestive activity, welfare, and partial cost-effectiveness of genetically improved farmed tilapia (Oreochromis niloticus) cultured in a recirculating aquaculture system and an indoor biofloc system. *Aquaculture, 422–423*, 1–7. https://doi.org/10.1016/J.AQUACULTURE.2013.11.023

Manam, V. K. (2023). Fish feed nutrition and its management in aquaculture. *International Journal of Fisheries and Aquatic Studies, 11*, 58–61. https://doi.org/10.22271/FISH.2023.V11.I2A.2791

Meena, D. K., Das, P., Kumar, S., Mandal, S. C., Prusty, A. K., Singh, S. K., Akhtar, M. S., Behera, B. K., Kumar, K., Pal, A. K., & Mukherjee, S. C. (2012). Beta-glucan: an ideal immunostimulant in aquaculture (a review). *Fish Physiology and Biochemistry, 39*, 431–457. https://doi.org/10.1007/S10695-012-9710-5

Mohammadiazarm, H., Maniat, M., Ghorbanijezeh, K., & Ghotbeddin, N. (2021). Effects of spirulina powder (Spirulina platensis) as a dietary additive on Oscar fish, Astronotus ocellatus: Assessing growth performance, body composition, digestive enzyme activity, immune-biochemical parameters, blood indices and total pigmentation. *Aquaculture Nutrition, 27*, 252–260. https://doi.org/10.1111/ANU.13182

Mohanty, B., Mahanty, A., Ganguly, S., Sankar, T. V., Chakraborty, K., Rangasamy, A., Paul, B., Sarma, D., Mathew, S., Asha, K. K., Behera, B., Aftabuddin, M., Debnath, D., Vijayagopal, P., Sridhar, N., Akhtar, M. S., Sahi, N., Mitra, T., Banerjee, S., Paria, P., Das, D., Das, P., Vijayan, K. K., Laxmanan, P. T., & Sharma, A. P. (2014). Amino acid compositions of 27 food fishes and their importance in clinical nutrition. *Journal of Amino Acids, 2014*, 269797. https://doi.org/10.1155/2014/269797

Oliva-Teles, A. (2012). Nutrition and health of aquaculture fish. *Journal of Fish Diseases, 35*, 83–108. https://doi.org/10.1111/J.1365-2761.2011.01333.X

Pailan, G. H., & Biswas, G. (2022). Advances in nutrient resource management for fisheries and aquaculture. *Agriculture, Livestock Production and Aquaculture: Advances for Smallholder Farming Systems, 2*(2), 291–311. https://doi.org/10.1007/978-3-030-93262-6_15

Prabu, E., Felix, N., & Uma, A. (2021). Optimizing amino acid balance in fish meal-free diets for GIFT strain of Nile tilapia (Oreochromis niloticus) by deletion method. *Aquaculture Nutrition, 27*, 1031–1041. https://doi.org/10.1111/ANU.13244

Qu, H., Ke, W., Wen, Z., Guo, B., Lu, X., Zhao, Y., & Yang, Y. (2022). Effects of dietary carbohydrate on growth, feed utilization, hepatic glucose and lipid metabolism in endangered Yangtze sturgeon (Acipenser dabryanus). *Aquaculture Reports, 26*, 101334. https://doi.org/10.1016/J.AQREP.2022.101334

Rahimnejad, S., Dabrowski, K., Izquierdo, M., Malinovskyi, O., Kolářová, J., & Policar, T. (2021). Effects of dietary protein and lipid levels on growth, body composition, blood biochemistry, antioxidant capacity and ammonia excretion of European grayling (Thymallus thymallus). *Frontiers in Marine Science, 8*, 715636. https://doi.org/10.3389/FMARS.2021.715636

Riaz, D., Hussain, S. M., Sarker, P. K., Ali, S., Naeem, A., Naeem, E., Nazish, N., Al-Anazi, K. M., & Farah, M. A. (2024). Use of protexin as a probiotic-supplemented feed additive: Assessment of growth, digestibility, serum antioxidant enzyme activity, and blood profile in Cirrhinus mrigala. *Frontiers in Sustainable Food Systems, 8*, 1449325. https://doi.org/10.3389/FSUFS.2024.1449325

Robinson, J. P. W., Nash, K. L., Blanchard, J. L., Jacobsen, N. S., Maire, E., Graham, N. A. J., MacNeil, M. A., Zamborain-Mason, J., Allison, E. H., & Hicks, C. C. (2022). Managing fisheries for maximum nutrient yield. *Fish and Fisheries, 23*, 800–811. https://doi.org/10.1111/FAF.12649

Roy, S., Kumar, V., Bossier, P., Norouzitallab, P., & Vanrompay, D. (2019). Phloroglucinol treatment induces transgenerational epigenetic inherited resistance against vibrio infections and thermal stress in a brine shrimp (Artemia franciscana) model. *Frontiers in Immunology, 10*, 467556. https://doi.org/10.3389/FIMMU.2019.02745

Schiemer, F., Amarasinghe, U. S., Simon, D., & Vijverberg, J. (2024). Sustainable aquatic resource management and inland fisheries in tropical Asia: Interdisciplinary and transdisciplinary approaches. *Ambio, 53*, 1050–1064. https://doi.org/10.1007/S13280-024-01996-8

Stone, D. A. J. (2003). Dietary carbohydrate utilization by fish. *Reviews in Fisheries Science, 11*, 337–369. https://doi.org/10.1080/10641260390260884

Tacon, A. G. J., & De Silva, S. S. (1997). Feed preparation and feed management strategies within semi-intensive fish farming systems in the tropics. *Aquaculture, 151*, 379–404. https://doi.org/10.1016/S0044-8486(96)01494-9

Tacon, A. G. J., Metian, M., & McNevin, A. A. (2022). Future feeds: Suggested guidelines for sustainable development. *Reviews in Fisheries Science & Aquaculture, 30*, 135–142. https://doi.org/10.1080/23308249.2020.1860474

Tocher, D. R. (2003). Metabolism and functions of lipids and fatty acids in teleost fish. *Reviews in Fisheries Science, 11*, 107–184. https://doi.org/10.1080/713610925

Turano, M. J., Davis, D. A., & Arnold, C. R. (2000). Observations and techniques for maturation, spawning, and larval rearing of the yellowtail snapper Ocyurus chrysurus. *Journal of the World Aquaculture Society, 31*, 59–68. https://doi.org/10.1111/J.1749-7345.2000.TB00698.X

Waagbø, R., & Remø, S. C. (2020). Functional diets in fish health management. In: *Aquaculture health management* (pp. 187–234). https://doi.org/10.1016/B978-0-12-813359-0.00007-5.

Wangkahart, E., Kersanté, P., Lee, P. T., Sanbut, O., Nontasan, S., & Chantiratikul, A. (2022). Effect of Kera-Stim®50, a feed additive containing free amino acid mix on growth, antioxidant and immune responses, digestive enzymes, and fatty acid composition in Nile tilapia (Oreochromis niloticus). *Aquaculture, 551*, 737874. https://doi.org/10.1016/J.AQUACULTURE.2021.737874

Wilson, R. P., & Halver, J. E. (1986). Protein and amino acid requirements of fishes. *Annual Review of Nutrition, 6*, 225–244. https://doi.org/10.1146/ANNUREV.NU.06.070186.001301

Xing, S., Liang, X., Zhang, X., Oliva-Teles, A., Peres, H., Li, M., Wang, H., Mai, K., Kaushik, S. J., & Xue, M. (2024). Essential amino acid requirements of fish and crustaceans, a meta-analysis. *Reviews in Aquaculture, 16*, 1069–1086. https://doi.org/10.1111/RAQ.12886

Yang, G., Zhang, Z., & Kumar, V. (2024). Editorial: Functional feed additives and intestinal health in aquatic animals. *Frontiers in Physiology, 15*, 1385046. https://doi.org/10.3389/FPHYS.2024.1385046

Zhang, C., Liu, Y., Yao, C., Zhang, J., Wang, Y., Liu, J., Hong, Y., Mai, K., & Ai, Q. (2023). Effects of supplemental fulvic acid on survival, growth performance, digestive ability and immunity of large yellow croaker (Larimichthyscrocea) larvae. *Frontiers in Physiology, 14*, 1159320. https://doi.org/10.3389/FPHYS.2023.1159320

Conclusion

A. K. Singh and Pavan Kumar

The book *Sustainable Agriculture Management in Semi-Arid Climates: Opportunities and Threats, Vol.-I* provides a holistic perspective on strategies and practices required to address the challenges of agricultural sustainability in semi-arid climates. This comprehensive compilation underscores the importance of adapting agricultural systems to climate variability, optimizing resource management, and fostering resilience in vulnerable ecosystems. Through its exploration of adaptation strategies, soil and water management techniques, livestock production systems, and value addition in meat and fish products, the book establishes a multidisciplinary framework for achieving sustainability in these fragile regions. Agriculture in semi-arid climates faces unique and pressing challenges, including erratic rainfall, rising temperatures, water scarcity, declining soil fertility, and increasing salinity. These regions are highly susceptible to the adverse impacts of climate change, which exacerbate existing vulnerabilities and threaten food security, livelihoods, and environmental health. The book emphasizes that addressing these challenges requires a multifaceted and integrated approach that combines innovative technologies, traditional knowledge, and policy interventions. By fostering climate-resilient practices, enhancing resource use efficiency, and promoting sustainable livelihoods, the book provides a roadmap for transforming the agricultural landscape in semi-arid regions.

One of the central themes of the book is climate resilience. The authors highlight the need for adaptive agricultural practices that can withstand the uncertainties of climate change. Techniques such as conservation agriculture, the use of drought-resistant crop varieties, and the adoption of smart irrigation systems are presented

A. K. Singh
Rani Lakshmi Bai Central Agricultural University, Jhansi, Uttar Pradesh, India

P. Kumar (✉)
Forest Resource Management, College of Horticulture and Forestry, Rani Lakshmi Bai Central Agricultural University, Jhansi, Uttar Pradesh, India

A. K. Singh et al. (eds.), *Sustainable Agriculture Management in Semi-Arid Climates*, https://doi.org/10.1007/978-3-031-94062-0_18

as essential strategies for enhancing resilience to climate extremes. Conservation agriculture, in particular, offers a sustainable approach that minimizes soil disturbance, maintains soil cover, and promotes crop rotation, thereby improving soil health and water retention.

Agroforestry is another key focus of the book, presented as a climate-smart agricultural practice with significant potential for semi-arid regions. By integrating trees, crops, and livestock, agroforestry systems provide multiple benefits, including carbon sequestration, enhanced soil fertility, improved water management, and biodiversity conservation. The book highlights the role of agroforestry in mitigating climate change impacts, fostering ecosystem stability, and providing economic benefits to farming communities. Moreover, the diversity of agroforestry systems allows for tailored approaches that address the specific needs and challenges of different semi-arid regions.

The book also explores the multifaceted applications of biochar, a carbon-rich material derived from biomass through pyrolysis. Beyond its well-known role in carbon sequestration, biochar is presented as a versatile tool for improving soil quality, enhancing nutrient retention, and increasing water-holding capacity. Its application in semi-arid agriculture offers a sustainable solution for addressing soil degradation and water scarcity while contributing to climate change mitigation. The discussion on biochar underscores the importance of adopting innovative and scientifically grounded practices to enhance the sustainability of agricultural systems. Soil and water management are foundational to agricultural sustainability, particularly in semi-arid climates where these resources are often limited and vulnerable to degradation. The book emphasizes the importance of soil conservation techniques, such as contour farming, terracing, and cover cropping, in preventing soil erosion and maintaining soil health. Adaptive management practices that respond to changing climatic conditions are also highlighted as critical for ensuring the long-term viability of soil resources. The integration of modern technologies, such as remote sensing and Geographic Information Systems (GIS), further enhances the ability to monitor and manage soil and water resources effectively. These technologies provide valuable insights into soil salinity, nutrient levels, and environmental variability, enabling precision agriculture and informed decision-making. Water quality is another critical aspect of sustainable agriculture in semi-arid regions. The book underscores the need for strategies to ensure the availability of clean and reliable irrigation water. Salinity management, water recycling, and the use of advanced filtration technologies are among the solutions proposed to address water quality challenges. The importance of monitoring and maintaining irrigation water quality is particularly evident in the context of climate-resilient agriculture, where water scarcity and contamination pose significant risks to productivity and sustainability. The role of soil microbiomes in enhancing agricultural resilience and sustainability is also explored in the book. The diversity and function of soil microbial communities are recognized as critical factors in nutrient cycling, soil structure, and plant health. In agroforestry systems, the microbiome contributes to climate resilience and tolerance to biotic stress, further enhancing the sustainability of these integrated systems. The book highlights the need for continued research and innovation to harness the potential of soil microbiomes for sustainable agriculture.

The discussion extends beyond crop production to include livestock systems, recognizing their integral role in semi-arid agricultural landscapes. Livestock nutrition and production systems face unique challenges in these regions, including limited feed resources, water scarcity, and vulnerability to climate change. The book presents innovative solutions to these challenges, such as the development of farmer-friendly technologies and the use of alternative feed sources. Capacity development and sustainable livelihood generation through livestock farming are also emphasized, with successful case studies demonstrating the potential for improving economic resilience and food security. For example, pig farming is highlighted as a viable and profitable enterprise that contributes to rural development while promoting sustainable practices.

Value addition in meat and fish products is another important aspect of the book, addressing both economic and nutritional dimensions of agricultural sustainability. By transforming raw products into value-added goods, farmers and producers can enhance their income and meet the growing demand for high-quality, nutrient-rich foods. The book explores various techniques for value addition, such as processing, packaging, and fortification, which not only improve the marketability of products but also contribute to public health and nutrition. The use of aquatic macrophytes as an alternative feed source in aquaculture is presented as an innovative approach to reducing reliance on conventional feed inputs while maintaining ecological balance.

The overarching themes of the book reflect the interconnectedness of environmental, economic, and social dimensions of agricultural sustainability. The authors emphasize the importance of integrating traditional knowledge with modern scientific advancements to create resilient and adaptive agricultural systems. The role of capacity building and knowledge dissemination is also highlighted as critical for ensuring the successful implementation of sustainable practices. By empowering farmers, researchers, and policymakers with the tools and knowledge needed to address the challenges of semi-arid agriculture, the book advocates for a collaborative and inclusive approach to sustainability.

In conclusion, *Sustainable Agriculture Management in Semi-Arid Climates: Opportunities and Threats, Vol.-I* serves as a comprehensive guide for addressing the complex challenges of agricultural sustainability in semi-arid regions. It provides a wealth of insights and practical solutions that are grounded in science and tailored to the unique needs of these vulnerable ecosystems. By fostering resilience, optimizing resource use, and promoting sustainable livelihoods, the book outlines a path toward a more sustainable and equitable agricultural future. Its multidisciplinary approach and emphasis on innovation and collaboration make it a valuable resource for researchers, practitioners, and policymakers committed to advancing sustainable agriculture in semi-arid climates. The strategies and practices presented in the book not only contribute to food security and environmental health but also enhance the well-being of farming communities, ensuring a resilient and prosperous future for generations to come.

Index

A. K. Singh et al. (eds.), *Sustainable Agriculture Management in Semi-Arid Climates*, https://doi.org/10.1007/978-3-031-94062-0

MIX
Papier aus verantwortungsvollen Quellen
Paper from responsible sources
FSC® C105338

If you have any concerns about our products,
you can contact us on
ProductSafety@springernature.com

In case Publisher is established outside the EU,
the EU authorized representative is:
Springer Nature Customer Service Center GmbH
Europaplatz 3, 69115 Heidelberg, Germany

Printed by Libri Plureos GmbH
in Hamburg, Germany